MODERN DICTIONARY OF BIOTECHNOLOGY

MODERN DICTIONARY OF BIOTECHNOLOGY

Comprehensive and Illustrated Encyclopaedic Dictionary

Edited by

DR. RANJANA ARYA
Ph.D. (J.N.U., New Delhi)
Post Doctorate : Harvard University, Boston, USA
University of North Carolina, Chappel Hill, USA

NTB

NuTech Books

from

DEEP & DEEP PUBLICATIONS PVT. LTD.
F-159, Rajouri Garden, New Delhi - 110 027

MODERN DICTIONARY OF BIOTECHNOLOGY

ISBN 81-7629-740-2

Printed in India at MAYUR ENTERPRISES,
WZ Plot No. 3, Gujjar Market, Tihar Village, New Delhi-110 018.

NuTech Books are Published by DEEP & DEEP PUBLICATIONS PVT. LTD.,
F-159, Rajouri Garden, New Delhi - 110 027 • Phone : 25435369, 25440916
E-mail : deep98@del3.vsnl.net.in • ddpbooks@yahoo.co.in
Showroom :
2/13, Ansari Road, Daryaganj, New Delhi - 110 002 • Telefax : 23245122

Preface

This is the first edition of the Modern Dictionary of Biotechnology which has been able to draw on the research and analysis of the language reflected in the Science of Biotechnology. This edition makes use of the innovative principles and methodology devised in the field of Biotechnology. It takes to heart the new approaches to presentation and understanding of meaning which forms such an integral part of Biotechnology modifying them to the needs of a smaller one and at the same time preserving the unique traditions of the concise. Definitions are sharp and crisp and the new open layout makes finding individual senses, sections, and entries easier than even before. In all respects this dictionary benefits from the extensive research carried out for producing the most accurate, up-to-date, and objective description of the Biotechnology.

The Dictionary of Biotechnology is designed to be straightforward and self explanatory as far as possible and the use of special dictionary symbols and conventions has been kept to a minimum. The aim is to help the user, not only to navigate the entries more easily and to find the relevant sense more readily but also to build up an understanding in general of the way in which different meanings of a word relate to each other.

I am grateful to all who have contributed in publishing this book. I acknowledge with thanks the help of various librarians and authors of various books from whom I have heavily drawn upon the material for this volume. I would also like to thank the publishers for taking up this book for publication and bringing it out in time.

New Delhi RANJANA ARYA

A

▫ abc model

Widely accepted model of flower organ identity that appears generally applicable to distantly related dicotyledonous, although less well to monocotyledonous plants. The model incorporates the *Arabidopsis* genes required for flower organ identity.

▫ abiotic

Absence of living organisms.

▫ abscisic acid

A phytohormone implicated in the control of many plant responses to abiotic stress, such as extent of stomatal opening under water deficit (i.e. drought) conditions.

▫ acaricide

A pesticide used to kill or control mites or ticks.

▫ acc synthase

Abbreviation for 1-aminocyclopropane-1-carboxylase. The enzyme catalyses the rate limiting step in the ethylene biosynthetic pathway, and is particularly significant in the fruit ripening process. Plants typically carry a number of distinct ACC synthase genes, which are differentially regulated in response to a variety of developmental, environmental and chemical factors.

▫ acceptor control

The regulation of the rate of respiration by the availability of ADP as a phosphate acceptor.

▫ acceptor junction site

The junction between the 3' end of an intron and the 5' end of an exon. *See*: donor junction site.

▫ accessory bud

A lateral bud occurring at the base of a terminal bud or at the side of an axillary bud.

❐ acclimatization

The adaptation of a living organism (plant, animal or micro-organism) to a changed environment that subjects it to physiological stress. Acclimatization should not be confused with adaptation.

❐ acellular

Tissues or organisms that are not made up of separate cells but often have more than one nucleus.

❐ acentric chromosome

Chromosome fragment lacking a centromere.

❐ acetyl co-enzyme A

(Abbreviation: acetyl CoA) A compound formed in the mitochondria when an acetyl group (CH_3CO-) - derived from breakdown of fats, proteins, or carbohydrates - combines with the thiol group (-SH) of co-enzyme A.

❐ acquired

Developed in response to the environment, not inherited, such as a character trait (acquired characteristic) resulting from environmental effect(s). *cf* acclimatization.

❐ acridine dyes

A class of positively charged polycyclic molecules that intercalate into DNA and induce frameshift mutations.

❐ acrocentric

A chromosome that has its centromere near the end.

❐ acropetal

Arising or developing in a longitudinal sequence beginning at the base and proceeding towards the apex.

❐ activated charcoal

Charcoal that has been treated to remove hydrocarbons and to increase its adsorptive properties. It acts by condensing and holding a gas or solute onto its surface; thus inhibitory substances in nutrient medium may be adsorbed to charcoal included in the medium.

❐ active transport

Active transport is the mediated transport of biochemicals, and other atomic/molecular substances, across membranes. Unlike passive transport, this process requires chemical energy. In this form of transport, molecules move against either an electrical or concentration gradient (collectively termed an electrochemical gradient).

This is achieved by either altering the affinity of the binding site or altering the rate at which the protein changes conformations.

Types

There are two main types, *primary* and *secondary*. In primary transport energy is directly coupled to movement of desired substance across a membrane, independent of any other species. Secondary transport concerns the diffusion of one species across a membrane to drive the transport of another.

Primary

Primary active transport directly uses energy to transport molecules across a membrane. Most of the enzymes that perform this type of transport are transmembrane ATPases. A primary ATPase universal to all cellular life is the sodium-potassium pump, which helps maintain the cell potential.

Secondary

In secondary active transport, there is no direct coupling of ATP; instead, the electrochemical potential difference created by pumping ions out of cells is used. The two main forms of this are counter-transport (antiport) and co-transport (symport).

Active transport

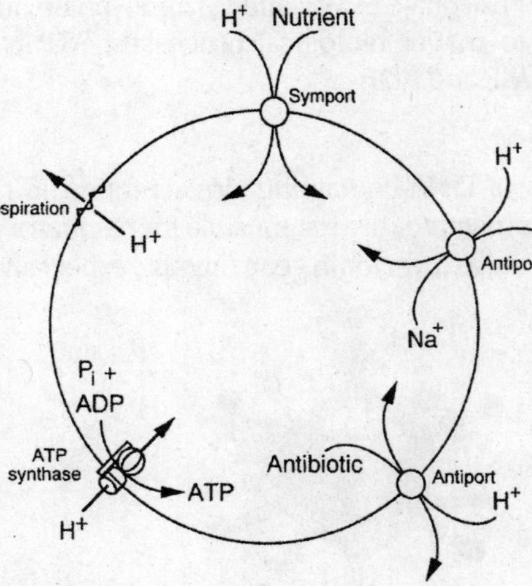

❐ acyl carrier protein

A class of molecules that bind acyl intermediates during the formation of long-chain fatty acids. ACPs are important because of their involvement in many of the reactions necessary for *in vivo* fatty acid synthesis.

❐ adaptation

Adjustment of a population to changes in environment over generations, associated (at least in part) with genetic changes resulting from selection imposed by the changed environment.

❐ additive genes

Genes whose net effect is the sum of their individual allelic effects, i.e. they show neither dominance nor epistasis.

❐ additive genetic variance

The net effect of the expresson of additive genes, and thus the chief cause of the resemblance between relatives. It represents the main determinant of the response of a population to selection. Formally, the variance of breeding values.

❐ adenosine

The (ribo)nucleoside resulting from the combination of the base adenine (A) and the sugar D-ribose. The corresponding deoxyribonucleoside is called deoxyadenosine.

❐ adenosine triphosphate (adenosine 5'-triphosphate)

A nucleotide of fundamental importance as the major carrier of chemical energy in all living organisms. It is also required for RNA synthesis since it is a direct precursor molecule. ATP consists of adenosine with three phosphate groups, linked together linearly. The phosphates are attached to adenosine through the 5'-hydroxyl of its ribose (sugar) portion. Upon hydrolysis, these bonds yield either one molecule of adenosine 5'-diphosphate (ADP) and the inorganic phosphate ion, or one molecule of adenosine 5'-monophosphate (AMP) and pyrophosphate; in both cases releasing energy that is used to power biological processes. ATP is regenerated by the phosphorylation of AMP and ADP.

❐ adenovirus

One of a group of DNA-containing viruses found in rodents, fowl, cattle, monkeys, and man. In man they are responsible for respiratory-tract infections, but they have been exploited as a vector in gene therapy, especially for genes targeted at the lungs.

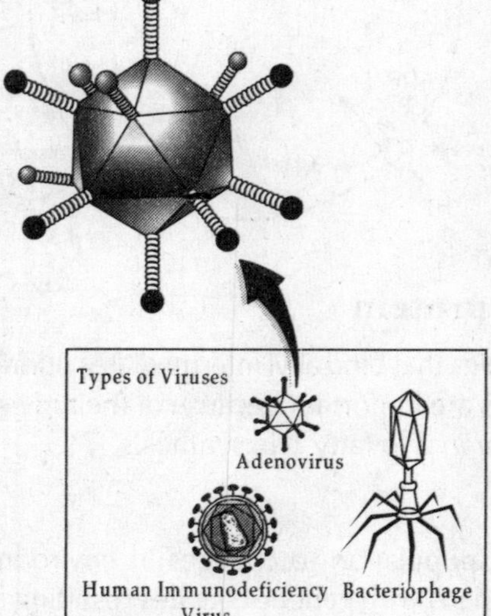

❐ adenylic acid

Synonym for adenosine monophosphate, a (ribo)nucleotide containing the nucleoside adenosine. The corresponding deoxyribonucleotide is called deoxyadenosine 5'-monophosphate or deoxyadenylic acid.

❐ adoptive immunization

The transfer of an immune state from one animal to another by means of lymphocyte transfusions.

❐ adp

Abbreviation for adenosine diphosphate.

❐ adventitious

A structure arising at sites other than the usual ones, e.g. shoots from roots or leaves, and embryos from any cell other than a zygote.

❐ aerobe

A micro-organism that grows in the presence of oxygen.

❐ aerobic

Active in the presence of free oxygen, e.g. aerobic bacteria that can live in the presence of oxygen.

❐ aerobic respiration

A type of respiration in which foodstuffs are completely oxidized to carbon dioxide and water, with the release of chemical energy, in a process requiring atmospheric oxygen.

❐ affinity chromatography

A method for purifying specific components in a solution by exploiting their specific binding to known molecule(s). The mixed solution is passed through a column containing a solid medium to which the binding molecule is covalently attached.

❐ affinity tag

An amino acid sequence that has been engineered into a protein to make its purification easier. The tag could be another protein or a short amino acid sequence, allowing purification by affinity chromatography.

❐ aflatoxins

A group of toxic compounds, produced by *Aspergillus flavus*, that bind to DNA and prevent replication and transcription. Aflatoxins can cause acute liver damage and cancer. A health hazard in certain stored foods or feed.

❐ aflp

Abbreviation for amplified fragment length polymorphism.

❐ ag

Abbreviation for antigen.

❐ agar

A polysaccharide gelifying agent used in nutrient media preparations and obtained from Rhodophyta (red algae). Both the type of agar and its concentration can affect the growth and appearance of cultured explants.

❐ agarose gel electrophoresis

A method to separate DNA and RNA molecules on the basis of their size, in which samples are subjected to an electric field applied to a gel made with agarose.

❐ agarose

The main functional constituent of agar.

❐ aggregate

1. A clump or mass formed by gathering or collecting units.
2. A body of loosely associated cells, such as a friable callus or cell suspension.
3. Coarse inert material, such as gravel, that is mixed with soil to increase its porosity.
4. A serological reaction in which the antibody and antigen react and precipitate.

❐ agonist

A drug, hormone or transmitter substance that forms a complex with a receptor site. The formation of the complex triggers an active response from a cell.

❐ agrobacterium

A genus of bacteria that includes several plant pathogenic species, causing tumour-like symptoms. *Agrobacterium rhizogenes* A bacterium that causes hairy root disease in some plants. Similar to the crown gall disease caused by *Agrobacterium tumefaciens*, this is achieved by the mobilization of the bacterial Ri plasmid with the transfer to the plant of some of the genetic material from the plasmid. This process has been used to insert foreign genes into plant cells, but to a lesser extent than the *Agrobacterium tumefaciens*-mediated transformation system, because regeneration of whole plants from hairy root cultures is problematic.

❐ agrobacterium tumefaciens

A bacterium that causes crown gall disease in some plants. The bacterium characteristically infects a wound, and incorporates a segment of Ti plasmid DNA into the host genome. This DNA causes the host cell to grow into a tumour-like structure which synthesizes specific opines that only pathogen can metabolize. This DNA-transfer mechanism is exploited in the genetic engineering of plants.

❐ agrobacterium tumefaciens-mediated transformation

The process of DNA transfer from *Agrobacterium tumefaciens* to plants, that occurs naturally during crown gall disease, and can be used as a method of transformation.

❐ ahg

Abbreviation for antihaemophilic globulin.

❐ ai

Abbreviation for artificial insemination.

❐ airlift fermenter

A cylindrical fermentation vessel in which the cells are mixed by air introduced at the base of the vessel and that rises through the column of culture medium. The cell suspension circulates around the column as a consequence of the gradient of air bubbles in different parts of the reactor.

❐ albinism

Hereditary absence of pigment in an organism. Albino animals have no colour in their skin, hair and eyes. Albino plants lack chlorophyll.

❐ albino

1. An organism lacking pigmentation, due to genetic factors. The condition is albinism
2. A conspicuous plastid mutant involving loss of chlorophyll.

❐ aleurone

The outermost layer of the endosperm in a seed, and the site of enzymes concerned with endosperm digestion during seedling growth.

❐ algal biomass

Single-celled plants (e.g. *Chlorella* spp. and *Spirulina* spp.) grown commercially in ponds to make feed materials for zooplankton, which are in turn harvested as feed for fish farms.

❐ alginate

Polysaccharide gelling agent.

❐ alkylating agent

A class of chemicals that transfer alkyl (methyl, ethyl, etc.) groups; for example to the bases in DNA. Some of these (especially ethyl methane sulphonate, abbreviated EMS) have been much used as mutagens.

❐ allele

A variant form of a gene. In a diploid cell there are two alleles of every gene (one inherited from each parent, although they could be identical). Within a population there may be many alleles of a gene. Alleles are symbolized with a capital letter to denote dominance, and lower case for recessive. In heterozygotes with co-dominant alleles, both are expressed.

❑ allele frequency

The relative number of copies of an allele in a population, expressed as a proportion of the total number of copies of all alleles at a given locus in a population.

❑ allele-specific amplification

The use of the polymerase chain reaction at a sufficiently high stringency that only one allele is amplified. A powerful means of genotyping for single-locus disorders that have been characterized at the molecular level.

❑ allelic exclusion

A phenomenon whereby only one functional allele of an antibody gene can be assembled in a given B lymphocyte.

❑ allelopathy

The secretion of chemicals, such as phenolic and terpenoid compounds, by a plant's roots, which inhibit the growth or reproduction of competitor plants.

❑ allergen

An antigen that provokes an immune response.

❑ allogenic

Differing at one or more loci, although belonging to the same species. Thus an organ transplant from one human donor to another is allogeneic, whereas a transplant from a baboon to a human would be xenogenic.

❑ allometric

When the growth rate of one part of an organism differs from that of another part or of rest of the body.

❑ allopatric

In the context of natural populations of animals or plants, inhabiting distinct and separate areas.

❑ allopatric speciation

Speciation occurring at least in part because of geographic isolation.

❑ allopolyploid

A polyploid organism with sets of chromosomes derived from different species.

❑ allosome

Synonym for sex chromosome.

❑ allosteric enzyme

An enzyme that has two structurally distinct forms, one of which is active and the other inactive. Active forms tend to catalyse the initial step in a pathway leading to the synthesis of molecules. The end product of this synthesis can act as a feedback inhibitor, converting the enzyme to the inactive form, thus controlling the amount of product synthesized.

❒ allosteric regulation

A catalysis-regulating process in which the binding of a small effector molecule to one site on an enzyme affects the activity at another site.

❒ allosteric site

That part of an enzyme molecule where the non-covalent binding of an effector molecule can affect the enzyme's catalytic activity.

❒ allosteric transition

A reversible interaction of a small molecule with a protein molecule, resulting in a change in the shape of the protein and consequent alteration of the interaction of that protein with a third molecule.

❒ allotetraploid

An allopolyploid having two different progenitor genomes.

❒ allotype

A classification of antibody molecules according to the antigenicity of the constant regions; a variation that is determined by a single allele.

❒ allozygote

A individual that is heterozygous for two different mutant alleles.

❒ alternative mRNA splicing

The inclusion or exclusion of different exons to form different mRNA transcripts from a single transcription unit.

❒ alu sequences

A highly repeated family of 300-bp long sequences dispersed throughout the human genome, so named because they are released by the digestion of genomic DNA with the restriction endonuclease AluI.

❒ amino acid

A molecule of the general formula NH2-CHR-COOH, where "R" is one of a number of different side chains. Amino acids are the building blocks of proteins. The sixty-four codons of the genetic code allow the use of twenty different amino acids (the primary amino acids) in the synthesis of proteins. Other nonprimary amino acids occur in proteins by enzymatic modification of amino acids in mature proteins, and as metabolic intermediates.

A compound containing both amino (-NH_2) and carboxyl (-COOH) groups. In particular, any of 20 basic building blocks of proteins having the formula NH_2-CR-COOH, where R is different for each specific amino acid.

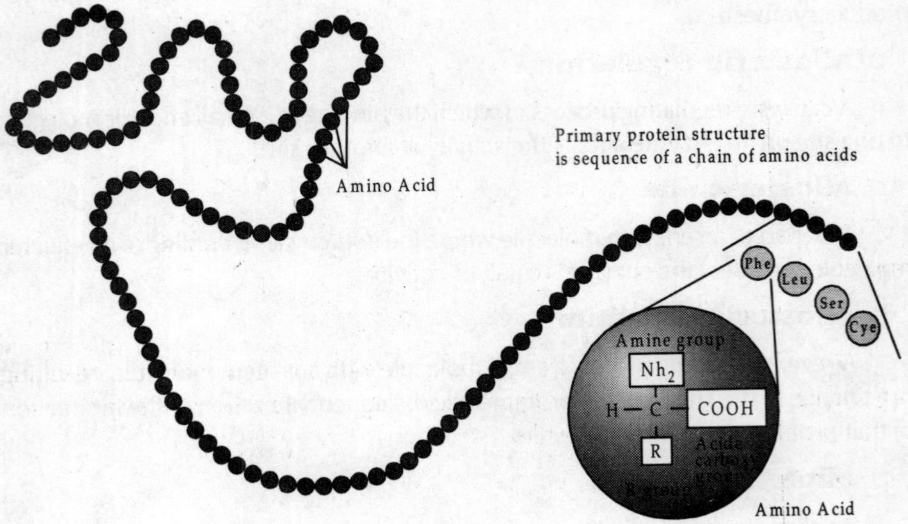

▢ amino acids found in proteins part 1

In chemistry, an amino acid is any molecule that contains both amino and carboxylic acid functional groups. In biochemistry, this shorter and more general term is frequently used to refer to alpha amino acids: those amino acids in which the amino and carboxylate functionalities are attached to the same carbon.

Amino acids are biochemical building blocks. They form short chemical chains called polypeptides or peptides which in turn form structures called proteins .

Twenty amino acids are encoded by the standard genetic code and are called proteinogenic. Rarer, more complicated ones are "made to order" by the body. Proline is the only cyclic proteinogenic amino acid. Other amino acids contained in proteins are usually formed by modification after translation (protein synthesis). These modifications are often essential for the function of the protein. At least two amino acids other than the standard 20 are sometimes incorporated into proteins during translation:

- Selenocysteine is incorporated into some proteins at a UGA codon, which is normally a stop codon.
- Pyrrolysine is used by some methanogens in enzymes that they use to produce methane. It is coded for similarly to selenocysteine but with the codon UAG instead.

Over 500 amino acids have been found in nature. Some of them have also been found in meteoritic material. Microorganisms and plants often produce very uncommon amino acids, which can be found in peptidic antibiotics (for example nisin or alamethicin). Lanthionine is a sulfide bridged alanine dimer which is found together with unsaturated amino acids in lantibiotics (antibiotic peptides from microbial origin). 1-Aminocycloproane-1-carboxylic acid (ACC) is a small disubstituted cyclic amino acid and a key intermediate in the production of the plant hormone ethylene.

In addition to amino acids for protein synthesis, there are other biologically important amino acids, such as the neurotransmitters glycine, GABA and glutamate, as well as carnitine (used in lipid transport within a cell), ornithine, citrulline, homocysteine, hydroxyproline, hydroxylysine, and sarcosine.

Some of the 20 amino acids in the genetic code are called essential amino acids, because they cannot be synthesized by the body from other compounds through chemical reactions, but instead must be taken in with food. In humans, the essential amino acids are lysine, leucine, isoleucine, methionine, phenylalanine, threonine, tryptophan, valine, and (in children) histidine and arginine.

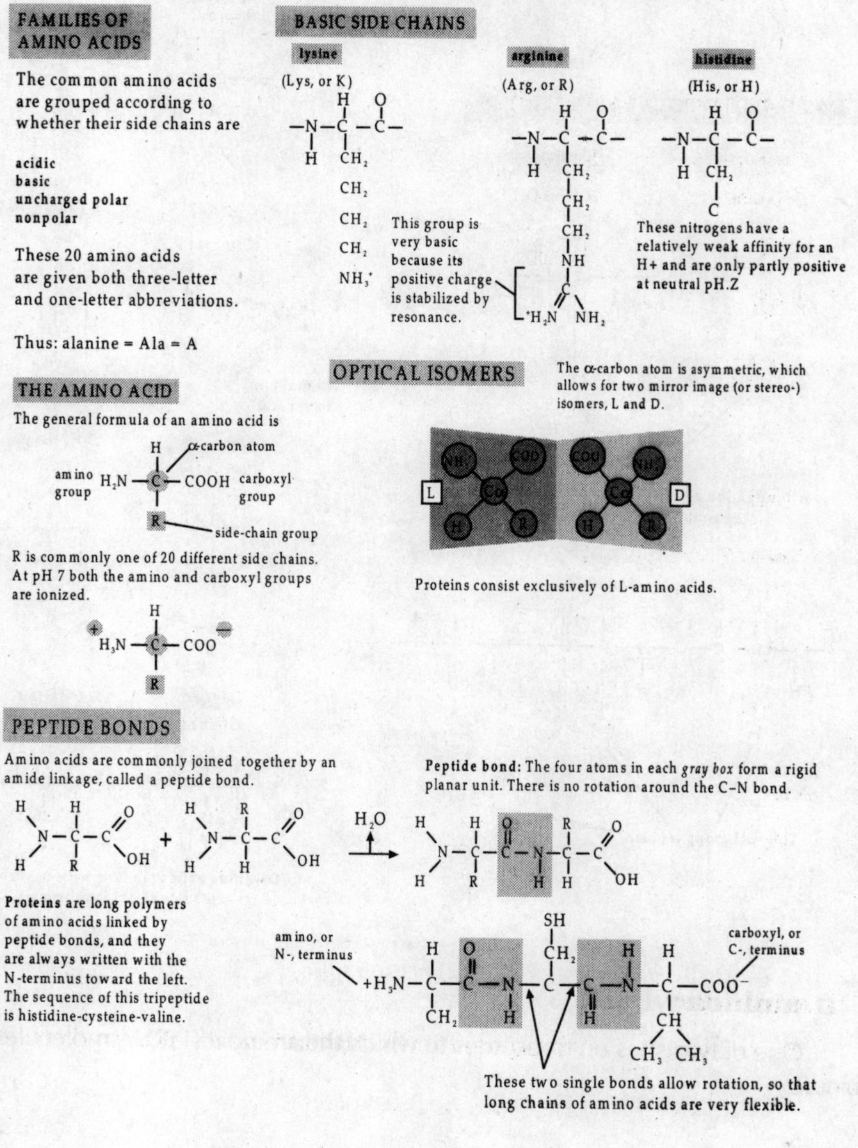

□ amino acids found in proteins part 2

The 20 amino acids found in proteins.

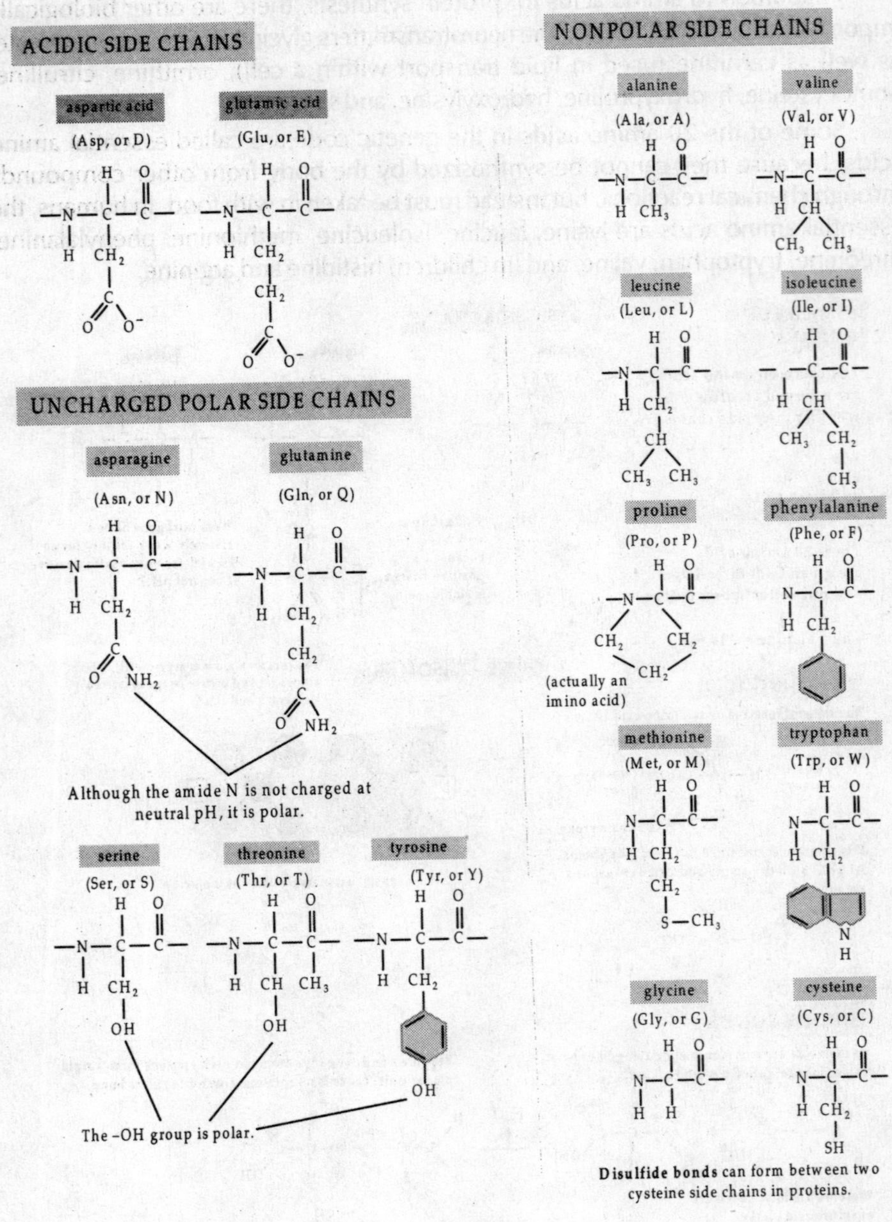

□ aminoacyl site

One of two sites on ribosomes to which the aminoacyl tRNA molecules can bind.

❐ aminoacyl tRNA synthetase

An enzyme that catalyses the attachment of an amino acid to its specific tRNA molecule.

❐ amitosis

A cell division (including nuclear division through constriction of the nucleus) that occurs without chromosome differentiation as in mitosis. The mechanism whereby the genetic integrity is maintained during amitosis is uncertain.

❐ amniocentesis

A procedure for obtaining foetal cells for prenatal diagnosis by sampling the amniotic fluid from a pregnant mammal. Cells are cultured, and the karyotype is checked for known irregularities (e.g. Down's syndrome and spina bifida in humans).

❐ amnion

The thin membrane that lines the fluid-filled sac in which the embryo develops in higher vertebrates, reptiles and birds.

❐ amniotic fluid

Liquid contents of the amniotic sac of higher vertebrates, containing foetal, but not maternal cells.

❐ amorph

A mutation that abolishes gene function.

❐ amp

Abbreviation for adenosine monophosphate.

❐ amphidiploid

A plant derived from doubling the chromosome number of an interspecific F_1 hybrid. Naturally found hybrids of this sort are referred to as allopolyploid.

❐ amphimixis

True sexual reproduction involving the fusion of male and female gametes and the formation of a zygote.

❐ ampicillin

A penicillin-type antibiotic that prevents bacterial growth by interfering with synthesis of the cell wall. Commonly used as a selectable marker in the creation of transgenic plants.

❐ amplicon

The product of a DNA amplification reaction.

❐ amplification

1. Creation of many copies of a segment of DNA by the polymerase chain reaction.
2. Treatment (e.g. use of chloramphenicol) designed to increase the proportion of plasmid DNA relative to that of bacterial (host) DNA.
3. Evolutionary expansion in copy number of a repetitive DNA sequence through a process of repeated duplication.

❐ amplified fragment length polymorphism

A type of DNA marker, generated by the PCR amplification of restriction endonuclease treated DNA. A small proportion of all restriction fragments is amplified in any one reaction, so that AFLP profiles can be analysed by gel electrophoresis. This has the important characteristic that many markers can be generated with relatively little effort.

❐ amplify

To increase the number of copies of a DNA sequence, either *in vivo* by inserting into a cloning vector that replicates within a host cell, or *in vitro* by polymerase chain reaction.

❐ amylase

Describing a wide class of enzymes that catalyse the hydrolysis of starch.

❐ amylolytic

The capability of enzymatically degrading starch into sugars.

❐ amylopectin

A polysaccharide comprising highly branched chains of glucose residues. The water-insoluble portion of starch.

❐ amylose

A polysaccharide consisting of linear chains of 100-1000 glucose residues. The water-soluble portion of starch.

❐ an antibody molecule

(A) Schematic drawing of a typical antibody molecule. As indicated on next page, this protein is Y-shaped and has two identical binding sites for its antigen, one on either arm of the 3Y.2 The protein is composed of four polypeptide chains (two identical heavy chains and two identical and smaller light chains) held together by disulfide bonds. Each chain is made up of several different domains. The antigen-binding site is formed where a heavy chain variable domain (VH) and a light chain variable domain (VL) come close together. These are the domains that differ most in their sequence and structure in different antibodies. (B) Ribbon drawing of a light chain showing the parts of the VL domain most closely involved in binding to the antigen these contribute half of the fingerlike loops that fold around each of the antigen molecules in (A).

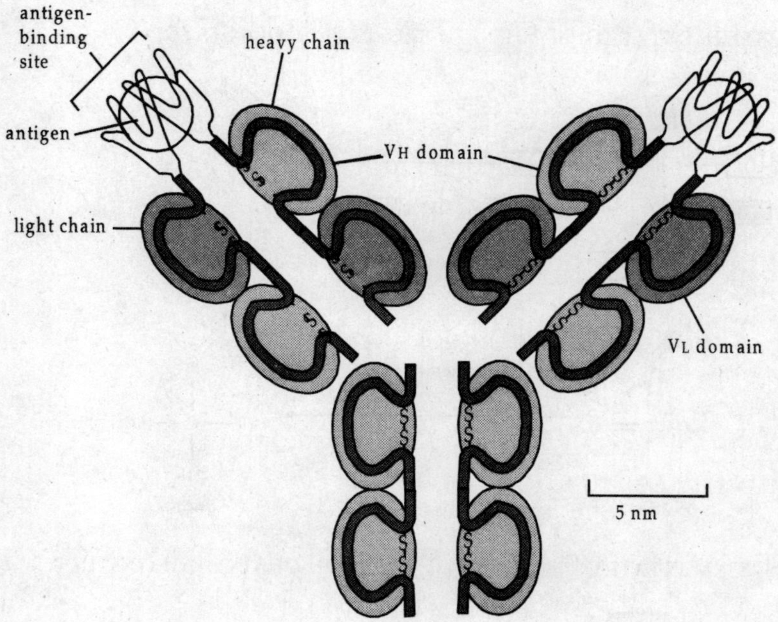

(A)

❒ anabolic pathway

A pathway by which a metabolite is synthesized; a biosynthetic pathway.

❒ anabolism

One of the two subcategories of metabolism, referring to the building up of complex organic molecules from simpler precursors.

❒ anaerobe

An organism that can grow in the absence of oxygen.

❒ anaerobic

An environment or condition in which molecular oxygen is not available for chemical, physical or biological processes.

❒ anaerobic breakdown of pyruvate

When inadequate oxygen is present, for example, in a muscle cell undergoing vigorous contraction, the pyruvate produced by glycolysis is converted to lactate as shown. This reaction restores the NAD^+ consumed in step 6 of glycolysis, but the whole pathway yields much less energy overall than complete oxidation. (B) In some organisms that can grow anaerobically, such as yeast, pyruvate is converted via acetaldehyde into carbon dioxide and ethanol. Again, this pathway regenerates NAD^+ from NADH, as required to enable glycolysis to continue. Both (A) and (B) are examples of fermentation.

(A) FERMENTATION LEADING TO EXCRETION OF LACTATE

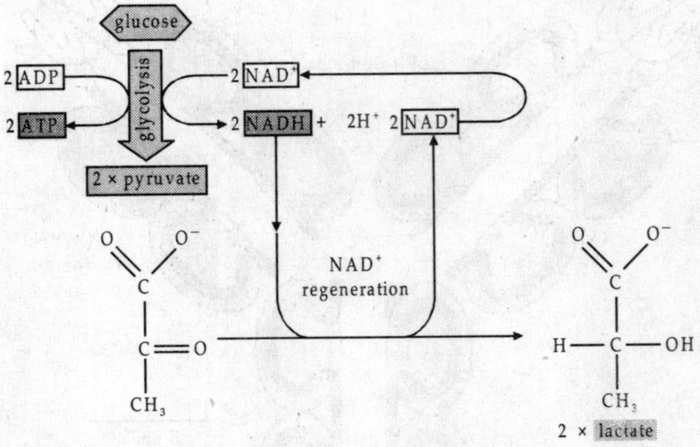

(B) FERMENTATION LEADING TO EXCRETION OF ALCOHOL AND CO_2

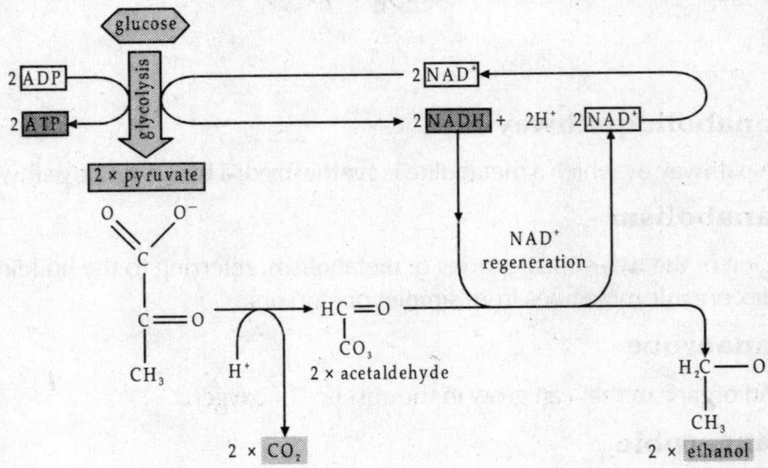

▫ anaerobic digestion

Digestion of materials in the absence of oxygen.

▫ anaerobic respiration

Respiration in which foodstuffs are partially oxidized, with the release of chemical energy, in a process not involving atmospheric oxygen. A notable example is in alcoholic fermentation, where sugar is metabolized into ethanol.

▫ analogous

Features of organisms or molecules that are superficially or functionally similar but have evolved in a different way or contain different compounds.

❐ anaphase

The stage of mitosis or meiosis during which the daughter chromosomes migrate to opposite poles of the cell (toward the ends of spindle). Anaphase follows metaphase and precedes telophase.

❐ anchor gene

A gene that has been positioned on both the physical map and the linkage map of a chromosome, and thereby allows their mutual alignment.

❐ androgen

Any hormone that stimulates the development of male secondary sexual characteristics, and contributes to the control of sexual activity in vertebrate animals. Usually synthesized in the testis.

❐ androgenesis

Male parthenogenesis, i.e. the development of a haploid embryo from a male nucleus. The maternal nucleus is eliminated or inactivated subsequent to fertilization of the ovum, and the haploid individual (referred to as androgenetic) contains the genome of the male gamete only in its cells.

❐ aneuploid

An organism or cell having a chromosome number other than the normal somatic number. Aneuploid gametes have a chromosome number other than the normal haploid number. The condition is *aneuploidy*.

❐ angiogenesis

The formation and development of new blood vessels in the body, stimulated by growth factors, such as angiogenin. The process is required for the spread of malignant tumours.

❐ angiogenin

One of the human angiogenic growth factors. In addition to stimulating (normal) blood vessel formation, angiogenin levels are correlated with placenta formation and tumour growth.

❐ angiosperm

A division of the plant kingdom that includes all flowering plants, i.e. vascular plants in which double fertilization occurs resulting in development of fruit containing seeds. Divided into two major groups, monocotyledons and dicotyledons.

❐ animal cell immobilization

Entrapment of animal cells in some solid material in order to produce some natural product or genetically engineered protein. Animal cells have the advantage that they already produce many proteins of pharmacological interest, and that genetically engineered proteins are produced by them with the post-translation modifications normal to animals. However, because animal cells are much more fragile than bacterial ones, they cannot tolerate a commercial fermentation process.

❐ anneal

The pairing of complementary DNA or RNA sequences, via hydrogen bonding, to form a double-stranded polynucleotide.

❐ annual 1

Taking one year, or occurring at intervals of one year. 2. A plant that completes its life cycle within one year.

❐ anonymous DNA marker

A DNA marker detectable by virtue of variation in its sequence. The function (if any) of the sequence is unknown. Microsatellites and AFLPs are typical anonymous DNA markers.

❐ antagonism

An interaction between two organisms (e.g. moulds or bacteria) in which the growth of one is inhibited by the other.

❐ antagonist

A compound that inhibits the effect of an agonist in such a way that the combined biological effect of the two becomes smaller than the sum of their individual effects.

❐ anther

The upper part of a stamen, containing pollen sacs within which the pollen develops and matures.

❐ anther culture

The aseptic culture of immature anthers to generate haploid plants from microspores via androgenesis.

❐ anthesis

The period during which anthers bear mature and functional pollen.

❐ anthocyanin

A water-soluble blue, purple or red flavonoid pigments found in vacuoles of cells of certain plants.

❐ antiauxin

A chemical that interferes with the auxin response, sometimes by the prevention of auxin transport. Some antiauxins may promote morphogenesis *in vitro* (e.g. 2,3,5-tri-iodobenzoate (TIBA) and 2,4,5-trichlorophenoxyacetate (2,4,5-T)) and are therefore used to stimulate the growth of some cultures.

❑ antibiosis

The prevention of growth or development of an organism by a substance or another organism.

❑ antibiotic

A class of natural and synthetic compounds that inhibit the growth of, or kill some micro-organisms. Antibiotics are widely used medicinally to control bacterial pathogens, but resistance in bacteria to particular antibiotics is often rapidly acquired through mutation.

❑ antibiotic resistance marker gene

Genes (usually of bacterial origin) used as selection markers in transgenesis, because their presence allows cell survival in the presence of normally toxic antibiotic agents. These genes were commonly used in the development and release of first generation transgenic organisms (particularly crop plants), but are no longer favoured because of perceived risks associated with the unintentional transfer of antibiotic resistance to other organisms.

❑ antibiotic resistance

The ability of a micro-organism to disable an antibiotic or prevent its transport into the cell.

❑ Antibodies

Proteins produced by certain white blood cells in response to foreign substances (antigens). Each antibody can bind only to a specific antigen. The purpose of this binding is to help destroy the antigen. Antibodies can work in several ways, depending on the nature of the antigen. Some antibodies disable antigens directly. Others make the antigen more vulnerable to destruction by white blood cells.

THE ANTIBODY MOLECULE

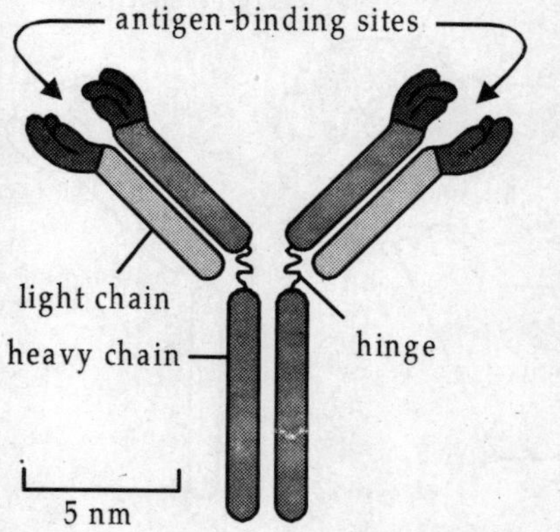

Antibodies are proteins that bind very tightly to their targets (antigens). They are produced in vertebrates as a defense against infection. Each antibody molecule is made of two identical light chains and two identical heavy chains, so the two antigen-binding sites are identical,

B CELLS

Antibodies are made by a class of white blood cells, called B lymphocytes, or B cells. Each resting B cell carries a different membrane-bound antibody molecule on its surface that serves as receptor for recognizing a specific antigen. When antigen binds to this receptor, the B cell is stimulated to divide and to secrete large amounts of the same antibody in a soluble form.

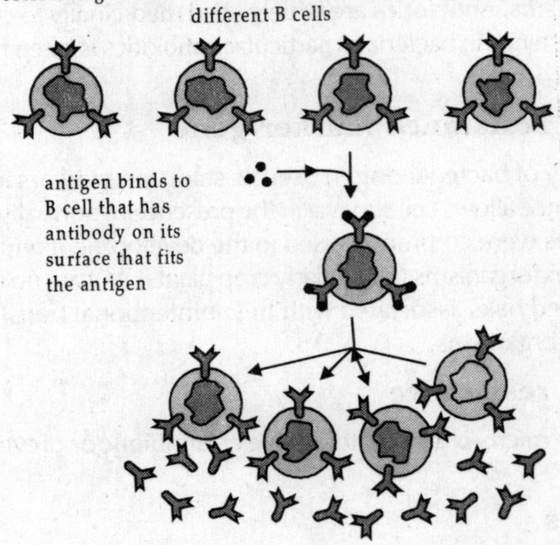

B cell is stimulated to make more of same antibody and secrete it

ANTIBODY SPECIFICITY

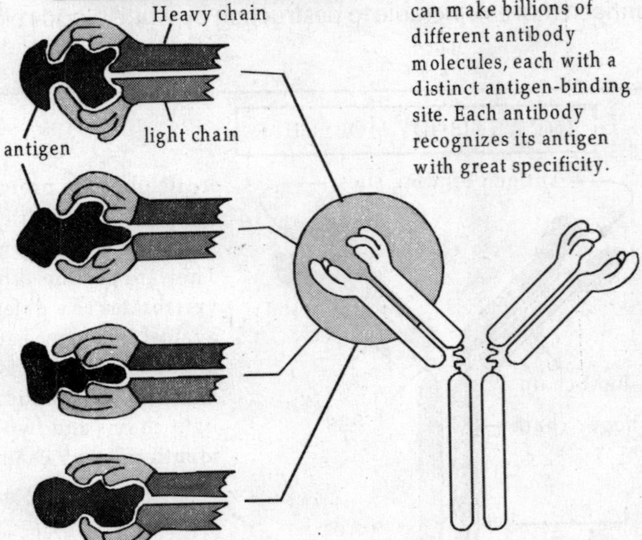

An individual animal can make billions of different antibody molecules, each with a distinct antigen-binding site. Each antibody recognizes its antigen with great specificity.

RAISING ANTIBODIES IN ANIMALS

Antibodies can be made in the laboratory by injecting an animal (usually a mouse, rabbit, sheep, or goat) with antigen A.

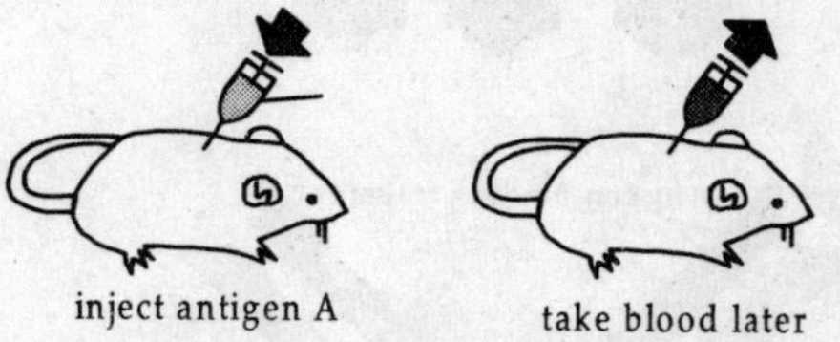

Repeated injections of the same antigen at intervals of several weeks stimulates specific B cells to secrete large amounts of anti-A antibodies into the bloodstream.

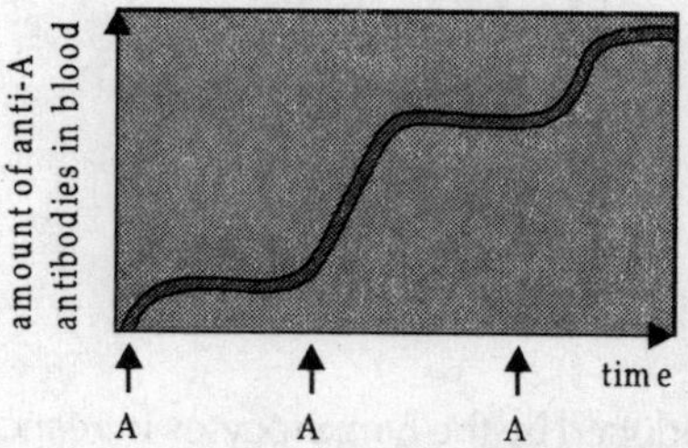

Because many different B cells are stimulated by antigen A, the blood will contain a variety of anti-A antibodies, each of which binds A in a slightly different way.

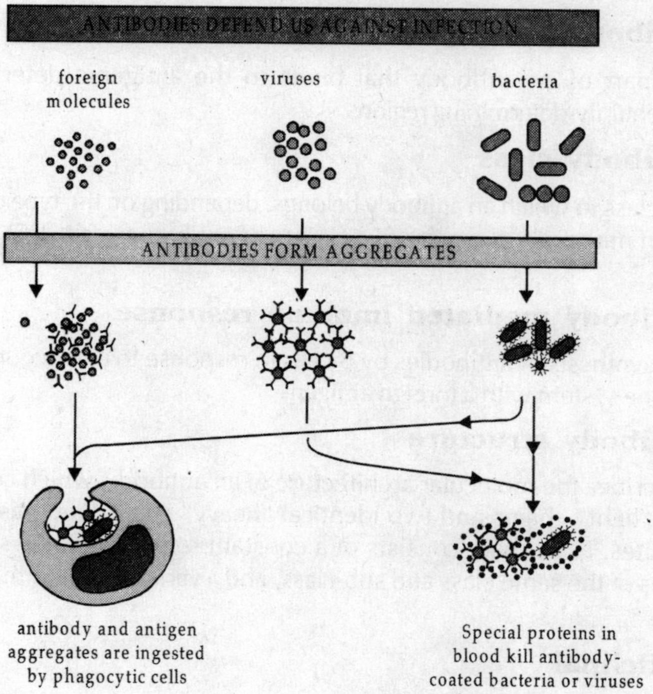

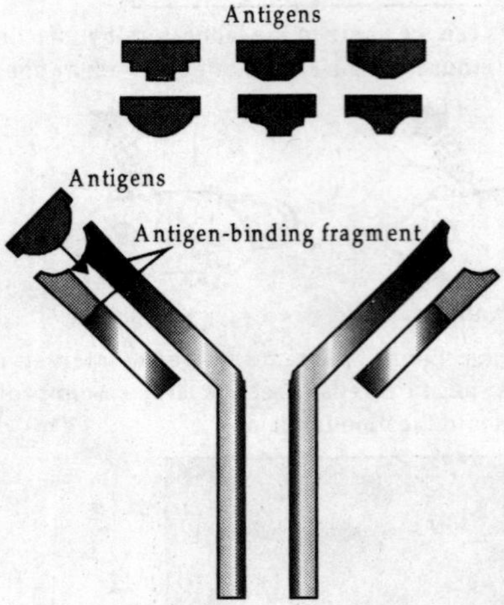

❒ antibody

An immunological protein produced by the lymphocytes in response to contact with an antigen. Each antibody recognizes just one antigenic determinant of one antigen and acts by specifically binding to it, thus rendering it harmless. Those from the IgG antibody class are found in the bloodstream and used in immunoassay. Synonym: immunoglobulin.

❒ antibody binding site

The part of an antibody that binds to the antigenic determinant. See: complementarity-determining regions.

❒ antibody class

The class to which an antibody belongs, depending on the type of heavy chain present. In mammals, there are five classes of antibodies: IgA, IgD, IgE, IgG, and IgM.

❒ antibody mediated immune response

The synthesis of antibodies by B cells in response to an encounter of cells of the immune system with a foreign antigen.

❒ antibody structure

Describes the molecular architecture of an antibody, which consists of two identical "light" chains and two identical "heavy" chains and has two antigen-binding sites. Each chain consists of a constant region which is same between antibodies of the same class and sub-class, and a variable region that is antibody-specific.

❒ anticlinal

The orientation of cell wall or plane of cell division perpendicular to the surface.

❐ anticoding strand

The DNA strand used as template for transcription. The resulting mRNA is complementary in sequence to that of the anticoding strand. Synonym: template strand.

❐ anticodon

A triplet of tRNA nucleotides that corresponds to a complementary codon in an mRNA molecule during translation.

❐ antigen

A macromolecule (usually a protein foreign to the organism), which elicits an immune response on first exposure to the immune system by stimulating the production of antibodies specific to its various antigenic determinants. During subsequent exposures, the antigen is bound and inactivated by these antibodies.

❐ antigenic determinant

The individual surface feature of an antigen, that elicits the production of a specific antibody in the course of immune response. Each antigenic determinant, typically a few amino acids in size, causes the synthesis of a different antibody and thus exposure to a single antigen may result in the expression of a number of antibodies.

❐ antigenic switching

The altering of a micro-organism's surface antigens through genetic re-arrangement, to elude detection by the host's immune system.

❐ antihaemophilic globulin

One of the blood clotting factors, a soluble protein that causes the fibrin matrix of a blood clot to form. Used as a treatment for haemophilia, AHG is usually obtained from genetically engineered cell cultures.

❐ anti-idiotype antibody

An antibody, produced by an organism, which specifically binds to the binding site of an antibody developed by that organism against a foreign antigen. Involved with the regulation of the immune response. Some allergic responses are in part due to the breakdown of this sort of regulation.

❐ antimicrobial agent

Any chemical or biological agent that inhibits the growth and/or survival of micro-organisms.

❐ antinutrient

Compounds that inhibit the normal uptake or utilization of nutrients.

❐ anti-oncogene

A gene whose product prevents the normal growth of tissue.

▫ antioxidant

Compounds that slow the rate of oxidation reactions.

▫ antiparallel orientation

The normal arrangement of the two strands of a double-stranded DNA molecule, and of other nucleic-acid duplexes (DNA-RNA, RNA-RNA), in which the two strands are oriented in opposite directions so that the 5'-phosphate end of one strand is aligned with the 3'-hydroxyl end of the complementary strand.

▫ antisense DNA

One of the two strands of double-stranded DNA, usually that which is complementary (hence "anti") to the mRNA, i.e. the non-transcribed strand. However, there is not universal agreement on this convention, and the preferred designations are coding strand for the strand whose sequence matches that of the mRNA, and non-coding strand or template strand for the complementary strand (i.e. the transcription template).

▫ antisense gene

A gene that produces an mRNA complementary to the transcript of a normal gene (usually constructed by inverting the coding region relative to the promoter).

▫ antisense RNA

An RNA sequence that is complementary to all or part of a functional mRNA molecule, to which it binds, blocking its translation.

▫ antisense therapy

The *in vivo* treatment of a genetic disease by blocking translation of a protein with a DNA or an RNA sequence that is complementary to a specific mRNA.

▫ antiseptic

Any substance that kills or inhibits the growth of disease-causing micro-organism (a micro-organism capable of causing sepsis), but is essentially non-toxic to cells of the body.

▫ antiserum

The fluid portion of the blood of an immunized animal (after coagulation of the blood), which retains any antibodies.

▫ anti-terminator

A protein which enables RNA polymerase to ignore certain transcriptional stop or termination signals and thereby produce longer than normal transcripts.

▫ antitranspirant

A compound designed to reduce plant transpiration. Applied to the leaves of newly transplanted trees, shrubs etc., or cuttings in lieu of misting. Can interfere with photosynthesis and respiration if the coating is too thick or is unbroken.

❑ antixenosis

The modification of the behaviour of an organism by a substance or another organism. Particularly used in the context of a plant's apparent resistance against insect feeding, when the insects are presented with a choice of plant genotypes.

❑ apex

The portion of a root or shoot containing the primary or apical meristem.

❑ apical cell

A meristematic initial in the apical meristem of shoots or roots of plants.

❑ apical dominance

The phenomenon where growth of lateral (axillary) buds in a plant is inhibited by the presence of the terminal (apical) bud on the branch. Explained by the export of auxins from the apical bud.

❑ apical meristem

A region of the tip of each shoot and root of a plant in which cell division is continually occurring to produce new stem and root tissue, respectively. Two regions are visible in the apical meristem: An outer 1-4-cell layered region (the *tunica*), where cell divisions are anticlinal; and below the tunica, (ii) the *corpus*, where the cells divide in all directions, and increase in volume.

❑ apoenzyme

Inactive enzyme that has to be associated with a co-enzyme in order to function. The apoenzyme/co-enzyme complex is called a holoenzyme.

❑ apomixis

The production of an embryo in the absence of meiosis. Apomictic higher plants produce asexual seeds, derived only from maternal tissue.

❑ apoptosis

The process of programmed cell death, which occurs naturally as a part of normal development, maintenance and renewal of tissue. Differs from necrosis, in which cell death is caused by external factors (stress or toxin).

❑ aptamer

A polynucleotide molecule that binds to a specific molecule, often a protein.

❑ aquaculture

Farming of aquatic organisms, including fish, molluscs, crustaceans and aquatic plants.

❑ arabidopsis

A genus of flowering plants in the *Cruciferae*. *A. thaliana* is used in research as a model plant because it has a small fully sequenced genome, can be cultured and transformed easily, and has a rapid generation time.

❑ arbitrarily primed polymerase chain reaction

An application of the polymerase chain reaction to generate DNA fingerprints. The technique uses arbitrary primers to amplify anonymous stretches of DNA.

❑ arbitrary primer

An oligonucleotide primer whose sequence is chosen at random, rather than one whose sequence matches that of a known locus. These primers therefore amplify DNA fragments which have not been pre-selected.

❑ Archaea

Single-celled life forms adapted to existence in high pressure, anaerobic, environments such as at extreme ocean depths. These organisms are seen as a promising source of enzymes robust enough for a number of demanding industrial processes.

❑ armg

Abbreviation for antibiotic resistance marker gene.

❑ ars

Abbreviation for autonomous(ly) replicating segment (or sequence).

❑ artificial inembryonation

Non-surgical transfer of embryo(s) to a recipient female. As in vitro embryo technology develops, artificial inembryonation may replace artificial insemination.

❑ artificial insemination

The deposition of semen, using a syringe, at the mouth of the uterus to make conception possible.

❑ artificial seed

Encapsulated or coated somatic embryos that are planted and treated like seed.

❑ artificial selection

The practice of choosing individuals from a population for reproduction, usually because these individuals possess one or more desirable traits.

❑ asa

Abbreviation for allele-specific amplification.

❑ ascites

Abnormal accumulation of fluid in the peritoneal cavity, occurring naturally as a complication of cirrhosis of the liver, among other conditions. In the context of monoclonal antibody production, hybridoma cells are injected into mice to induce their proliferation in the resulting ascites. This method has been largely superseded by *in vitro* culture of hybridomas.

❒ ascospore

One of the spores contained in the ascus of certain fungi.

❒ ascus

Reproductive sac in the sexual stage of a type of fungi (*Ascomycetes*) in which ascospores are produced.

❒ aseptic

Sterile, free of contaminating organisms (bacteria, fungi, algae but not generally including viruses, and particularly not internal symbionts).

❒ asexual

Reproduction not involving meiosis or the union of gametes.

❒ asexual propagation

Vegetative, somatic, non-sexual reproduction of a plant without fertilization.

❒ asexual reproduction

Reproduction that does not involve the formation and union of gametes from the different sexes or mating types. It occurs mainly in lower animals, micro-organisms and plants. In plants, asexual reproduction is by vegetative propagation (e.g. bulbs, tubers, corms) and by formation of spores.

❒ a-site

Abbreviation for aminoacyl site.

❒ assay

1. To test or evaluate.
2. The procedure for measuring the quantity of a given substance in a sample (chemically or by other means).

❒ assortative mating

Mating in which the partners are chosen on the basis of phenotypic similarity.

❒ asymmetric hybrid

A hybrid formed, usually via protoplast fusion, between two donors, where the chromosome complement of one of the donors is incomplete. This chromosome loss can be induced by irradiation or chemical treatment, or can occur naturally.

❒ asynapsis

The failure or partial failure in the pairing of homologous chromosomes during the first meiotic prophase.

❒ atp

Abbreviation for adenosine triphosphate.

❒ atpase

An enzyme that brings about the hydrolysis of adenosine triphosphate, by the cleavage of either one phosphate groups with the formation of ADP and inorganic phosphate, or of two phosphate groups, with the formation of AMP and pyrophosphate.

❒ attenuated vaccine

A virulent organism that has been modified to produce a less virulent form, but nevertheless retains the ability to elicit antibodies against the virulent form.

❒ attenuation

A mechanism for controlling gene expression in prokaryotes that involves premature termination of transcription.

❒ attenuator

A nucleotide sequence in the 5' region of a prokaryotic gene (or in its RNA) that causes premature termination of transcription, possibly by forming a secondary structure.

❒ aureofacin

An antifungal antibiotic produced by a strain of *Streptomyces aureofaciens*. A possible candidate for the transgenic control of plant fungal disease.

❒ authentic protein

A recombinant protein that has all the properties - including any post-translational modifications - of its naturally occurring counterpart.

❒ autocatalysis

Catalysis in which one of the products of the reaction is a catalyst for the reaction.

❒ autoclave

An enclosed chamber in which materials can be heated under pressure to sterilize utensils, liquids, glassware, etc., using steam.

❒ autogenous control

The action of a gene product to inhibit (negative autogenous control) or enhance (positive autogenous control) the expression of the gene that codes for it.

❒ auto-immune disease

Disorder in which the immune system of affected individuals produce antibodies against molecules that are normally produced by those individuals (called self antigens).

❒ auto-immunity

A disorder in the body's defence mechanism in which an immune response is elicited against its own (self) tissues.

❐ autologous cells

Cells taken from an individual, cultured (or stored), and, possibly, genetically manipulated before being transferred back into the original donor.

❐ autolysis

The process of self destruction of a cell, cell organelle, or tissue, through the action of lysosomal enzymes.

❐ autonomous

A term applied to any biological unit that can function on its own, i.e. without the help of another unit, such as a transposable element that encodes an enzyme for its own transposition.

❐ autonomous(ly) replicating segment (or sequence)

Any eukaryotic DNA sequence that initiates and supports chromosomal replication; they have been isolated in yeast cells.

❐ autopolyploid

A polyploid whose constituent genomes are derived from the same or nearly the same progenitor. In an autotetraploid, each chromosome is present in four copies, so meiotic configurations may include many (or exclusively) quadrivalents (four paired chromosomes), and the inheritance of alleles will be quadruplex. Quadrivalents do not always segregate normaly at meiosis, resulting in lowered fertility, so some established autotetraploid species that reproduce sexually have restricted quadrivalent formation.

❐ autoradiograph

A technique for visualizing the presence, location and intensity of radioactivity in histological preparations, paper chromatograms or electrophoretic gel separations, obtained by overlaying the surface with X-ray film and allowing the radiation to form an image on the film.

❐ autosome

A chromosome not involved in sex determination. The diploid human genome consists of 46 chromosomes, 22 pairs of autosomes, and one pair of sex chromosomes (the X and Y chromosomes).

The next chromosome identifies the sex of the infant which is explained in the figure on the next page.It has 22 number of chromosomes as an autosome and the last 23rd identifies the sex of an infant.

1. 22 chromosomes are autosomes
2. 1 chromosome sex chromosome

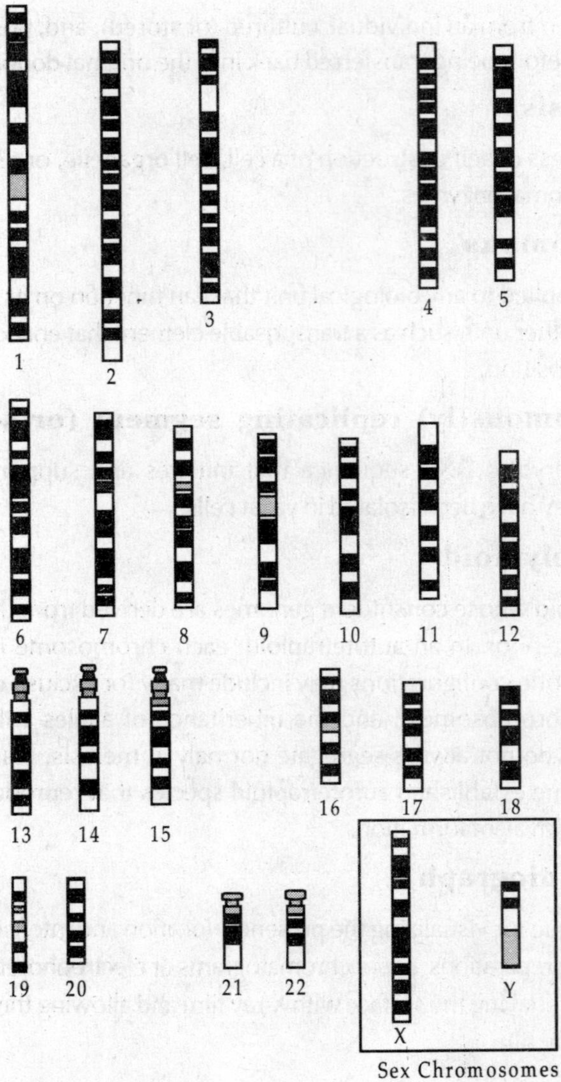

□ autotroph

Organism capable of self-nourishment utilizing carbon dioxide or carbonates as the sole source of carbon and obtaining energy from radiant energy or from the oxidation of inorganic elements, or compounds such as iron, sulphur, hydrogen, ammonium and nitrites.

□ auxin

A group of plant growth regulators (natural or synthetic) which stimulate cell division, enlargement, apical dominance, root initiation, and flowering.

❒ auxin-cytokinin ratio

The relative proportion of auxin to cytokinin present in plant tissue culture media. Varying the relative amounts of these two hormones affects the proportional growth of shoots and roots.

❒ auxotroph

A mutant cell or micro-organism lacking one metabolic pathway present in the parental strain, and that consequently will not multiply on a minimal medium, but requires for growth the addition of a specific compound, such as an amino acid or a vitamin.

❒ availability

A reflection of the form and location of nutritional elements and their suitability for absorption.

❒ avidin

A glycoprotein present in egg white, which has a strong affinity to biotin. Can lead to biotin deficiency if given in large quantities. Used as a biological reagent in the same way as streptavidin.

❒ avidity

A measure of the binding strength of an antibody to its antigen.

❒ avirulence gene

Many plants contain *R* genes, which confer simply-inherited resistance to a specific pathogen race. The plants are able to recognize the presence of the pathogen by an interaction between their *R* gene and the matching pathogen's avirulence gene. Successful recognition triggers a cascade of further genes, often leading to a hypersensitive response.

❒ axenic culture

Free of external contaminants and internal symbionts; generally not possible with surface sterilization alone, sometimes used incorrectly to indicate aseptic culture.

❒ axillary bud

A bud found at the axil of a leaf.

❒ axillary bud proliferation

Propagation of plant tissue *in vitro* to promote axillary growth, to generate large numbers of plantlets in culture.

B

□ b cell

An important class of lymphocytes that mature in bone marrow (in mammals) and the *Bursa of Fabricius* (in birds) and produce antibodies. Largely responsible for the antibody-mediated or humoral immune response, giving rise to the antibody-producing plasma cells and some other cells of the immune system. *Synonym*: B lymphocyte.

□ b chromosome

A supernumerary chromosome present in some individuals (both plant and animal). They are smaller than the normal chromosomes, behave abnormally in both mitosis and meiosis, can vary in number between somatic cells and are not thought to have any significant gene content.

□ babs

Abbreviation for biosynthetic antibody binding sites.

□ bac

Abbreviation for bacterial artificial chromosome.

□ bacillus

A rod-shaped bacterium.

□ bacillus thuringiensis

A bacterium that produces a toxin against certain insects, particularly Coloeoptera and Lepidoptera; a major means of insecticide for organic farming. Some of the toxin genes are important for transgenic approaches to crop protection.

□ back mutation

A second mutation at the same site in a gene as the original mutation. The second mutation restores the wild-type protein sequence.

□ backcross

Crossing an individual with one of its parents or with the genetically equivalent organism. The offspring of such a cross are referred to as the backcross generation or backcross progeny.

❑ **bacterial artificial chromosome**

A plasmid vector that can be used to clone large inserts of DNA (up to 500 kb).

❑ **bacterial toxin**

A toxin produced by a bacterium, such as Bt toxin of Bacillus thuringiensis.

❑ **bacteriocide**

A chemical or drug that kills bacterial cells.

❑ **bacteriocin**

A protein produced by bacteria of one strain and active against those of a closely related strain.

❑ **bacteriophage**

(Abbreviation: phage). A virus that infects bacteria. Altered forms are used as cloning vectors.

❑ **bacteriostat**

A substance that inhibits or slows down growth and reproduction of bacteria.

❑ **bacterium**

Unicellular prokaryotic organisms, without a distinct nucleus. Major distinctive groups are defined by Gram staining. Also classified on the basis of oxygen requirement (aerobic vs anaerobic) and shape (spherical = coccus; rodlike = bacillus; spiral = spirillum; comma-shaped = vibrio; corkscrew-shaped = spirochaete; filamentous).

❑ **baculovirus**

A class of insect virus used to make DNA cloning vectors for gene expression in eukaryotic cells. Production of a target protein can be up to 50% of the cells' protein content, and several proteins can be made simultaneously, so that multi-sub-unit enzymes can be made by this system.

❑ **baculovirus expression vector**

A method for the *in vitro* production of complex recombinant eukaryotic proteins. A genetically engineered baculovirus (a virus that infects certain types of insects) is introduced into appropriate cultured insect cells, which then expresses the recombinant protein.

❑ **balanced lethal system**

A system for maintaining a recessive lethal allele at each of two loci on the same pair of chromosomes. In a closed population with no crossing-over between the loci, only the double heterozygotes for the lethal mutations survive.

❑ **balanced polymorphism**

Two or more phenotypes maintained in the same breeding population.

□ barnase

A bacterial ribonuclease, which, when transformed into plants and expressed in the anthers, generates a male sterile phenotype. Thus it is a technology applicable to F_1 hybrid seed production, which relies on the ability to genetically sterilize genotypes to ensure that all seed borne on the plant are the result of out crossing. The sterility phenotype is suppressed by the barstar protein, which can therefore be used to reverse the sterility where this is necessary.

□ barr body

A condensed mass of chromatin found in the nuclei of female mammals. It is a late-replicating, inactive X-chromosome.

□ barstar protein

A polypeptide inhibitor of barnase.

□ basal

1. Located at the base of a plant or a plant organ.
2. A fundamental formulation of a tissue culture medium containing nutrients but no growth promoting agents.

□ base

One of the components of nucleosides, nucleotides and nucleic acids. Four different bases are found in naturally occurring DNA - the purines A (adenine) and G (guanine); and the pyrimidines C (cytosine) and T (thymine, the common name for 5-methyluracil). In RNA, T is replaced by U (uracil).

□ base analogue

A non-natural purine or pyrimidine base that differs slightly in structure from the normal bases, but can be incorporated into nucleic acids. They are often mutagenic.

□ base pair

Two bases which form a "rung of the DNA ladder." A DNA nucleotide is made of a molecule of sugar, a molecule of phosphoric acid, and a molecule called a base. The bases are the "letters" that spell out the genetic code. In DNA, the code letters are A, T, G, and C, which stand for the chemicals adenine, thymine, guanine, and cytosine, respectively. In base pairing, adenine always pairs with thymine, and guanine always pairs with cytosine.

Base Pair

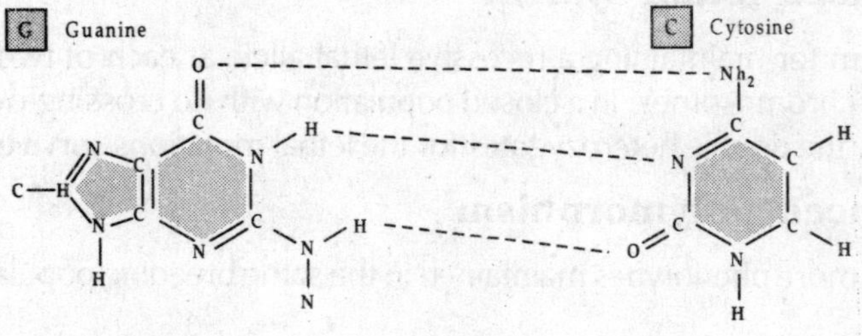

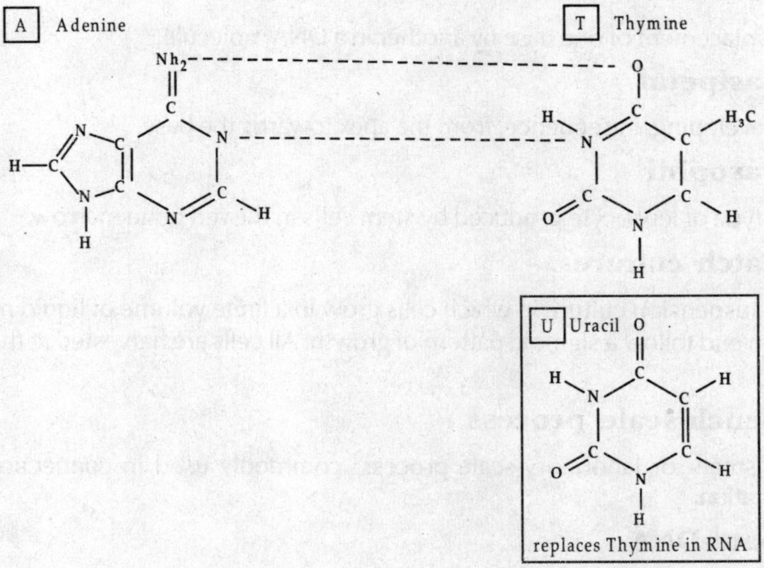

Deoxyribonucleic Acid (DNA)

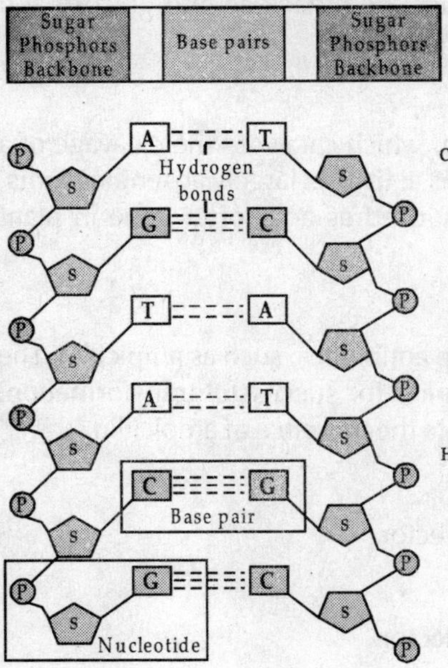

Nitrogenous Bases

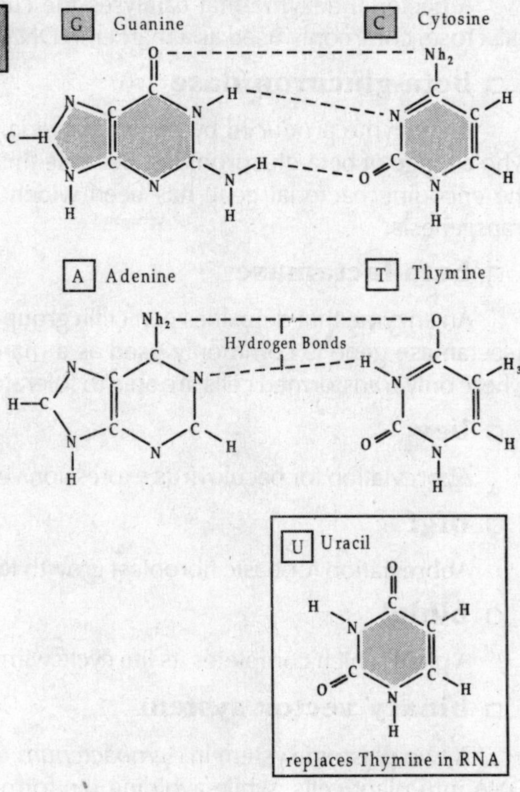

❐ base substitution

Replacement of one base by another in a DNA molecule.

❐ basipetal

Developing, in sequence, from the apex towards the base.

❐ basophil

A type of leukocyte produced by stem cells in the red bone marrow.

❐ batch culture

A suspension culture in which cells grow in a finite volume of liquid nutrient medium and follow a sigmoid pattern of growth. All cells are harvested at the same time.

❐ bench-scale process

A small- or laboratory-scale process; commonly used in connection with fermentation.

❐ beta-DNA

The form of DNA generally found in nature. A right-handed helix.

❐ beta-galactosidase

A bacterial enzyme that catalyses the cleavage of lactose into glucose and galactose, commonly used as a marker in DNA cloning.

❐ beta-glucuronidase

An enzyme produced by certain bacteria, which catalyses the cleavage of a whole range of beta-glucuronides. Because this activity is largely absent in plants, the encoding bacterial gene has been widely used as a reporter gene in plant transgenesis.

❐ beta-lactamase

An enzyme that detoxifies penicillin group antibiotics, such as ampicillin. The a-lactamase gene is commonly used as a marker for successful transformation, where only transformed cells are able to tolerate the presence of ampicillin.

❐ bev

Abbreviation for baculovirus expression vector.

❐ bfgf

Abbreviation for basic fibroblast growth factor.

❐ binial

A plant which completes its life cycle within two years and then dies.

❐ binary vector system

A two plasmid system in *Agrobacterium tumefaciens* designed to transfer T-DNA into plant cells, while avoiding the formation of crown gall tumours. One plasmid contains the virulence gene (responsible for transfer of the T-DNA), and the other the T-DNA borders, the selectable marker and the DNA to be transferred.

❐ binding

The ability of molecules to bind each other non-covalently because of the exact shape and chemical nature of parts of their surfaces. A common biological phenomenon, e.g. an enzyme to its substrate; an antibody to its antigen; a DNA strand to its complementary strand.

❐ binding Site of a Protein

(A) The folding of the polypeptide chain typically creates a crevice or cavity on the protein surface. This crevice contains a set of amino acid side chains disposed in such a way that they can make noncovalent bonds only with certain ligands. (B) Close-up view of an actual binding site showing the hydrogen bonds and ionic interactions formed between a protein and its ligand (in this example, cyclic AMP is the bound ligand).

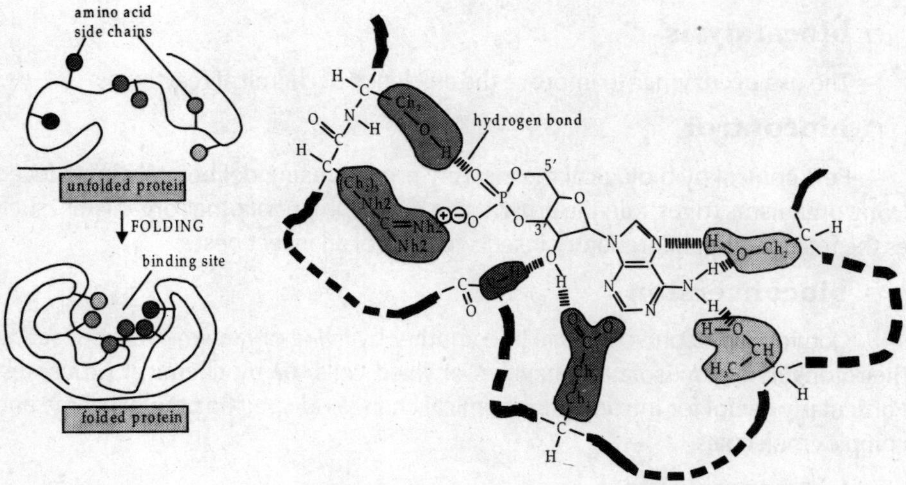

❐ bio-

A prefix used in scientific words to associate the concept of "living organisms." Usually written with a hyphen before vowels, for emphasis or in neologisms.

❐ bio-accumulation

A problem that can arise when a stable chemical such as a heavy metal or DDT is introduced into a natural environment. Where there are no agents present able to biodegrade it, its concentration can increase as it passes up the food chain and higher organisms may suffer toxic effects. This phenomenon may be employed beneficially for the removal of toxic metals from wastewater, and for bioremediation.

❐ bio-assay

1. The assessment of a substance's activity on living cells or on organisms. Animals have been used extensively in drug research in bio-assays in the

pharmaceutical and cosmetics industries. Current trends are to develop bio-assays using bacteria or animal or plant cells, as these are easier to handle than whole animals or plants, are cheaper to make and keep, and avoid the ethical problems associated with testing of animals.

2. An indirect method to detect sub-measurable amounts of a specific substance by observing a sample's influence on the growth of live material.

❐ bio-augmentation

Increasing the activity of bacteria that decompose pollutants; a technique used in bioremediation.

❐ bioavailability

The proportion of a nutrient or administered drug etc. that can be taken up by an organism in a biologically effective form. For example, some soils high in phosphorus have a low level of P availability because the pH of the soil renders much of the P insoluble.

❐ biocatalysis

The use of enzymes to improve the efficiency of chemical reactions.

❐ biocontrol

Pest control by biological means. Any process using deliberately introduced living organisms to restrain the growth and development of other organisms, such as the introduction of predatory insects to control an insect pest.

❐ bioconversion

Conversion of one chemical into another by living organisms, as opposed to their conversion by isolated enzymes or fixed cells, or by chemical processes. Particularly useful for introducing chemical changes at specific points in large and complex molecules.

❐ biodegradable

Capable of being biodegraded.

❐ biodegrade

The breakdown by micro-organisms of a compound to simpler chemicals. Materials that are easily biodegraded are colloquially termed biodegradable.

❐ biodesulphurization

The removal of organic and inorganic sulphur from coal by bacterial and soil micro-organisms. Certain bacteria can oxidize insoluble sulphur compounds into soluble sulphates, which can be washed away with the bacteria.

❐ biodiversity

The variability among living organisms from all sources, including, inter alia, terrestrial, marine and other ecosystems and the ecological complexes of which they are part; this includes diversity within species, between species and of ecosystems.

❐ bio-energetics

The study of the flow and the transformation of energy that occur in living organisms.

❐ bio-engineering

The use of artificial tissues, organs and organ components to replace parts of the body that are damaged, lost or malfunctioning.

❐ bio-enrichment

Adding nutrients or oxygen to increase microbial breakdown of pollutants.

❐ bio-ethics

The branch of ethics that deals with the life sciences and their potential impact on society.

❐ biofilms

A layer of micro-organisms growing on a surface, in a bed of polymeric material which they themselves have made. Biofilms tend to form wherever a surface on which bacteria can grow is exposed to some suitable medium and a supply of bacteria.

❐ biofuel

A gaseous, liquid or solid fuel derived from a biological source, e.g. ethanol, rapeseed oil or fish liver oil.

❐ biogas

A mixture of methane and carbon dioxide resulting from the anaerobic decomposition of waste such as domestic, industrial and agricultural sewage.

❐ bio-informatics

The use and organization of information of biological interest. In particular, concerned with organizing bio-molecular databases (particularly DNA sequences), utilizing computers for analysing this information, and integrating information from disparate biological sources.

❐ bioleaching

The recovery of metals from their ores, using the action of micro-organisms, rather than chemical or physical treatment. For example, *Thiobacillus ferroxidans* has been used to extract gold from refractory ores.

❐ biolistics

A technique to generate transgenic cells, in which DNA-coated small metal particles (tungsten or gold) are propelled by various means fast enough to puncture target cells. Provided that the cell is not irretrievably damaged, the DNA is frequently taken up by the cell. The technique has been successfully used to transform animal, plant and fungal cells, and even mitochondria inside cells. *Synonym*: microprojectile bombardment.

▫ biological containment

Restricting the movement of organisms from the laboratory. Can take two forms: making the organism unable to survive in the outside environment, or making the outside environment inhospitable to the organism. For micro-organisms, the favoured approach is to engineer organisms to require a supply of a specific nutrient that is usually available only in the laboratory. For higher organisms (plants and animals), it is more possible to ensure that the outside environment is unsuited to growth, spread and reproduction.

▫ biological oxygen demand

The dissolved oxygen required for the respiration of a population of aerobic organisms present in water. Expressed in terms of the oxygen consumed in water at a temperature of 20°C per unit time. The BOD is used as an indication of the degree to which the sample of water is polluted, particularly by inorganic nutrients for plants.

▫ biologics

Agents, such as vaccines, that give immunity to diseases or harmful biotic stresses.

▫ bioluminescence

The enzyme-catalyzed production of light by a number of diverse organisms (e.g. fireflies and many deep ocean marine organisms). Utilized as a reporter gene in plant transgenesis, and for the detection of food-borne pathogenic bacteria.

▫ biomass

1. The cell mass produced by a population of living organisms.
2. The organic matter that can be used either as a source of energy or for its chemical components.
3. All the organic matter that derives from the photosynthetic conversion of solar energy.

▫ biomass concentration

The amount of biological material in a specific volume.

▫ biome

A major ecological community or complex of communities, extending over a large geographical area and characterized by a dominant type of vegetation.

▫ biometry

The application of statistical methods to the analysis of continuous variation in biological systems.

❐ biomimetic materials

Employed to describe synthetic analogues of natural materials with advantageous properties. For instance, some synthetic molecules act chemically like natural proteins, but are not as easily degraded by the digestive system. Other systems such as reverse micelles and/or liposomes exhibit certain properties that mimic certain aspects of living systems.

❐ biopesticide

A compound that kills organisms by virtue of specific biological effects rather than as a broader chemical poison. Differ from biocontrol agents in being passive agents, whereas biocontrol agents actively seek the pest. The rationale behind replacing conventional pesticides with biopesticides is that the latter are more likely to be selective and biodegradable.

❐ biopharming

The use of genetically transformed crop plants and livestock animals to produce valuable compounds, especially pharmaceuticals.

❐ biopiracy

The patenting of genetic stocks, and the subsequent privatization of genetic resources collections. The term implies a lack of consent on the part of the originator.

❐ biopolymer

Any large polymer (protein, nucleic acid, polysaccharide) produced by a living organism. Includes some material (such as polyhydroxybutyrate) suitable for use as plastics.

❐ bioprocess

Any process that uses complete living cells or their components (e.g. enzymes, chloroplasts) to effect desired physical or chemical changes.

❐ bioreactor

A tank in which cells, cell extracts or enzymes carry out a biological reaction. Often refers to a fermentation vessel for cells or micro-organisms.

❐ biorecovery

The use of micro-organisms for the recovery of valuable materials (metals or particular organic compounds) from complex mixtures.

❐ bioremediation

A process that uses living organisms to remove contaminants, pollutants or unwanted substances from soil or water. *See*: remediation, bio-accumulation, bio-augmentation.

❐ biosafety

Referring to the avoidançe of risk to human health and safety, and to the conservation of the environment, as a result of the use for research and commerce of infectious or genetically modified organisms.

❒ biosafety protocol

An internationally agreed protocol set up to protect biological diversity from the potential risks posed by the release of genetically modified organisms. It establishes a procedure for ensuring that countries are provided with the information necessary to make informed decisions before agreeing to the import of such organisms into their territory.

❒ biosensor

A device that uses an immobilized biologically-related agent (such as an enzyme, antibiotic, organelle or whole cell) to detect or measure a chemical compound. Reactions between the immobilized agent and the molecule being analysed are converted into an electric signal.

❒ biosilk

A biomimetic fibre produced by the expression of the relevant orb-weaving spider genes in yeast or bacteria, followed by the spinning of the expressed protein into a fibre.

❒ biosorbents

Micro-organisms which, either by themselves or in conjunction with a substrate are able to extract and/or concentrate a desired molecule by means of its selective retention.

❒ biosphere

The part of the earth and its atmosphere that is inhabited by living organisms.

❒ biosynthesis

Synthesis of compounds by living cells, which is the essential feature of anabolism.

❒ biotechnology

1. "Any technological application that uses biological systems, living organisms, or derivatives thereof, to make or modify products or processes for specific use" (Convention on Biological Diversity).
2. "Interpreted in a narrow sense, ... a range of different molecular technologies such as gene manipulation and gene transfer, DNA typing and cloning of plants and animals".

❒ biotechnology - the present and the future

In 1885, a scientist named Roux demonstrated embryonic chick cells could be kept alive outside an animal's body. For the next hundred years, advances in **cell tissue culture** have provided fascinating glimpses into many different areas such as biological clocks and cancer therapy.

Monoclonal antibodies are new tools to detect and localize specific biological molecules. In principle, monoclonal antibodies can be made against any macromolecule and used to locate, purify or even potentially destroy a molecule as for example with anticancer drugs.

Molecular biology is useful in many fields. DNA technology is utilized in solving crimes. It also allows researchers to produce banks of DNA, RNA and proteins, while mapping the human genome. Tracers are used to synthesize specific DNA or RNA probes, essential to localize sequences involved in genetic disorders.

With genetic engineering, new proteins are synthesized. They can be introduced into plants or animal genomes, producing a new type of disease resistant plants, capable of living in inhospitable environments (i.e. temperature and water extremes,...). When introduced into bacteria, these proteins have also produced new antibiotics and useful drugs.

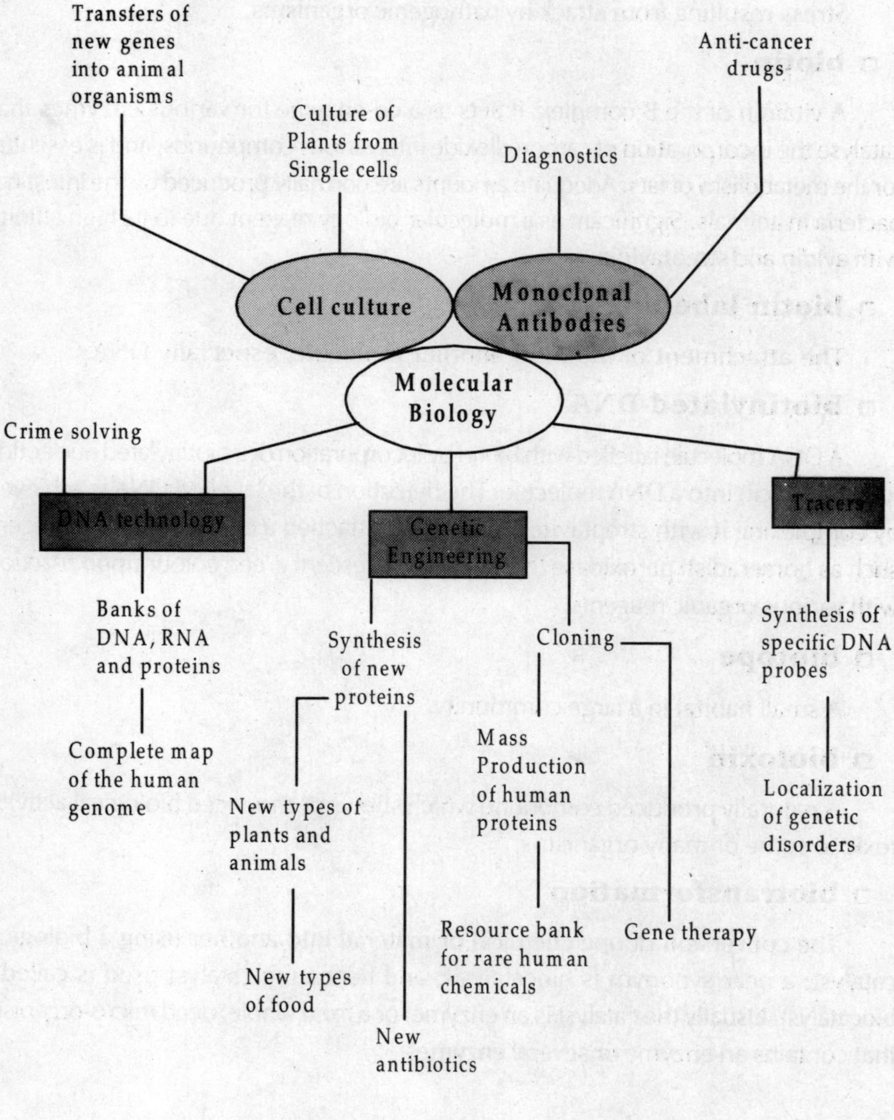

Techniques of cloning generate large quantities of pure human proteins, which are used to treat diseases like diabetes. In the future, a resource bank for rare human proteins or other molecules is a possibility. For instance, DNA sequences which are modified to correct a mutation, to increase the production of a specific protein or to produce a new type of protein can be stored . This technique will be probably play a key role in gene therapy.

❑ biotic factor

Other living organisms that are a component of an organism's environment, and form the biotic environment, affecting the organism in many ways.

❑ biotic stress

Stress resulting from attack by pathogenic organisms.

❑ biotin

A vitamin of the B complex, it acts as a co-enzyme for various enzymes that catalyse the incorporation of carbon dioxide into various compounds, and is essential for the metabolism of fats. Adequate amounts are normally produced by the intestinal bacteria in animals. Significant as a molecular biology reagent due to its high affinity with avidin and streptavidin.

❑ biotin labelling

The attachment of biotin to another molecule, especially DNA.

❑ biotinylated-DNA

A DNA molecule labelled with biotin by incorporation of a biotinylated nucleotide (usually uracil) into a DNA molecule. The detection of the labelled DNA is achieved by complexing it with streptavidin to which is attached a colour-generating agent such as horseradish peroxidase that gives a fluorescent green colour upon reaction with various organic reagents.

❑ biotope

A small habitat in a large community.

❑ biotoxin

A naturally produced compound which shows pronounced biological activity, toxic to some or many organisms.

❑ biotransformation

The conversion of one chemical or material into another using a biological catalyst: a near synonym is biocatalysis, and hence the catalyst used is called a biocatalyst. Usually the catalyst is an enzyme, or a fixed whole, dead micro-organism that contains an enzyme or several enzymes.

❐ bivalent

Two paired homologous chromosomes (one of maternal origin; the other of paternal origin) at prophase to anaphase of the first meiotic division. Because DNA is replicated in prophase, each duplicated chromosome comprises two chromatids, and thus a bivalent comprises four chromatids.

❐ blast cell

A large, rapidly dividing cell that develops from a B cell in response to an antigenic stimulus. The blast cell then becomes an antibody-producing plasma cell.

❐ blastocyst

Early stage of an embryo produced by cleavage of an ovum; a liquid-filled sphere whose wall is composed of a single layer of cells; during this stage (about eight days after fertilization) implantation in the wall of the uterus occurs.

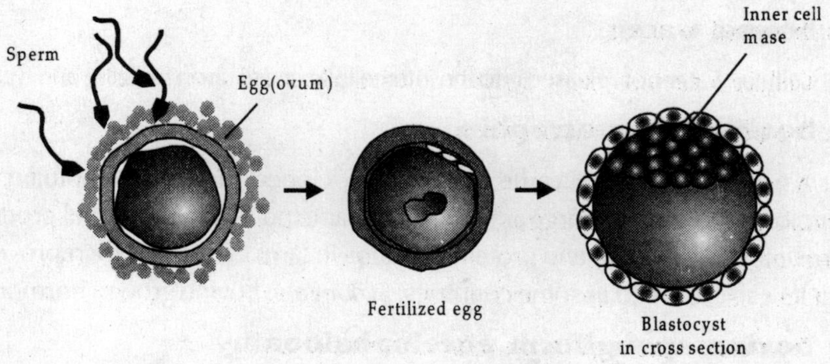

❐ blastomere

Any one of the cells formed from the first few cleavages in animal embryology. The embryo usually divides into two, then four, then eight blastomeres, and so on.

❐ blastula

In animals, an early embryo form that follows the morula stage; typically, a single-layered sheet (blastoderm) or ball of cells (blastocyst).

❐ bleeding

1. Collection of blood from immunized animals.
2. Used to describe the occasional purplish-black colouration of media due to phenolic products given off by (usually fresh) transfers.

❐ blot

As a verb, to transfer DNA, RNA or protein to an immobilizing matrix. As a noun, the immobilizing matrix carrying DNA, RNA or protein. The various types of blot are named according to the probe and/or the probed molecules: Southern blot (DNA/DNA), northern blot (DNA/mRNA), western blot (antibody/protein), southwestern blot (DNA/protein). Only "Southern" is written with an initial capital, as it is named after Ed Southern, the inventor of the technique.

❐ blunt end

The end of a double-stranded DNA molecule in which neither strand extends beyond the other. blunt-end cut To cut a double-stranded DNA with a restriction endonuclease which generates blunt ends. blunt-end ligation The joining of two blunt-ended double-stranded DNA molecules.

❐ bod

Abbreviation for biological oxygen demand.

❐ boring platform

Sterile bottom half of a petri dish used for preparing explants with a cork borer.

❐ bound water

Cellular water not released into the intercellular space upon freezing and thawing.

❐ bovine somatotrophin

A natural protein in cattle. It has been cloned, using recombinant DNA technology, expressed in large amounts and marketed as an agricultural product to improve the growth rate and protein:fat ratios in farm cattle, and to improve milk yield. Its use is banned in some countries. *Synonyms*: bovine growth hormone.

❐ bovine spongiform encelophalopathy

Cattle disease (colloquially called mad cow disease) caused by proteinaceous infectious particles.

❐ bract

A modified leaf that subtends flowers or inflorescences and may appear to be a petal.

❐ breed

1. a sub-specific group of domestic livestock with definable and identifiable external characteristics that enable it to be separated by visual appraisal from other similarly defined groups within the same species.
2. a group of domestic livestock for which geographical and/or cultural separation from phenotypically similar groups has led to acceptance of its separate identity.

❐ breed at risk

An animal breed that is in danger of becoming extinct because its population has fallen below a critical number.

❐ breeding

The process of sexual reproduction and production of offspring.

❐ breeding value

A quantitative genetics term, describing that part of the deviation of an individual phenotype from the population mean that is due to the additive effects of alleles. Thus, if an individual is mated with a random sample of individuals from a population, its breeding value for a given trait is twice the average deviation of its offspring from the population mean for that trait.

❐ brewer's yeast

Strains of the yeast *Saccharomyces cerevisiae* that are used for the production of beer.

❐ bridge

A filter paper or other substrate used as a wick and support structure for a plant tissue in culture when a liquid medium is used.

❐ broad-host-range plasmid

A plasmid that can replicate in a number of different bacterial species.

❐ broad-sense heritability

The proportion of the total phenotypic variation which results from genetic variation or interaction between the genotype and the environment.

❐ broodstock

The group of males and females from which fish are bred.

❐ browning

Discolouration of freshly cut surfaces of plant tissue due to phenolic oxidation. In plant tissue culture, it may indicate a nutritional or pathogenic problem, generally leading to necrosis.

❐ bsa

Abbreviation for bovine serum albumin.

❐ bse

Abbreviation for bovine spongiform encephalopathy. *See* proteinaceous infectious particle.

❐ bst

Abbreviation for bovine somatotrophin.

❐ bt

Abbreviation for *Bacillus thuringiensis*.

❐ bubble column fermenter

A bioreactor in which the cells or micro-organisms are kept suspended in a tall cylinder by rising air, which is introduced at the base of the vessel.

❐ bud

A region of meristematic tissue with the potential for developing into leaves, shoots, flowers or combinations of these; generally protected by modified scale leaves.

❐ budding

1. A method of asexual reproduction in which a new individual is derived from an outgrowth (bud) that becomes detached from the body of the parent.
2. Among fungi, budding is characteristic of the brewers yeast *Saccharomyces cerevisiae*.
3. A form of graft in which a single vegetative bud is taken from one plant and inserted into stem tissue of another plant so that the two will grow together. The inserted bud develops into a new shoot.

❐ bud sport

A somatic mutation arising in a bud, generating a genetically different shoot. Includes changes due to gene mutation, chromosomal mutation or polyploidy.

❐ bulked segregant analysis

A method to obtain markers linked to a target trait, in which DNA samples, prepared from a number of individuals of each of two contrasting phenotypes, are separately pooled and used to generate contrasting DNA fingerprints. DNA fragments unique to one pool become candidates for a marker linked to the gene controlling the trait.

❐ buoyant density

The intrinsic density which a molecule, virus or sub-cellular particle has when suspended in an aqueous solution of a salt, such as CsCl, or a sugar, such as sucrose. DNA from different species has different characteristic buoyant densities, determined largely by the relative proportion of the base-pairs G+C, to A+T.

C

❐ c

Abbreviation for cytosine.

❐ caat

A conserved DNA sequence found within the promoter region of the protein-encoding genes of many eukaryotic organisms. So-called because of its consensus sequence GGCCAATCT, it occurs around 75 bases prior to the transcription initiation site; and is one of several sites for the recognition and binding of transcription factors.

❐ callipyge

An inherited trait in livestock (e.g. sheep) that results in thicker, meatier hind-quarters, and hence a higher meat yield per animal.

❐ callus

A protective tissue, consisting of parenchyma cells, that develops over a cut or damaged plant surface. 2. Mass of undifferentiated, thin-walled parenchyma cells induced by hormone treatment. 3. Actively dividing non-organized masses of undifferentiated and differentiated cells often developing from injury (wounding) or in tissue culture in the presence of growth regulators.

❐ callus culture

A technique of plant tissue culture, usually on solidified medium and initiated by inoculation of small explants. Used as the basis for organogenic (shoot or root forming) cultures, cell cultures or proliferation of embryoids. Callus cultures can be indefinitely maintained through regular sub-culturing.

❐ cambial zone

Region in stems and roots consisting of the cambium and its recent derivatives.

❐ cambium

A one or two cells thick layer of plant meristematic tissue, between the xylem and phloem tissues, which gives rise to secondary tissues, thus resulting in an increase in the diameter of the stem or root. The two most important cambia are the vascular (fascicular) cambium and the cork cambium.

❐ camp

Abbreviation for cyclic adenosine monophosphate.

❐ camv 35S

Abbreviation for cauliflower mosaic virus 35S ribosomal DNA promoter.

❐ camv

Abbreviation for cauliflower mosaic virus.

❐ candidate gene

A gene whose deduced function (on the basis of DNA sequence) suggests that it may be involved in the genetic control of an aspect of phenotype.

❐ candidate-gene strategy

An experimental approach in which knowledge of the biochemistry and/or physiology of a trait is used to identify candidate genes.

❐ canola

A specific subgroup of oilseed rape cultivars; canola oil is the highly mono-unsaturated fatty acid and low in erucic acid product produced in the seed of these cultivars.

❐ cap

Abbreviation for catabolite activator protein.

❐ capacitation

The final stage, inside the female genital tract, in the maturation process of a spermatozoan, as it penetrates the ovum.

❐ capillary electrophoresis

A form of electrophoresis used widely in current large-scale DNA sequencing facilities, where the sample is passed through a long, very-narrow-bore tube containing a re-usable matrix.

❐ capsid

The protein coat of a virus. The capsid often determines the shape of the virus.

❐ cap site

The site on a DNA template where transcription begins. It corresponds to the nucleotide at the 5' end of the RNA transcript which accepts the G cap.

❐ capsule

Carbohydrate coverings that have antigenic specificity, present on some types of bacteria and other micro-organisms. The capsule is usually composed of polysaccharides, polypeptides, or polysaccharide-protein complexes. These materials are arranged in a compact manner around the cell surface.

❐ carbohydrade recognition at infection site

The recognition of cell-surface carbohydrate on neutrophils is the first stage of their migration out of the blood at sites of infection. Specialized transmembrane proteins (called lectins) are made by the cells lining the blood vessel (called endothelial cells) in response to chemical signals emanating from the site of infection. These proteins recognize particular groups of sugars carried by glycolipids and glycoproteins on the surface of neutrophils circulating in the blood. The neutrophils consequently stick to the blood vessel wall. This association is not very strong, but it leads to another, much stronger protein-protein interaction (not shown) that helps the neutrophil migrate out of the bloodstream between the endothelial cells into the tissue at the site of infection.

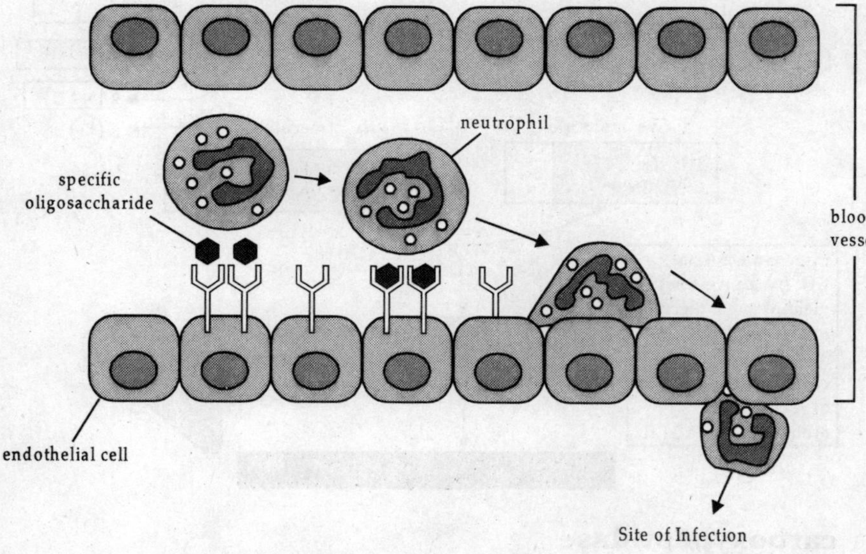

❐ carbon-fixation cycle

The carbon-fixation cycle, which forms organic molecules from CO_2 and H_2O. The number of carbon atoms in each type of molecule is indicated in the white box. There are many intermediates between glyceraldehyde 3-phosphate and ribulose 5-phosphate, but they have been omitted here for clarity. The entry of water into the cycle is also not shown.

Photosynthesis is the synthesis of organic molecules using the energy of light. For the sugar glucose (one of the most abundant products of photosynthesis) the equation is:

$$6CO_2 + 12H_2O \text{ -> } C_6H_{12}O_6 + 6H_2O + 6O_2$$

ATP and **NADPH** provide the energy and electrons to reduce carbon dioxide (CO_2) to organic molecules.

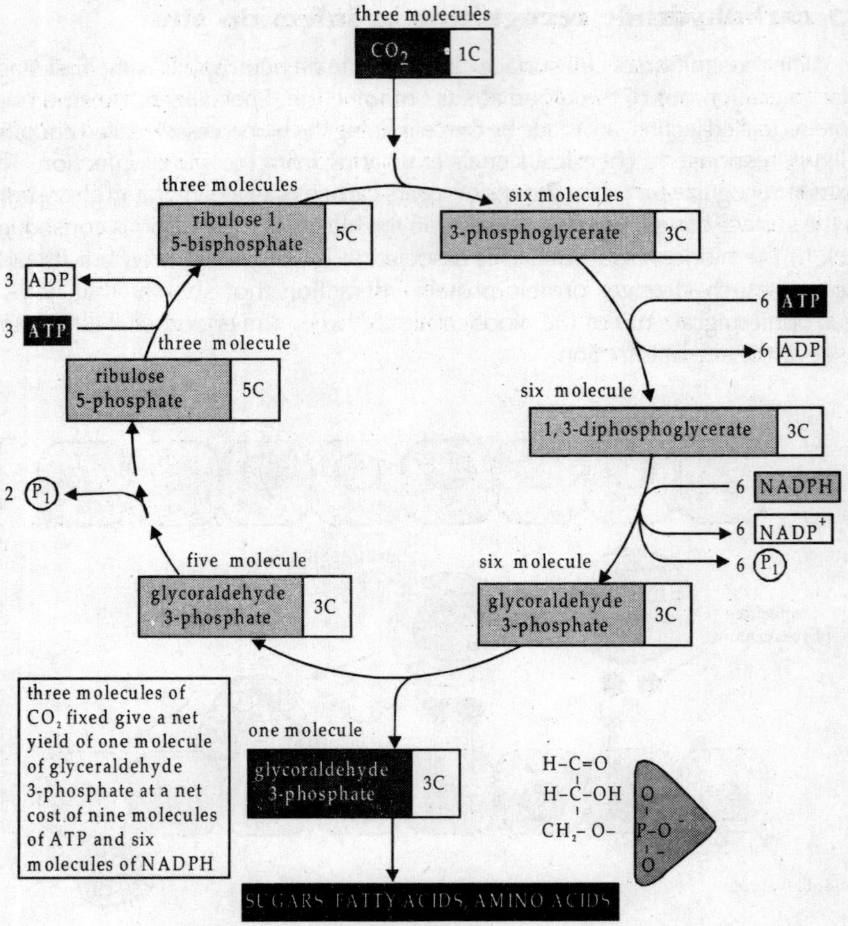

❒ carboxypeptidase

A class of enzymes which catalyse the cleavage of peptide bonds, requiring a free carboxyl group in the substrate. The peptide bond adjacent to this group is cleaved and a free amino acid is released. Used for deriving the amino acid sequence of peptides.

❒ carcinogen

A substance capable of inducing cancer in an organism.

❒ carcinoma

A malignant tumour derived from epithelial tissue, which forms the skin and the outer cell layers of internal organs.

❒ carotene

A reddish-orange plastid pigment involved in photosynthesis. A carotenoid and precursor of vitamin A.

❑ carotenoid

A group of chemically similar red to yellow pigments responsible for the characteristic colour of many plant organs or fruits, such as tomatoes, carrots, etc. Oxygen-containing carotenoids are called xanthophylls. Carotenoids serve as light-harvesting molecules in photosynthetic assemblies and also play a role in protecting prokaryotes from the deleterious effects of light.

❑ carpel

Female reproductive organ of flowering plants, consisting of stigma, style and ovary.

❑ carrier

A heterozygous individual bearing a recessive mutant allele for a defective condition that is "masked" by the presence of the dominant normal allele; the phenotype is normal, but the individual passes the defective (recessive) allele to half of its offspring.

❑ carrier DNA

DNA of undefined sequence which is added to the transforming (plasmid) DNA used in physical DNA-transfer procedures. This additional DNA increases the efficiency of transformation in electroporation and chemically-mediated DNA delivery systems. The mechanism responsible is not known.

❑ carrier molecule

1. A molecule that plays a role in moving electrons through the electron transport chain. They are usually proteins bound to non-protein groups and able to undergo oxidation and reduction relatively easily, thus allowing electrons to flow.
2. A lipid-soluble molecule that can bind to lipid-insoluble molecules and transport them across membranes. Carrier molecules have specific sites that interact with the molecules they transport. The efficiency of carrier molecules may be modified by changing the interacting sites through genetic engineering.

❑ casein

A group of milk proteins.

❑ casein hydrolysate

The mixture of amino acids and peptides produced by enzymatic or acid hydrolysis of casein.

❑ catabolic and anabolic pathways

Schematic representation of the relationship between catabolic and anabolic pathways in metabolism. As suggested here, since a major portion of the energy stored in the chemical bonds of food molecules is dissipated as heat, the mass of food required by any organism that derives all of its energy from catabolism is much greater than the mass of the molecules that can be produced by anabolism.

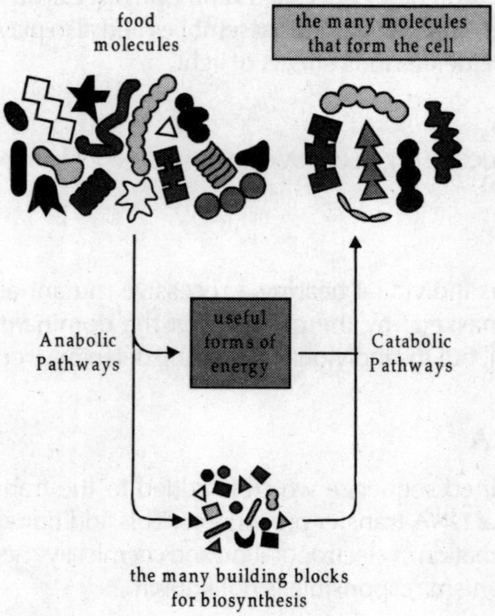

▫ catabolic pathway

A pathway by which an organic molecule is degraded in order to release energy for growth and other cellular processes.

▫ catabolism

The breakdown of large molecules in living organisms, with the accompanying release of energy.

▫ catabolite activator protein

A protein which combines with cyclic AMP. The cAMP-CAP complex binds to the promoter regions of *E. coli* and stimulates transcription of the relevant operon.

▫ catabolite repression

Glucose-mediated reduction in the rates of transcription of genes that encode enzymes involved in catabolic pathways (e.g. the *lac* operon).

▫ catalase

A metalloenzyme, present in both plants and animals, that catalyzes the decomposition of hydrogen peroxide to water and oxygen. This activity is important in the detoxification of reactive oxygen generated as part of the response to stress.

❐ catalysis

The process of increasing the rate of a chemical reaction by the addition of a substance that is not itself changed by the reaction (the catalyst).

❐ catalyst

A substance that promotes a chemical reaction by lowering the activation energy of a chemical reaction, without itself undergoing any permanent chemical change.

❐ catalytic antibody

An antibody selected for its ability to catalyse a chemical reaction by binding to and stabilizing the transition-state intermediate.

❐ catalytic site

The part of the surface of an enzyme molecule (usually only a small portion of the total) necessary for the catalytic process.

❐ cauliflower mosaic virus

A DNA virus affecting cauliflower and many other dicot species. Its importance is due to the promoter of its 35S ribosomal DNA, which is constitutively active in most plant tissues, and has therefore been widely used as a promoter for the expression of transgenes.

❐ cauliflower mosaic virus 35S promoter

A promoter sequence isolated from the ribosomal gene of the cauliflower mosaic virus

❐ caulogenesis

Stem organogenesis; induction of shoot development from callus.

❐ cbd

Abbreviation for Convention on Biological Diversity.

❐ ccc dna

Abbreviation for covalently-closed circle DNA.

❐ cd molecules

Abbreviation for: cluster of differentiation molecules. Any group of surface antigens associated with a specific sub-population of T cells.

❐ cdna

Abbreviation for complementary DNA.

❐ cdna clone

A double-stranded cDNA molecule propagated in a vector, and used as a probe in RFLP analyses, as template for the production of EST sequences, and for gene expression studies.

□ cdna cloning

A method of cloning the coding sequence of a gene, starting with its mRNA transcript.

□ cdna library

An collection of cDNA clones.

□ cdr

Abbreviation for complementarity-determining regions.

□ cell

Fundamental structural unit of all life. The cell consists primarily of an outer plasma membrane, which separates it from the environment; the genetic material (DNA), which encodes heritable information for the maintenance of life; and the cytoplasm, a heterogeneous assemblage of ions, molecules, and fluid.

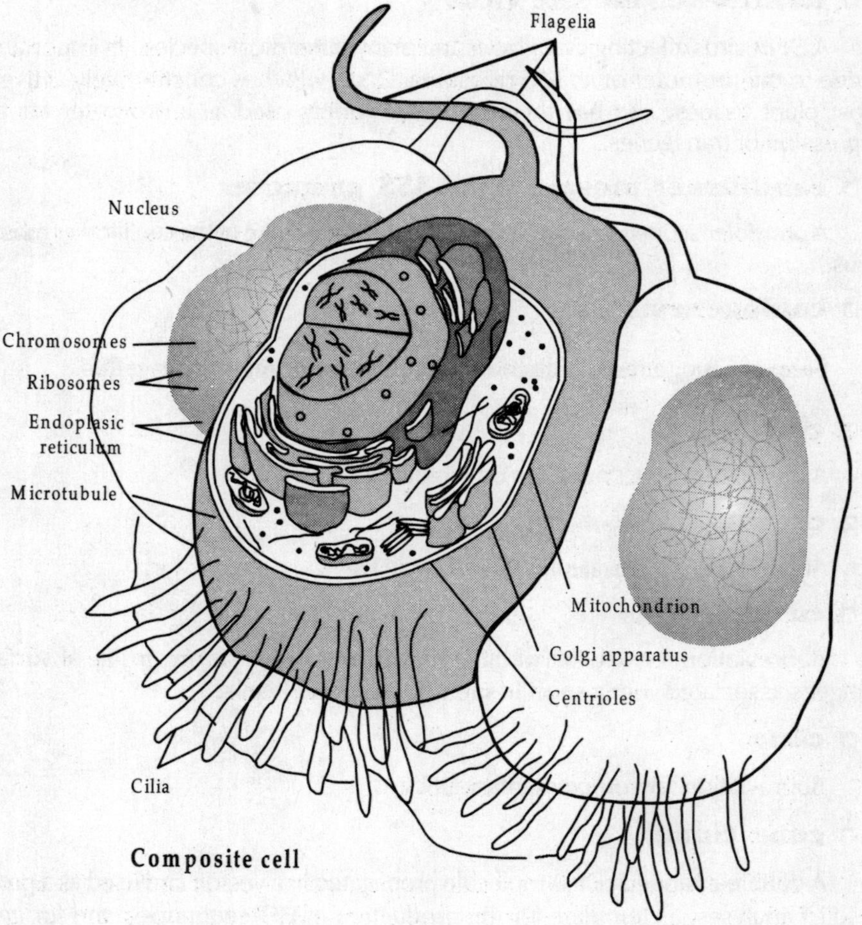

Composite cell

❑ cell breakage and fractionation

The basic principle for all microscopes is that the cell is composed of smaller physical units, the organelles. Definition of the organelles is possible with microscopy, but the function of individual organelles is often beyond the ability of observations through a microscope. We are able to increase our chemical knowledge of organelle function by isolating organelles into reasonably pure fractions.

A host of fractionation procedures are employed by cell biologists. Each organelle has characteristics (size, shape and density for example) which make it different from other organelles within the same cell. If the cell is broken open in a gentle manner, each of its organelles can be subsequently isolated. The process of breaking open cells is homogenization and the subsequent isolation of organelles is fractionation. Isolating the organelles requires the use of physical chemistry techniques, and those techniques can range from the use of simple sieves, gravity sedimentation or differential precipitation, to ultra centrifugation of fluorescent labeled organelles in computer generated density gradients.

Cell wall fractionation was performed basically with some modifications. Yeast cells and hyphae were collected by centrifugation and filtration, respectively, and washed five times with lysis buffer (10 mM Tris-HCl, pH 7.4, 1 mM phenylmethylsulfonyl fluoride). Subsequently cells were resuspended in ice-cold lysis buffer and lysed mechanically with an equal volume of glass beads in a cell homogenizer (Braun, MSK). This procedure was carried out until complete cell breakage, verified beforehand by phase-contrast microscopic examinations and *a posteriori* by the failure of cells to grow on YPD-chloramphenicol plates.

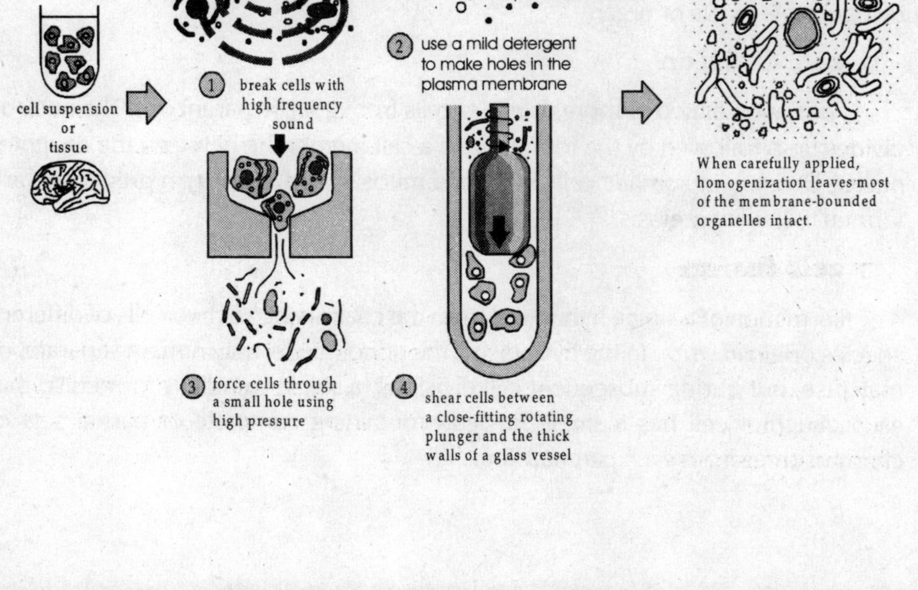

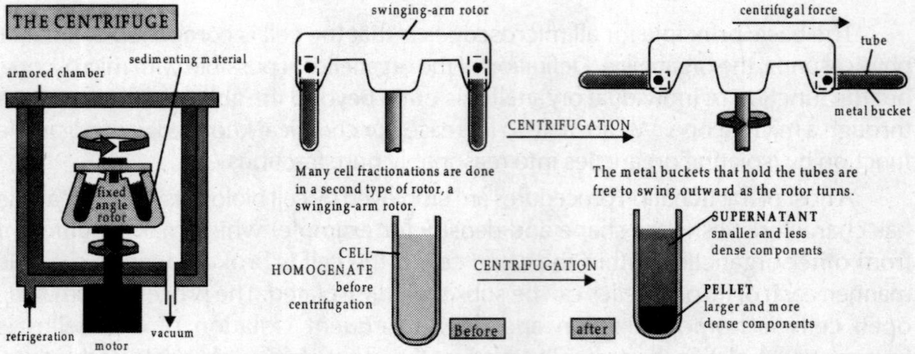

Centrifugation is the most widely used procedure to separate homogenate into different parts, or fractions. The homogenate placed in test tubes and rotated at high speed in a centrifuge (sometimes called an ultracentrifuge). Present-day ultracentrifuges rotate at speeds up to 100,000 revolutions per minute and oduce enormous forces, as high as 600,000 times gravity. At such speeds, centrifuge chambers must be refrigerated and evacuated so that friction does not heat up the homogenate. The centrifuge is surrounded by thick armor plating, since an unbalanced rotor can shatter with an explosive release of energy. A fixed-angle rotor can hold larger volumes than a swinging-arm rotor, but the pellet forms less evenly.

□ cell culture

The *in vitro* growth of cells isolated from multi-cellular organisms.

□ cell cycle

The sequence of stages that a cell passes through between one division and the next. The cell cycle oscillates between mitosis (M) and the interphase, which is divided into the G_1 phase (involving a high rate of biosynthesis and growth), the S phase (in which the DNA content is doubled as a consequence of chromosome replication), and the G_2 phase (preparatory for cell division).

□ cell differentiation

The transition of cells (by the programmed activation and de-activation of the necessary genes) from an tissue-unspecific type, in which daughter cells are similarly undifferentiated, to a committed type in which the cell line specializes to become a recognizable tissue or organ.

□ cell division

Formation of two or more daughter cells from a single parent cell. The nucleus divides first, followed by the formation of a cell membrane between the daughter nuclei. Division of somatic cells is termed mitosis; egg and sperm precursors are formed following meiosis.

□ cell fusion

Formation of a single hybrid cell from the coalescence of two cells of different species origin *in vitro* . In the hybrid cell, the donor nuclei may remain separate, or may fuse, but during subsequent cell divisions, a single spindle is formed so that each daughter cell has a single nucleus containing complete or partial sets of chromosomes from each parental line.

▫ cell generation time

The interval between the beginning of consecutive divisions of a cell, equivalent to the time that it takes for a population of single-celled organisms to double its cell number.

▫ cell line

A cell lineage that can be maintained *in vitro*. Significant genetic changes can occur during lengthy periods in culture, so that the genotype of long-term cell lines may not be the same as that of the starter cell. A cell lineage that can be recognized *in vivo*.

▫ cell metabolism

Simplified diagram of the three stages of cellular metabolism that lead from food to waste products in animal cells. This series of reactions produces ATP, which is then used to drive biosynthetic reactions and other energy-requiring processes in the cell. Stage 1 occurs outside cells. Stage 2 occurs mainly in the cytosol, except for the final step of conversion of pyruvate to acetyl groups on acetyl CoA, which occurs in mitochondria. Stage 3 occurs in mitochondria.

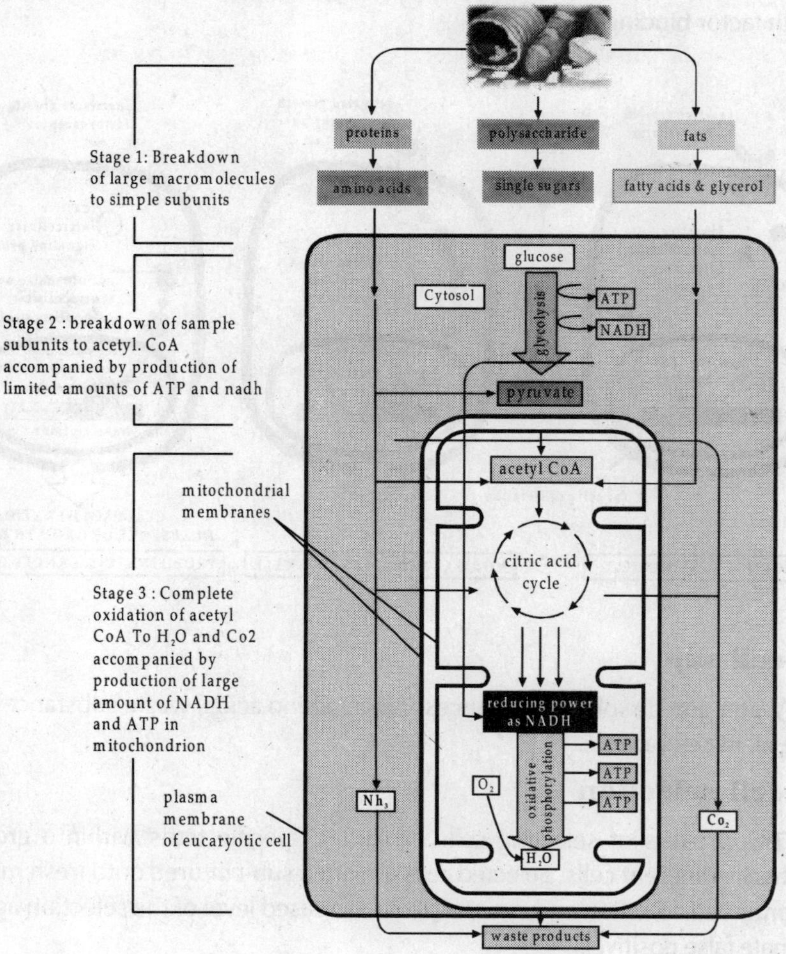

▢ cell number

The number of cells per unit volume of a culture.

▢ cell plate

The precursor of the cell wall, formed at the beginning of cell division. The cell plate develops in the region of the equatorial plate and arises from membranes in the cytoplasm.

▢ cell proliferation

In a normal resting cell the intracellular signaling proteins and genes that are normally activated by extracellular growth factors are inactive. (B) When the normal cell is stimulated by an extracellular growth factor, these signaling proteins and genes become active and the cell proliferates. (C) In this cancer cell, a mutation in a proto-oncogene that encodes an intracellular signaling protein that is normally activated by extracellular growth factors has created an oncogene. The oncogene encodes an altered form of the signaling protein that is active even in the absence of growth factor binding.

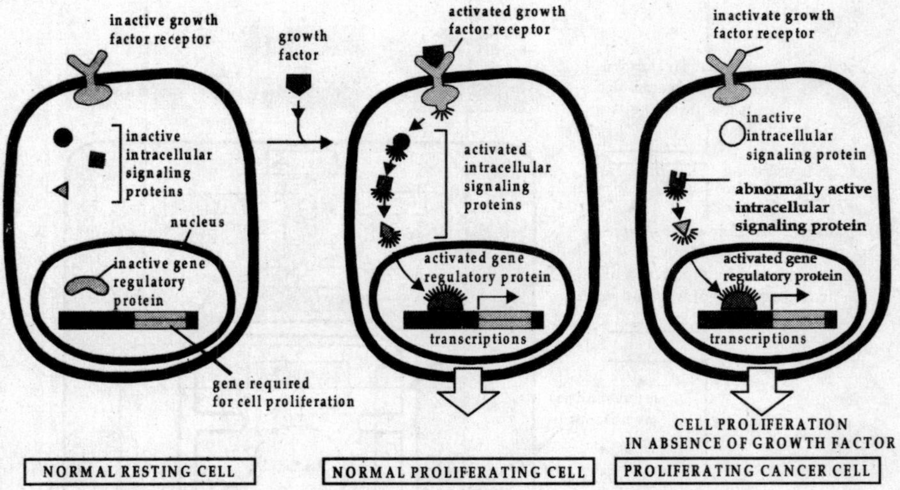

▢ cell sap

Water and dissolved substances, sugar, amino acids, waste substances, etc., in the plant cell vacuole.

▢ cell selection

The process of selecting cells exhibiting specific traits within a group of genetically different cells. Selected cells are often sub-cultured onto fresh medium for continued selection and exposed to an increased level of the selection agent to eliminate false positives.

❐ cell strain

An *in vitro* culture initiated by asexual reproduction from a single cell. Such cell lines should represent a population of genetically homogenous cells. Strains are defined by specific properties or markers used for their selection.

❐ cell structure

The fundamental level of structural organization in complex organisms. Eukaryotic cells contain a nucleus (with chromosomes) and cytoplasm with the protein synthesis machinery, bounded by a membrane. Prokaryotic cells have no nucleus.

❐ cell suspension

Cells in culture in moving or shaking liquid medium, often used to describe suspension cultures of single cells and cell aggregates.

❐ cell wall

A rigid external structure which surrounds plant cells. It is formed outside the plasmalemma and consists primarily of cellulose.

❐ cellular oxidation *vs.* burning of sugar

Schematic representation of the controlled stepwise oxidation of sugar in a cell, compared with ordinary burning. In the cell, enzymes catalyze oxidation via a series of small steps in which free energy is transferred in conveniently sized packets to carrier molecules.

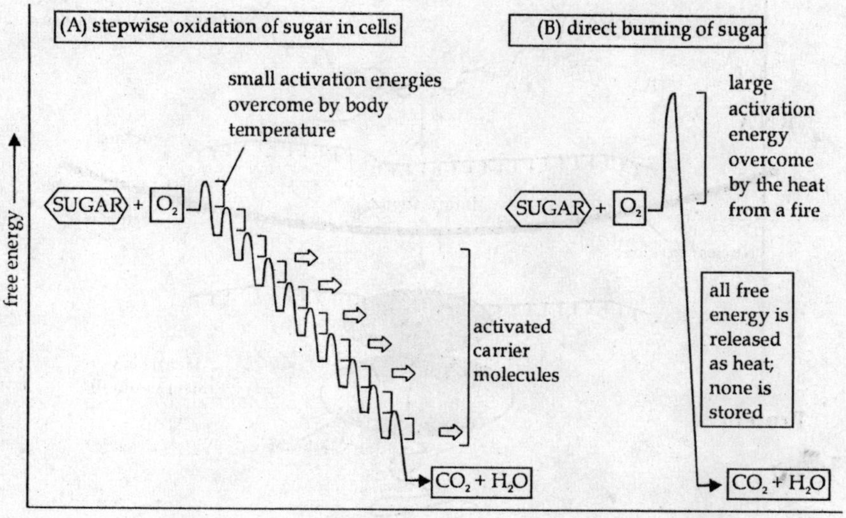

❐ cellulase

Enzyme catalysing the breakdown of cellulose.

❐ cellulose

A complex polysaccharide composed of long linear chains of glucose residues. It comprises 40% to 55% by weight of the plant cell wall.

❐ cellulosome

A multi-protein aggregate present in some micro-organisms which degrade cellulose. It contains multiple copies of the enzymes necessary for this process, and is often found on the outer surface of the micro-organism cell.

❐ centiMorgan

Unit of map distance. For small recombination fractions, cM and % recombination frequency are equivalent.

❐ central dogma of molecular biology

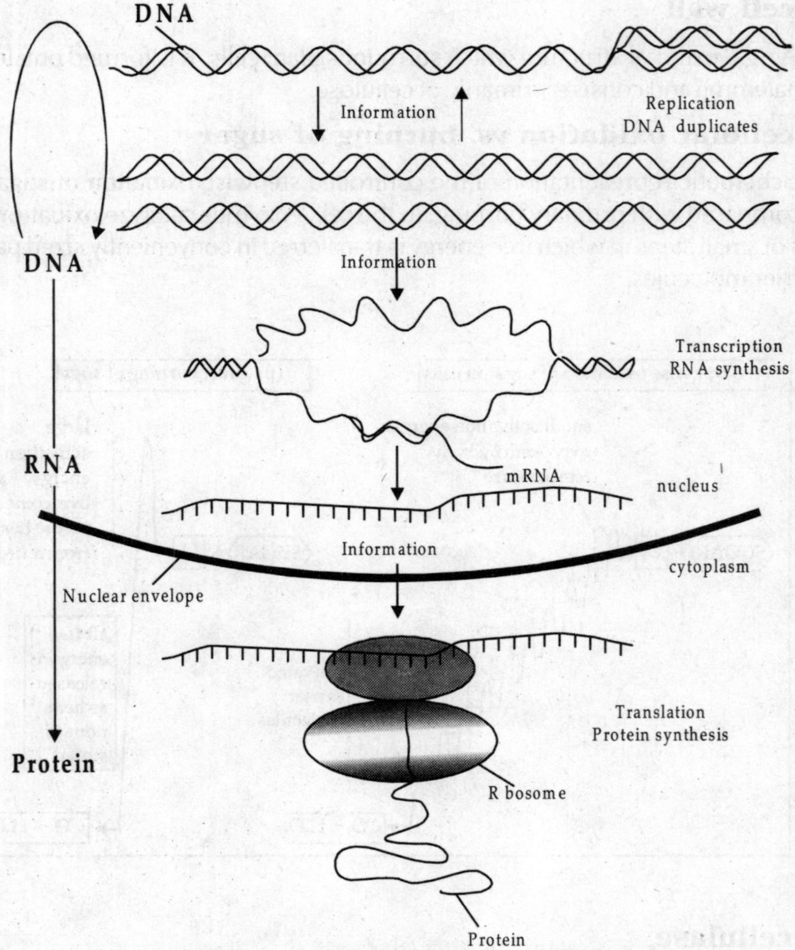

The Central dogma of Molecular Biology

Transcription of DNA to RNA to protein: This dogma forms the backbone of molecular biology and is represented by four major stages.

1. The DNA replicates its information in a process that involves many enzymes: replication.
2. The DNA codes for the production of messenger RNA (mRNA) during transcription.
3. In eucaryotic cells, the mRNA is processed (essentially by splicing) and migrates from the nucleus to the cytoplasm.
4. Messenger RNA carries coded information to ribosomes. The ribosomes "read" this information and use it for protein synthesis. This process is called translation.

Proteins do not code for the production of protein, RNA or DNA. They are involved in almost all biological activities, structural or enzymatic.

❐ central dogma

The basic concept that, in nature, genetic information generally flows from DNA to RNA to protein. However, information contained in the RNA molecules of retroviruses can also flow back to DNA.

❐ central mother cell

A subsurface cell located in a plant apical meristem and characterized by a large vacuole.

❐ centre of origin

The geographic locations where a particular domesticated plant species originated. These areas are the likeliest source of natural genetic variation, and represent ideal targets for *in situ* conservation.

❐ centrifugation

Separating molecules by size or density using centrifugal forces generated by a spinning rotor. G-forces of several hundred thousand times gravity are generated in ultracentrifugation.

❐ centrifuge

A mechanical device which delivers the centrifugal forces necessary for centrifugation.

❐ centriole

An organelle in many animal cells that appears to be involved in the formation of the spindle during mitosis. During cell division, the two centrioles move to opposite sides of the nucleus to form the ends of the spindle.

❐ centromere

The eukaryotic chromosome structure, which appears as a constriction in karyotype analysis, to which the spindle fibres attach during mitotic and meiotic division. Composed of highly repetitive DNA.

▫ centrosome

A specialized region of a living cell, situated next to the nucleus, where microtubules are assembled and broken down during cell division. The centrosome of most animal cells contains a pair of centrioles.

▫ cephem-type antibiotic

An antibiotic that shares the basic chemical structure of cephalosporin.

▫ chain terminator

In the Sanger method of DNA sequencing, refers to the labelled di-deoxynucleotide triphosphates which are added to disrupt DNA polymerase extension.

▫ chakrabarty decision

A landmark legal case in the U.S.A., in which it was held that the inventor of a new micro-organism whose invention otherwise met the legal requirements for obtaining a patent, could not be denied a patent solely because the invention was alive. This has set the precedent for the patenting of life forms.

▫ chaperone

A family of proteins that ensure the correct assembly and conformation of other polypeptides *in vivo* as they emerge from the ribosome, but are not themselves components of the functional assembled structures. The prokaryotic equivalents are known as chaperonins.

▫ characterization

Description of the essential properties of an organism or system.

▫ charcoal

The black porous residue of partly burnt wood, bones, etc; a form of carbon.

▫ chelate

A cation bound to an organic molecule through electron pair donation from nitrogen and/or oxygen atoms in its structure. Ethylenediamine tetraacetic acid is a typical and frequently employed chelating agent. Soluble chelates can supply plants with micronutrients which would otherwise be unavailable because of precipitation.

▫ chemical bonds and groups part 1

Chemical bonds and groups commonly encountered in biological molecules. Organic chemistry encompasses a very large number of compounds (many millions), and focused on their structural characteristics. We turn to the roles they are inclined to play in the scientific drama staged by the multitude of chemical reactions that define organic chemistry. The three kinds of chemical bonds are the covalent, ionic, and hydrogen bonds. The covalent bonds include single, double, and triple bonds. Additionally, some covalent bonds formed by the reaction of two functional groups are given special names. "Peptide bond" refers to the covalent bond between amino acids in a peptide.

CARBON SKELETONS

Carbon has a unique role in the cell because of its ability to form strong covalent bonds with other carbon atoms. Thus carbon atoms can join to form chains.

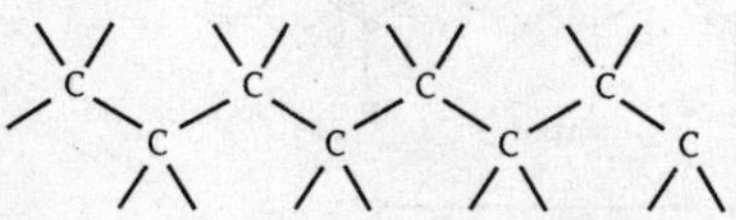

also written as

or branched trees

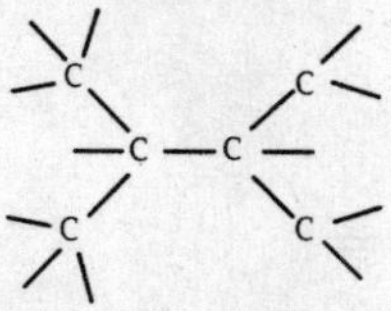

also written as

or rings

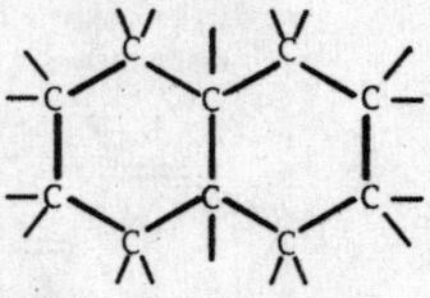

also written as

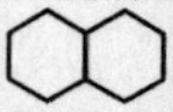

COVALENT BONDS

A covalent bond forms when two atoms come very close together and share one or more of their electrons. In a single bond one electron from each of the two atoms is shared; in a double bond a total of four electrons are shared.
Each atom forms a fixed number of covalent bonds in a defined spatial arrangement. For example, carbon forms four single bonds arranged tetrahedrally, whereas nitrogen forms three single bonds and oxygen forms two single bonds arranged as shown below.

Double bonds exist and have a different spatial arrangement.

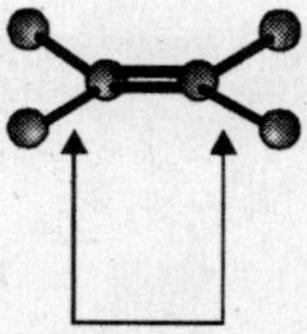

Atoms joined by two or more covalent bonds cannot rotate freely around the bond axis.
This restriction is a major influence on the three-dimensional shape of many macromolecules.

C–H COMPOUNDS

Carbon and hydrogen together make stable compounds (or groups) called hydrocarbons. These are nonpolar, do not form hydrogen bonds, and are generally insoluble in water.

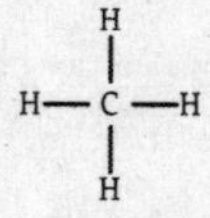

methane

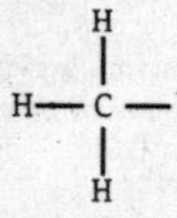

methyl group

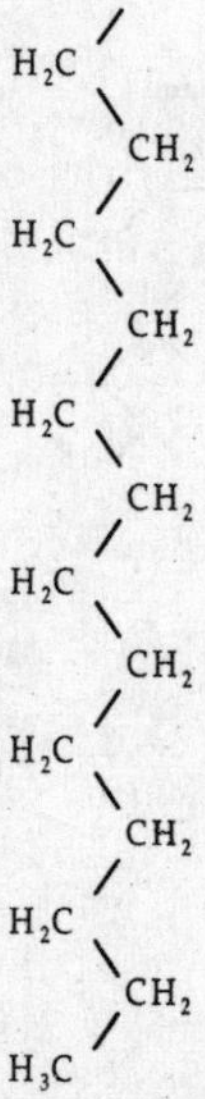

part of the hydrocarbon "tail" of a fatty acid molecule

ALTERNATING DOUBLE BONDS

The carbon chain can include double bonds. If these are on alternate carbon atoms, the bonding electrons move within the molecule, stabilizing the structure by a phenomenon called resonance.

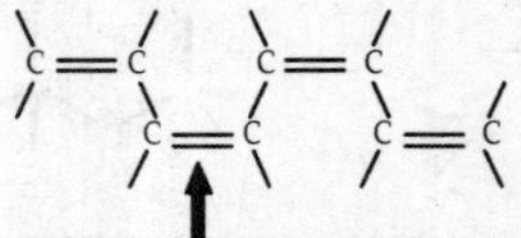

the truth is somewhere between these two structures

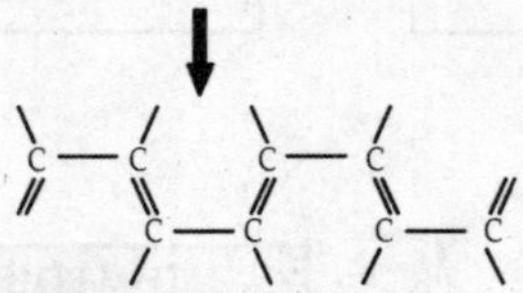

Alternating double bonds in a ring can generate a very stable structure.

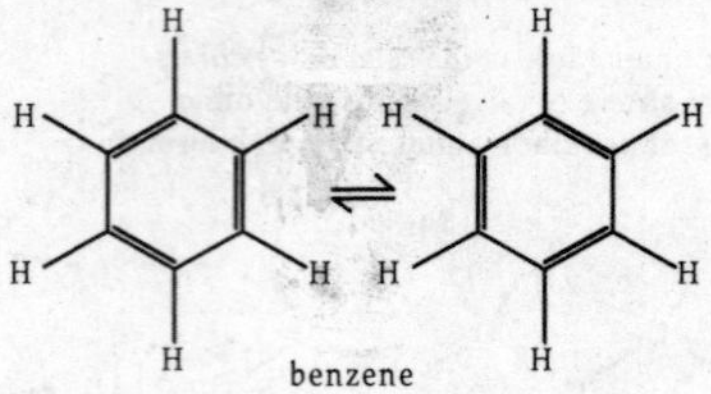

benzene

often written as 

C-O COMPOUNDS

Many biological compounds contain a carbon bonded to an oxygen. For example,

alcohol

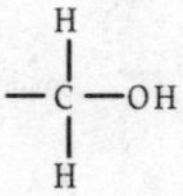

The -OH is called a **hydroxyl** group.

aldehyde

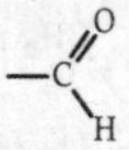

ketone

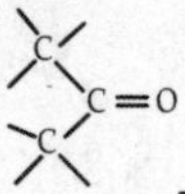

The C—O is called a **carbonyl** group.

carboxylic acid

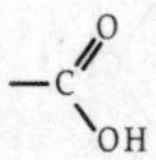

The -COOH is called a **carboxyl** group. In water this loses an H+ ion to become -COO_.

esters Esters are formed by combining an acid and an alcohol.

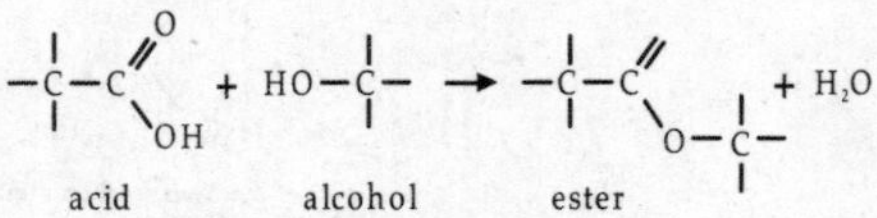

acid alcohol ester

C-N COMPOUNDS

Amines and amides are two important examples of compounds containing a carbon linked to a nitrogen.

Amines in water combine with an H^+ ion to become positively charged.

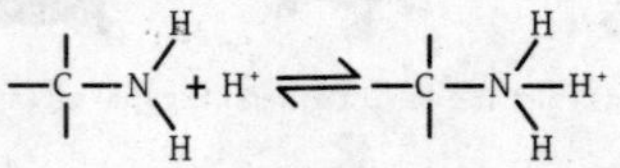

Amides are formed by combining an acid and an amine. Unlike amines, amides are uncharged in water. An example is the peptide bond that joins amino acids in a protein.

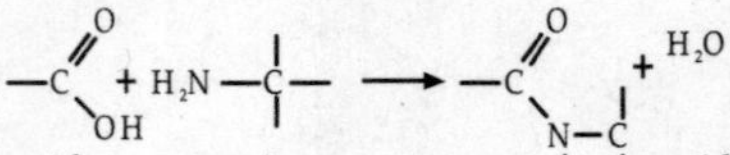

acid amine amide

Nitrogen also occurs in several ring compounds, including important constituents of nucleic acids: purines and pyrimidines.

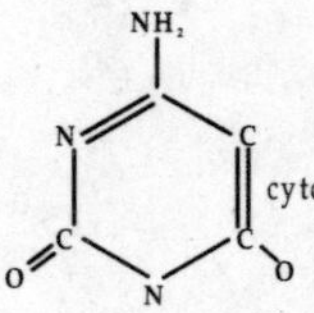

cytosine (a pyrimidine)

PHOSPHATES

Inorganic phosphate is a stable ion formed from phosphoric acid, H_3PO_4. It is often written as P_i.

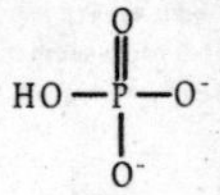

Phosphate esters can form between a phosphate and a free hydroxyl group. Phosphate groups are often attached to proteins in this way.

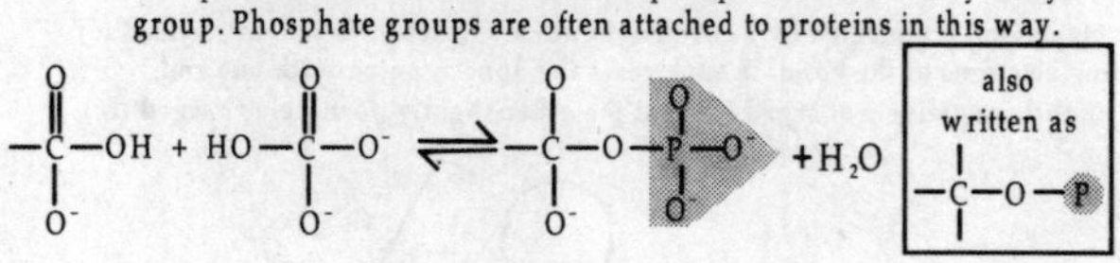

The combination of a phosphate and a carboxyl group, or two or more phosphate groups, gives an acid anhydride.

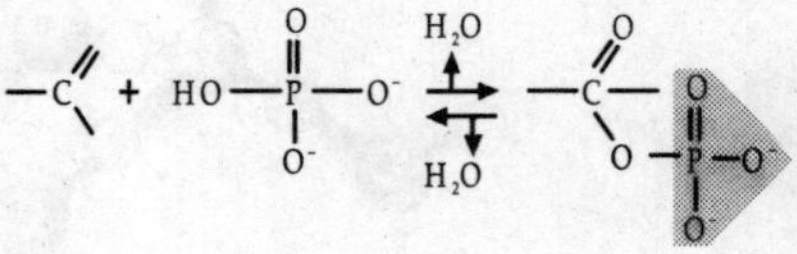

high-energy acyl phosphate bond (carboxylic–phosphoric acid anhydride) found in some metabolites

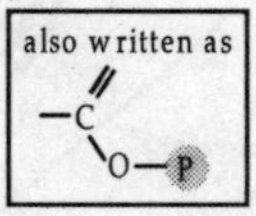

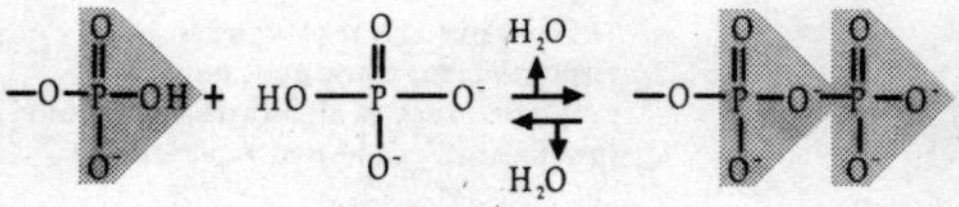

phosphoanhydride — a high energy bond found in molecules such as ATP

▫ chemical mutagen

A chemical capable of inducing mutations in living organisms.

▫ chemical properties of water part 1

The chemical properties of water and their influence on the behavior of biological molecules.

HYDROGEN BONDS

Because they are polarized, two adjacent H_2O molecules can form a linkage known as a **hydrogen bond**. Hydrogen bonds have only about 1/20 the strength of a covalent bond.

Hydrogen bonds are strongest when the three atoms lie in a straight line.

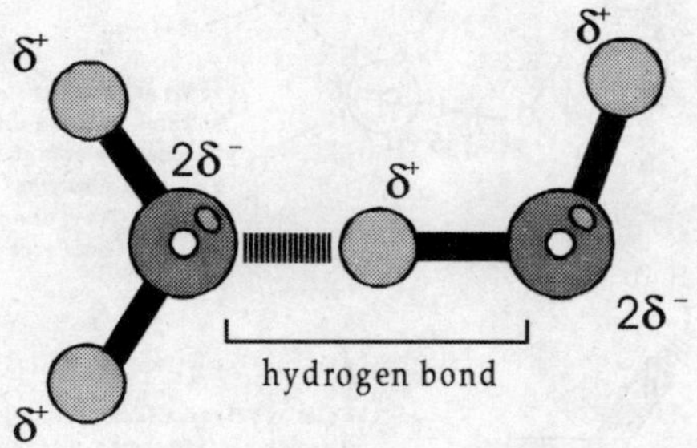

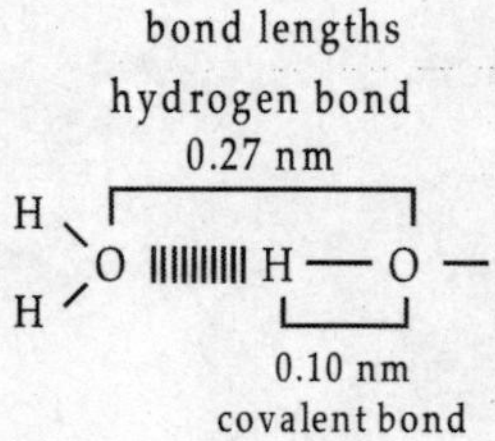

WATER

Two atoms, connected by a covalent bond, may exert different attractions for the electrons of the bond. In such cases the bond is polar, with one end slightly negatively charged (δ^-) and the other slightly positively charged (δ^+).

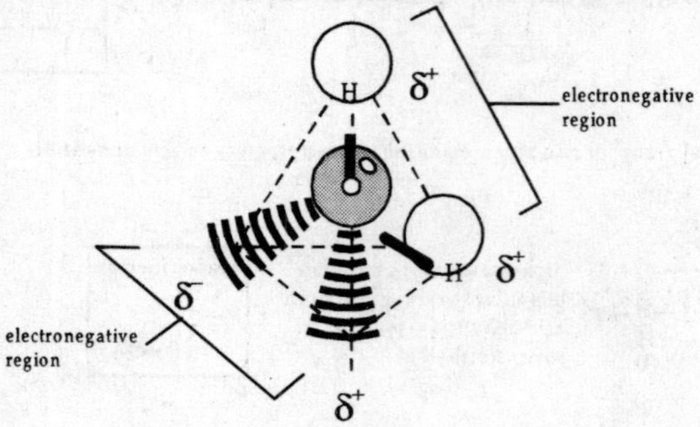

WATER STRUCTURE

Molecules of water join together transiently in a hydrogen-bonded lattice. Even at 37oC, 15% of the water molecules are joined to four others in a short-lived assembly known as a "flickering cluster."

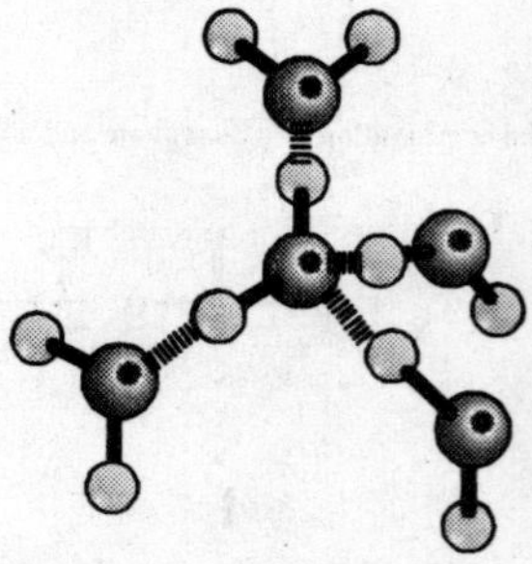

The cohesive nature of water is responsible for many of its unusual properties, such as high surface tension, specific heat, and heat of vaporization.

HYDROPHILIC MOLECULES

Substances that dissolve readily in water are termed **hydrophilic.** They are composed of ions or polar molecules that attract water molecules through electrical charge effects. Water molecules surround each ion or polar molecule on the surface of a solid substance and carry it into solution.

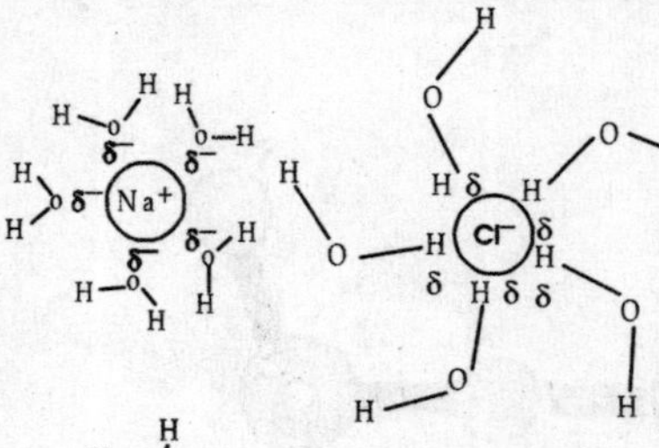

Ionic substances such as sodium chloride dissolve because water molecules are attracted to the positive (Na+) or negative (Cl_) charge of each ion.

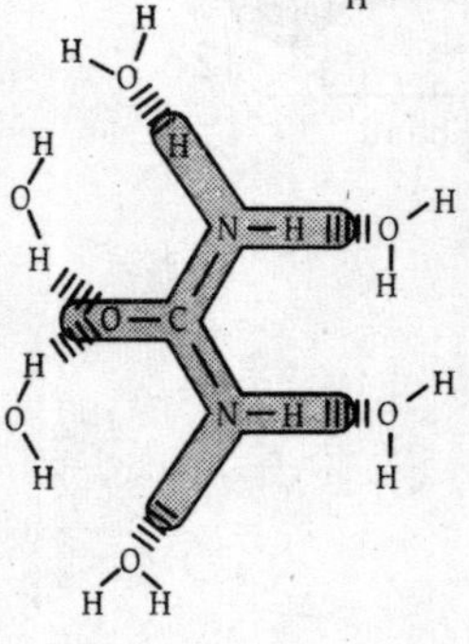

Polar substances such as urea dissolve because their molecules form hydrogen bonds with the surrounding water molecules

HYDROPHOBIC MOLECULES

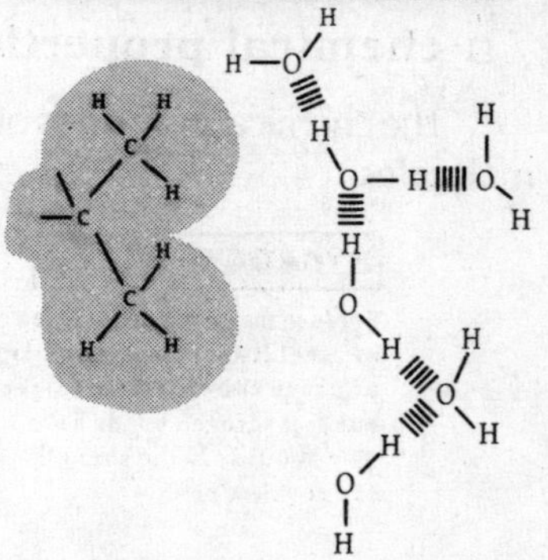

Molecules that contain a preponderance of nonpolar bonds are usually insoluble in water and are termed **hydrophobic.** This is true, especially, of hydrocarbons, which contain many C–H bonds. Water molecules are not attracted to such molecules and so have little tendency to surround them and carry them into solution.

❑ chemical properties of water part 2

The chemical properties of water and their influence on the behavior of biological molecules.

WATER AS A SOLVENT

Many substances, such as household sugar, dissolve in water. That is, their molecules separate from each other, each becoming surrounded by water molecules.

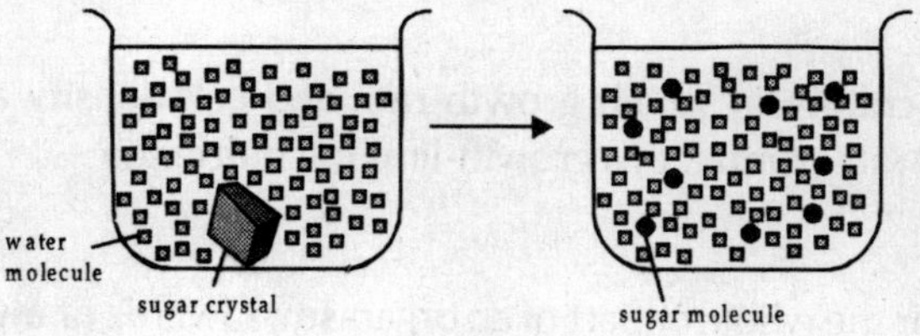

When a substance dissolves in a liquid, the mixture is termed a solution. The dissolved substance (in this case sugar) is the solute, and the liquid that does the dissolving (in this case water) is the solvent. Water is an excellent solvent for many substances because of its polar bonds.

ACIDS

Substances that release hydrogen ions into solution are called **acids.**

$$HCl \longrightarrow H^+ + Cl^-$$

hydrochloric acid (strong acid) → hydrogen ion + chloride ion

Many of the acids important in the cell are only partially dissociated, and they are therefore **weak acids**—for example, the carboxyl group (–COOH), which dissociates to give a hydrogen ion in solution

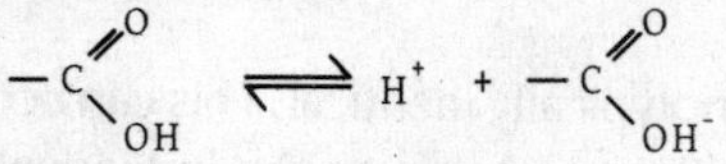

Note that this is a reversible reaction.

HYDROGEN ION EXCHANGE

Positively charged hydrogen ions (H+) can spontaneously move from one water molecule to another, thereby creating two ionic species.

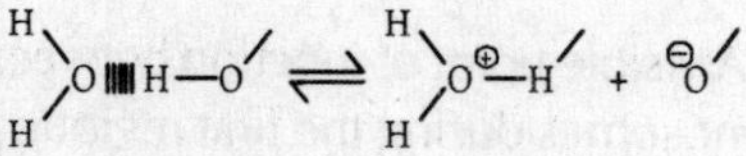

often written as: $H_2O \rightleftharpoons H^+ + OH^-$

hydrogen ion + hydroxyl ion

Since the process is rapidly reversible, hydrogen ions are continually shuttling between water molecules. Pure water contains a steady state concentration of hydrogen ions and hydroxyl ions (both 10-7 M).

pH

The acidity of a solution is defined by the concentration of H+ ions it possesses. For convenience we use the pH scale, where

$$pH = -\log_{10}[H^+]$$

For pure water

$[H^+] = 10^{-7}$ moles/liter

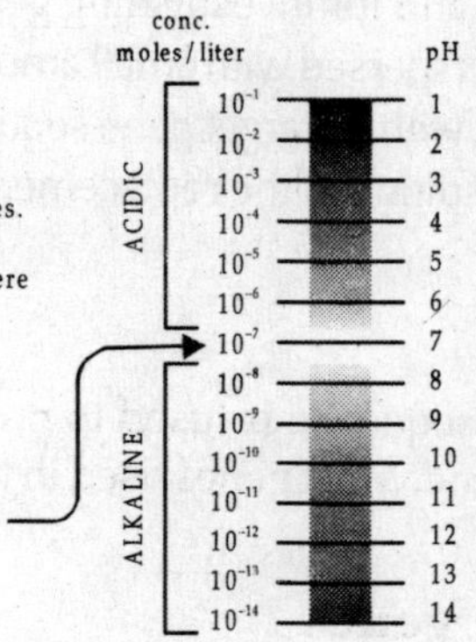

BASES

Substances that reduce the number of hydrogen ions in solution are called **bases.** Some bases, such as ammonia, combine directly with hydrogen ions.

$$NH_3 + H^+ \longrightarrow NH_4^+$$

ammonia + hydrogen ion → ammonium ion

Other bases, such as sodium hydroxide, reduce the number of H^+ ions indirectly, by making OH^- ions that then combine directly with H^+ ions to make H_2O.

$$NaOH \longrightarrow Na^+ + OH^-$$

sodium hydroxide (strong base) → sodium ion + hydroxyl ion

Many bases found in cells are partially dissociated and are termed **weak bases.** This is true of compounds that contain an amino group (–NH2), which has a weak tendency to reversibly accept an H+ ion from water, increasing the quantity of free OH- ions.

$$-NH_2 + H^+ \rightleftharpoons -NH_3^+$$

▫ chemically-defined medium

When all of the chemical components of a culture medium are fully known and defined.

▫ chemiluminescence

The emission of light during the course of a chemical reaction.

▫ chemostat

A continuous and open culture in which growth rate and cell density are maintained constant by a fixed rate of input of a growth-limiting nutrient.

▫ chemotaxis

The movement of a cell, or the whole or part of an organism, towards or away from an increasing concentration of a particular substance.

▫ chemotherapy

The treatment of disease, especially infections or cancer, by means of chemicals.

▫ chiasma

A visible point of junction between two non-sister chromatids of homologous chromosomes during the first meiotic prophase.

▫ chimera (or chimaera)

1. An organism whose cells are not all genotypically identical. This can occur as a result of: somatic mutation; grafting; or because the individual is derived from two or more embryos or zygotes.
2. A recombinant DNA molecule that contains sequences from different organisms.

▫ chimeraplasty

A method designed to create defined alterations in DNA sequence at a target locus, with potential both for gene therapy and for investigating gene function. A synthetic nucleic acid that contains DNA interspersed with small amounts of RNA is introduced into the target cell, where it pairs with its target gene sequence and then triggers the cell's DNA repair machinery, resulting in the replacement of the native sequence by the synthetic one.

▫ chimeric gene

An engineered gene, where a coding sequence is fused to promoter and/or other sequences derived from a different gene. Most genes used in transformation are chimeric.

▫ chimeric selectable marker gene

A gene that is constructed from parts of two or more different genes and allows the host cell to survive under conditions where it would otherwise die.

❐ chitin

A nitrogenous polysaccharide that gives structural strength to the exoskeleton of insects and the cell walls of fungi.

❐ chitinase

An enzyme which breaks down chitin.

❐ chloramphenicol

An antibiotic that interferes with protein synthesis.

❐ chlorenchyma

Plant tissue (leaf mesophyll and other parenchyma cells) containing chloroplasts.

❐ chlorophyll

One of the two pigments responsible for the green colour of most plants. It is an essential component of the machinery to absorb light energy for photosynthesis.

❐ chloroplast

Specialized plastid that contains chlorophyll. Lens-shaped and bounded by a double membrane, chloroplasts contain membranous structures (thylakoids) piled up into stacks, surrounded by a gel-like matrix (stroma). They are the site of solar energy transfer and some important reactions involved in starch or sugar synthesis. Chloroplasts have their own DNA; these genes are inherited only through the female parent, and are independent of nuclear genes.

❐ chloroplast dna

The DNA present in the chloroplast. Although the chloroplast has a small genome, the large number of chloroplasts per cell ensures that chloroplast DNA is a significant proportion of the total DNA in a plant cell.

❐ chloroplast transit peptide

A transit peptide that, when fused to a protein, acts to transport that protein into plant chloroplasts. Once inside the chloroplast, the transit peptide is cleaved off the protein used to target transgene expression to the chloroplast, where this is appropriate.

❐ chlorosis

The appearance of yellow colour in plants, due to the failure of development or the breakdown of chlorophyll. This is generally a symptom of either nutritional disturbance or of pathogen infection.

❐ chromatid

Each of the two strands of chromatin comprising a duplicated chromosome. The term is applied only while the two chromatids are joined at the centromere. As soon as the centromere divides, setting the two chromatids adrift (during anaphase of mitosis; and during anaphase II of meiosis), they are called chromosomes.

❐ chromatin

Substance of which eukaryotic chromosomes are composed. It consists of a complex of DNA, histone and non-histone chromosomal proteins (mainly histones), and a small amount of RNA.

❐ chromatin fibre

The standard structural conformation of chromatin in strands of 30 nm average diameter.

❐ chromatography

A method for separating the components of mixtures of molecules by partitioning them between two phases, one stationary and the other mobile. Appropriate selection of partitioning mechanism can produce separation of very closely-related molecules.

❐ chromocentre

Body produced by fusion of the heterochromatic regions of the chromosomes in the polytene tissues (e.g. the salivary glands) of certain *Diptera*.

❐ chromogenic substrate

A compound or substance that contains a colour-forming group.

❐ chromomeres

Small dense bodies identified by their characteristic size and linear arrangement along a chromosome.

❐ chromonema

An optically single thread forming an axial structure within each chromosome.

❐ chromoplast Plastid

containing pigments other than chlorophyll.

❐ chromosomal aberration

An abnormal change in chromosome structure or number, including deficiency, duplication, inversion, translocation, aneuploidy, polyploidy, or any other change from the normal pattern.. Although it can be a mechanism for enhancing genetic diversity, most alterations are fatal or debilitating, especially in animals.

❐ chromosomal integration site

A chromosomal location where foreign DNA can be integrated, often without impairing any essential function in the host organism.

❐ chromosomal polymorphism

The occurrence of one to several chromosomes in two or more alternative structural forms within a population; the structurally changed chromosomes are the result of chromosome mutations (i.e. any structural change involving the gain, loss or re-location of chromosome segments).

▢ Chromosome

Chromosome a structure of compact intertwined molecules of DNA found in the nucleus of cells. Chromosome contain the cell's genetic information. Humans normally have 46 chromosomes.

It is found in the nucleus of the cell and made of the [2: nucleic] acid [1: DNA] and [2: protein] called chromatin. It contains the [2: genetic] code that is transmitted when the sperm and egg combine to form a new organism. Small units on the chromosomes, called [2: genes], determine the hereditary characteristics, such as eye colour, leaf shape etc. of the organism. [2: Genetics chapter on chromosomes]; [3: chromosome number of various species]; [3: what are telomeres?]; [3: see human chromosmes (left) and chimpanzee chromosomes (right) (ref)] [3: chromosomes of human, chimpanzee, gorilla, and organgutan.

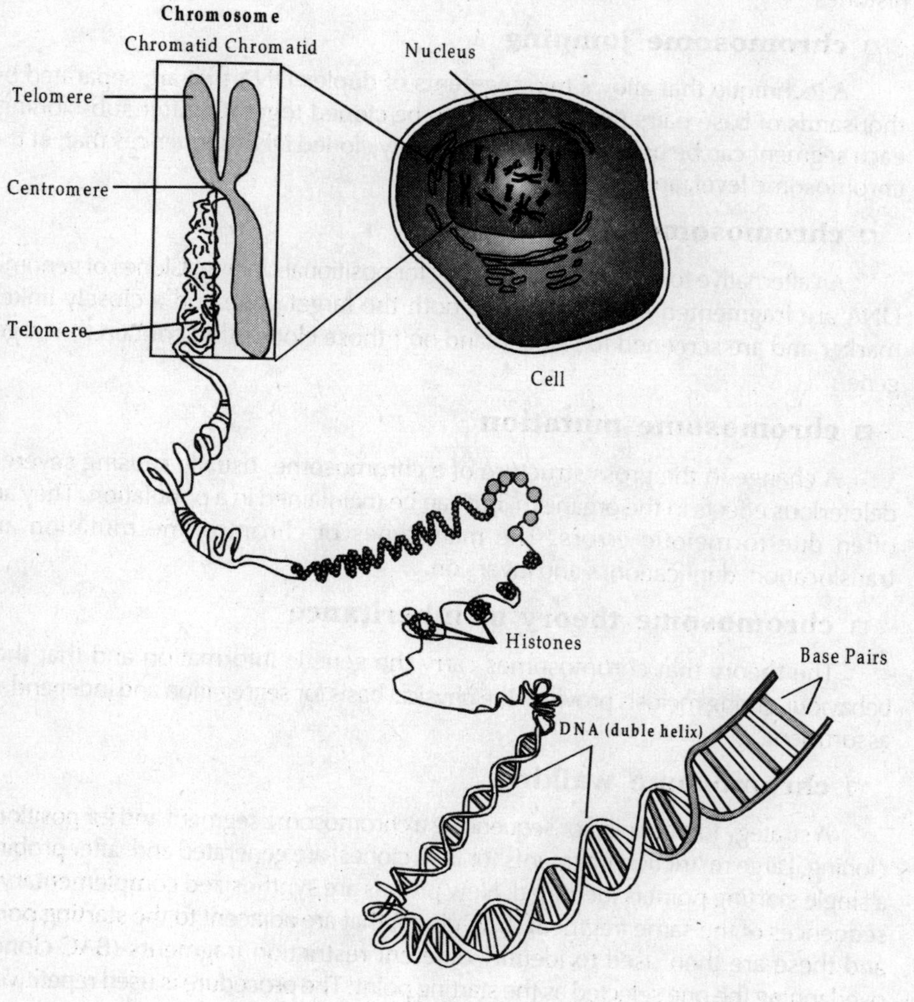

❒ chromosome banding

Differential staining of chromosomes in such a way that light and dark areas occur along the length of the chromosomes in repeatable patterns. Identical banding pattern implies chromosome homology.

❒ chromosome cells

In eukaryotic cells, chromosomes are the nuclear bodies containing most of the genes largely responsible for the differentiation and activity of the cell. Chromosomes are most easily studied in their contracted state, which occurs around the metaphase of mitosis or meiosis; they contain most of the cell's DNA in the form of chromatin. Each eukaryotic species has a characteristic number of chromosomes. Bacterial and viral cells contain only one chromosome, which consists of a single or double strand of DNA or, in some viruses, RNA, without histones.

❒ chromosome jumping

A technique that allows two segments of duplex DNA that are separated by thousands of base pairs (about 200 kb) to be cloned together. After sub-cloning, each segment can be used as a probe to identify cloned DNA sequences that, at the chromosome level, are roughly 200 kb apart.

❒ chromosome landing

An alternative to chromosome walking for positional cloning. Clones of genomic DNA are fragmented so as to include both the target gene and a closely linked marker and are screened to select ('land on') those clones that contain the target gene.

❒ chromosome mutation

A change in the gross structure of a chromosome, usually causing severely deleterious effects in the organism, but can be maintained in a population. They are often due to meiotic errors. The main types of chromosome mutation are translocation, duplication, s and inversion.

❒ chromosome theory of inheritance

The theory that chromosomes carry the genetic information and that their behaviour during meiosis provides the physical basis for segregation and independent assortment.

❒ chromosome walking

A strategy for mapping or sequencing a chromosome segment and for positional cloning. Large restriction fragments (or BAC clones) are generated and, after probing, a single starting point is identified. New probes are synthesized complementary to sequences of the same fragment (BAC clone) that are adjacent to the starting point, and these are then used to identify different restriction fragments (BAC clones) overlapping the one selected as the starting point. The procedure is used repetitively, walking away from the starting point.

❐ chymosin

An enzyme that clots milk, used in the manufacture of cheese.

❐ cilium

Hairlike locomotor structure on certain cells; a locomotor structure on a ciliate protozoan.

❐ circadian

Cycle of physiological activity, etc.: recurring at approximately 24-hour intervals.

❐ circularization

The self-ligation of a linear DNA fragment having complementary ends, generally generated by digestion with a restriction endonuclease. Successful ligation produces a molecule in the form of a covalently-closed circle. Plastid DNA and plasmids are examples of naturally circularized DNA.

❐ *cis* heterozygote

A double heterozygote that contains two mutations arranged in a *cis* configuration (e.g. *a*+ *b*+ / *a b)*.

❐ *cis*-acting protein

A protein with the particular property of acting only on the molecule of DNA from which it was expressed.

❐ *cis*-acting sequence

A nucleotide sequence that only affects the expression of genes located on the same chromosome.

❐ cistron

A DNA sequence that codes for a specific polypeptide; a gene.

❐ citric acid cycle

(also called the Kreb's Cycle and the Tricarboxylic Acid cycle) - this cycle stores energy, released by the oxidation of fats, proteins, and carbohydrates in foodstuffs, in high energy phosphate bonds of ATP. About 90 percent of the energy released from food occurs in the Citric Acid Cycle. In the process, a series of acids are oxidized to release the energy used in forming high-energy. ATP phosphate bonds, plus carbon dioxide and water. ATP is life's universal energy supply.

The reaction of acetyl CoA with oxaloacetate starts the cycle by producing citrate (citric acid). In each turn of the cycle, two molecules of CO_2 are produced as waste products, plus three molecules of NADH, one molecule of GTP, and one molecule of $FADH_2$. The number of carbon atoms in each intermediate is shown in a yellow box.

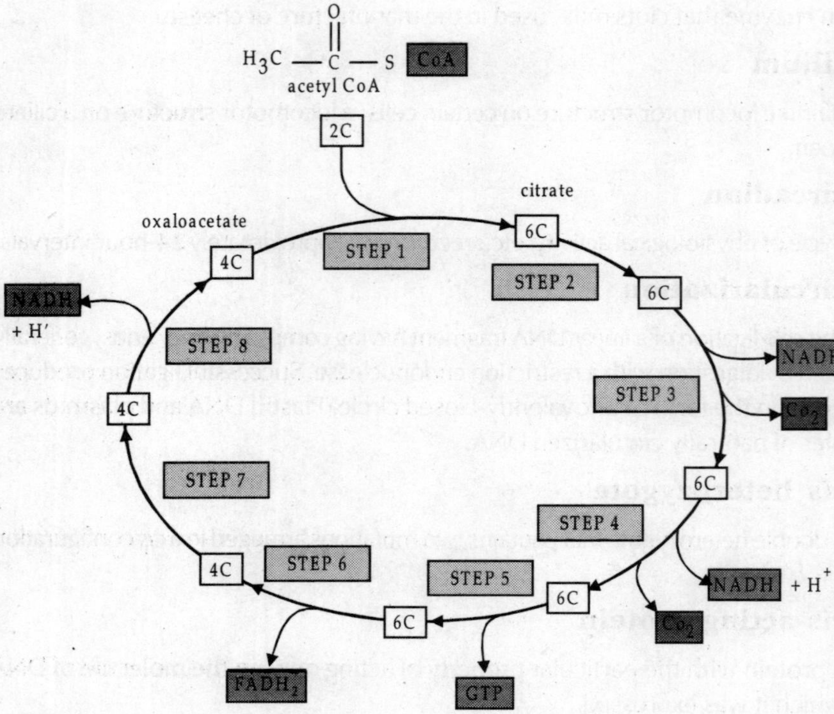

Net Result: One turn of the cycle produces three NADH, ONE GTP, AND ONE FADH$_2$, AND RELEASES TWO MOLECULES OF CO

❑ citric acid cycle - part 1 (overview and steps 1 to 8)

It also called the Kreb's Cycle and the Tricarboxylic Acid cycle.This cycle stores energy, released by the oxidation of fats, proteins, and carbohydrates in foodstuffs, in high energy phosphate bonds of ATP. About 90 percent of the energy released from food occurs in the Citric Acid Cycle. In the process, a series of acids are oxidized to release the energy used in forming high-energy. ATP phosphate bonds, plus carbon dioxide and water. ATP is life's universal energy supply.

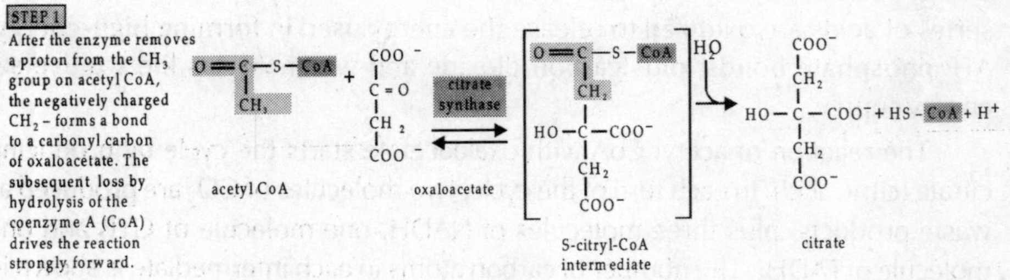

STEP 2

An isomerization reaction, in which water is first removed and then added back, moves the hydroxyl group from one carbon atom to its neighbor.

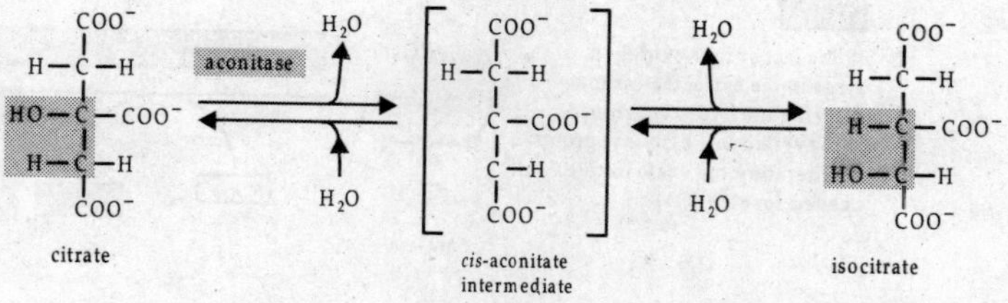

STEP 3

In the first of four oxidation steps in the cycle, the carbon carrying the hydroxyl group is converted to a carbonyl group. The immediate product is unstable, losing CO_2 while still bound to the enzyme.

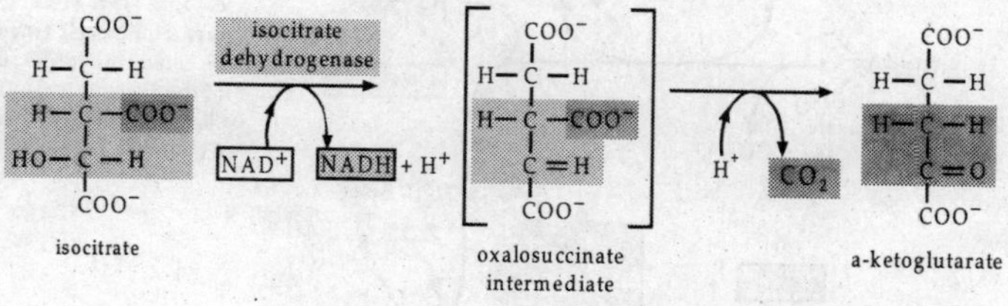

STEP 4

The *aketoglutarate dehydrogenase complex* closely resembles the large enzyme complex that converts pyruvate to acetyl CoA (*pyruvate dehydrogenase*). It likewise catalyzes an oxidation that produces NADH, CO_2, and a high-energy thioester bond to coenzyme A (CoA).

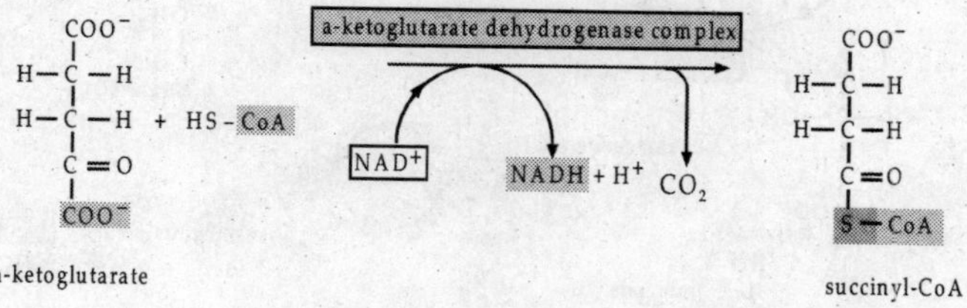

STEP 5

A phosphate molecule from solution displaces the CoA, forming a high-energy phosphate linkage to succinate. This phosphate is then passed to GDP to form GTP. (In bacteria and plants, ATP is formed instead.

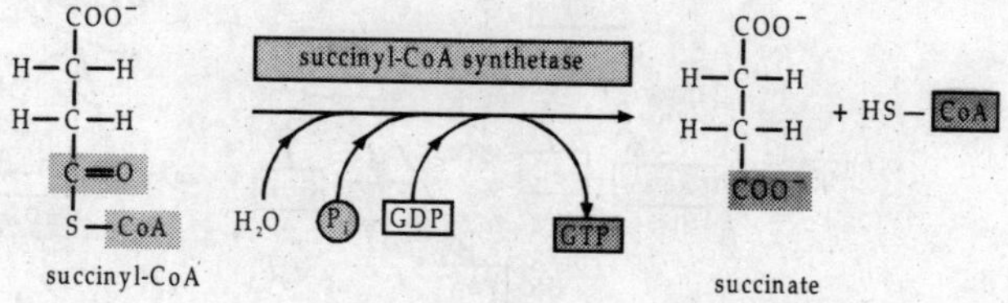

STEP 6

In the third oxidation step in the cycle, FAD removes two hydrogen atoms from succinate.

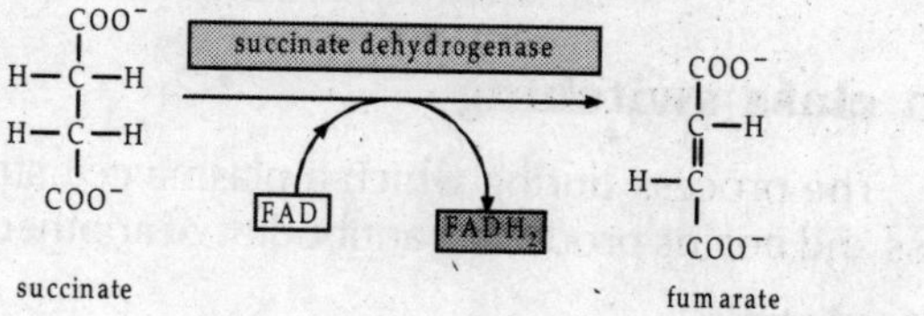

STEP 7

The addition of water to fumarate places a hydroxyl group next to a carbonyl carbon.

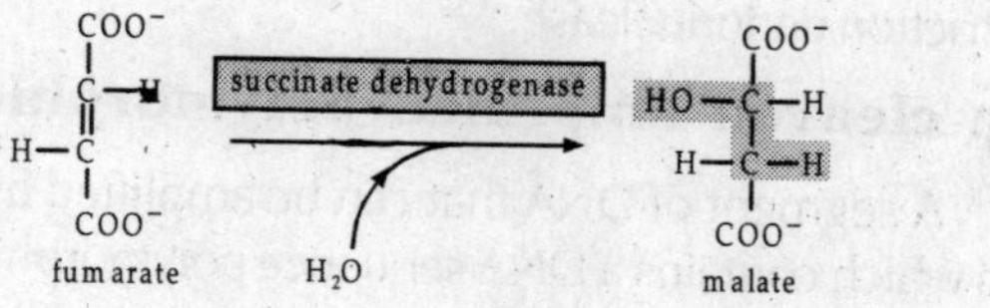

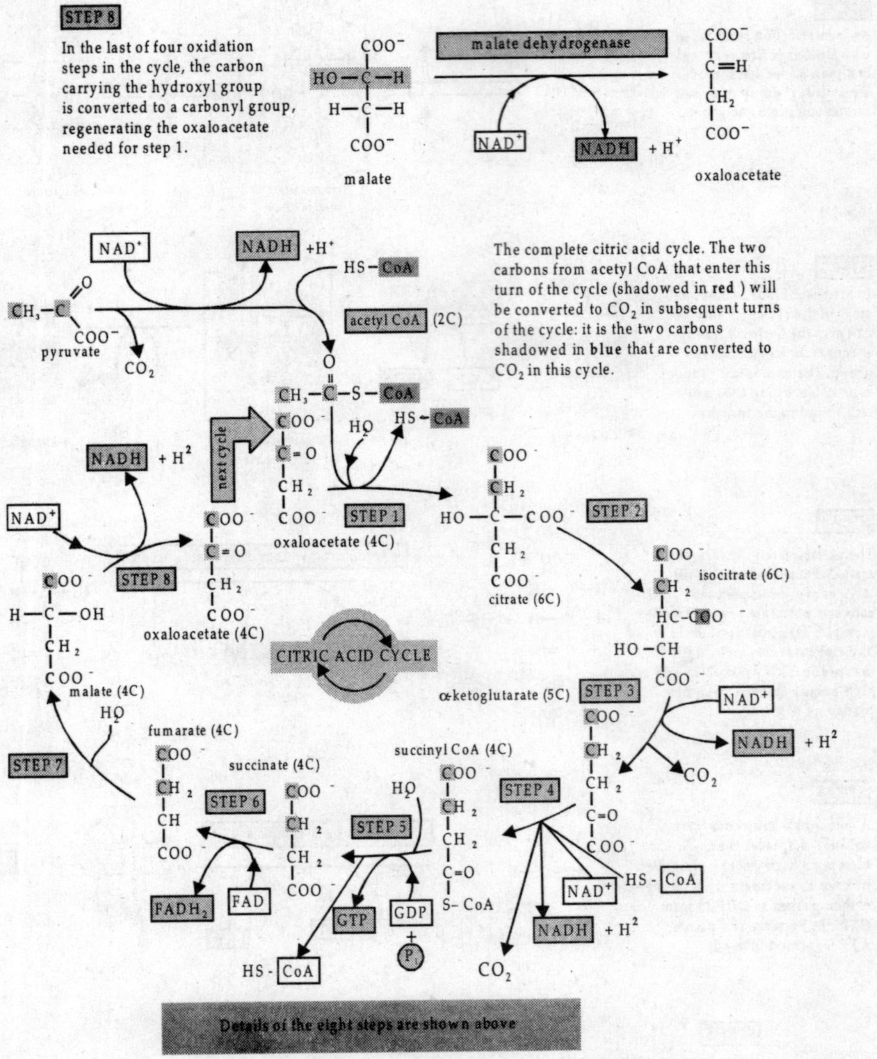

□ class switching

The process during which a plasma cell stops producing antibodies of one class and begins producing antibodies of another class.

□ cleave

To break phosphodiester bonds of double-stranded DNA, usually with a type II restriction endonuclease.

□ cleaved amplified polymorphic sequence

A segment of DNA that can be amplified by polymerase chain reaction (PCR) and which contains a DNA sequence polymorphism. Following PCR amplification of a locus, the amplicon is treated with a restriction endonuclease. If the recognition site for this enzyme is present in the amplicon, two or more restriction fragments are generated. Thus sequence variation between individuals at the recognition site(s) can be detected by electrophoresis.

❐ cline

Variation in one or more phenotypic characters or allele frequencies across a geographical gradient.

❐ clonal propagation

Asexual propagation of many new plants (ramets) from an individual (ortet); all have the same genotype.

❐ clonal selection

The production of a population of plasma cells all producing the same antibody in response to the interaction between a B lymphocyte producing that specific antibody and the antigen bound by that antibody.

❐ clone

1. A group of cells or individuals that are genetically identical as a result of asexual reproduction, breeding of completely inbred organisms, or forming genetically identical organisms by nuclear transplantation.
2. Group of plants genetically identical in which all are derived from one selected individual by vegetative propagation.
3. Verb: to clone. To insert a DNA segment into a vector or host chromosome.

❐ cloned strain or line

A strain or line descended directly from a clone.

❐ cloning from adult vertebrate cells

In nature, cloning is common among plants and is used extensively in plant propagation. The offspring produced by cloning (and other methods of asexual reproduction) simply develop from cells produced by the parent. Because the offspring have genetic information identical to the parent, they develop very similar characteristics to their parent (and to one another) as they mature. This means that an agriculturalist who grows new plants from pieces of an older plant ensures that the new crop will be a fairly uniform one. In contrast, an individual formed by sexual reproduction develops from a cell produced by the union of two cells, usually from different parents. Offspring produced in this way are not genetically identical to each other or to their parents, unlike offspring formed by asexual methods of reproduction.

The only clones produced naturally in vertebrate animals are identical twins. These are formed when cells produced by the early divisions of the fertilised egg separate and independently develop into new individuals. They are therefore genetically identical to each other but not identical to their parents.

The nucleus of a differentiated cell from the skin of an adult frog contains all the instructions, encoded in DNA, to control the formation of an entire tadpole. A similar experiment was recently performed in sheep.

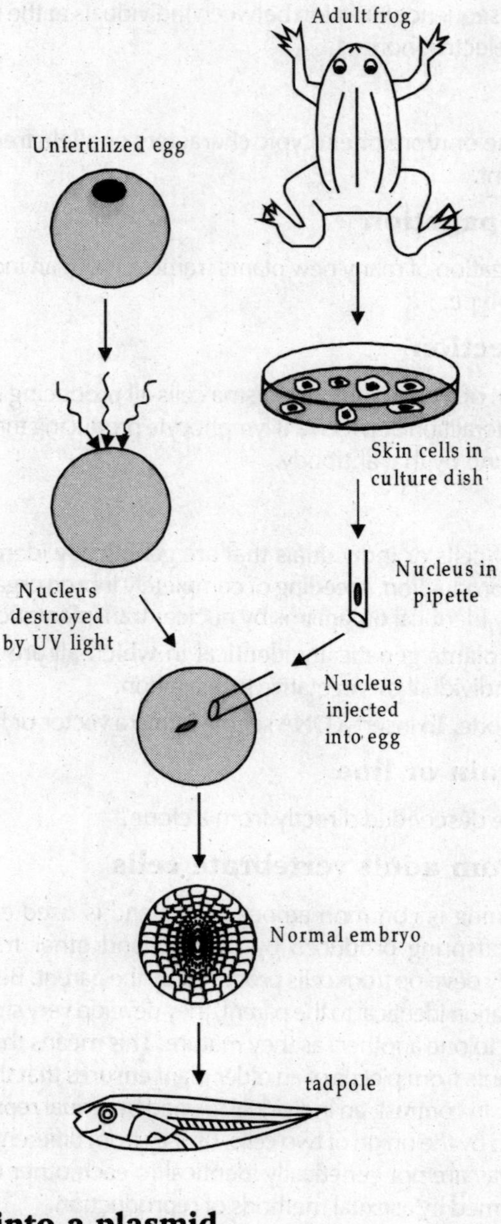

◻ cloning into a plasmid

Process by which a plasmid is used to import recombinant DNA into a host cell for cloning.

Many diseases are caused by gene alterations. Our understanding of genetic diseases was greatly increased by information gained from DNA cloning. In DNA cloning, a DNA fragment that contains a gene of interest is inserted into a cloning vector or plasmid.

The plasmid carrying genes for antibiotic resistance, and a DNA strand, which contains the gene of interest, are both cut with the same restriction endonuclease. The plasmid is opened up and the gene is freed from its parent DNA strand. They have complementary "sticky ends." The opened plasmid and the freed gene are mixed with DNA ligase, which reforms the two pieces as recombinant DNA.

Plasmids + copies of the DNA fragment produce quantities of recombinant DNA.

This recombinant DNA stew is allowed to transform a bacterial culture, which is then exposed to antibiotics. All the cells except those which have been encoded by the plasmid DNA recombinant are killed, leaving a cell culture containing the desired recombinant DNA.

DNA cloning allows a copy of any specific part of a DNA (or RNA) sequence to be selected among many others and produced in an unlimited amount. This technique is the first stage of most of the genetic engineering experiments: production of DNA libraries, PCR, DNA sequencing, et al.

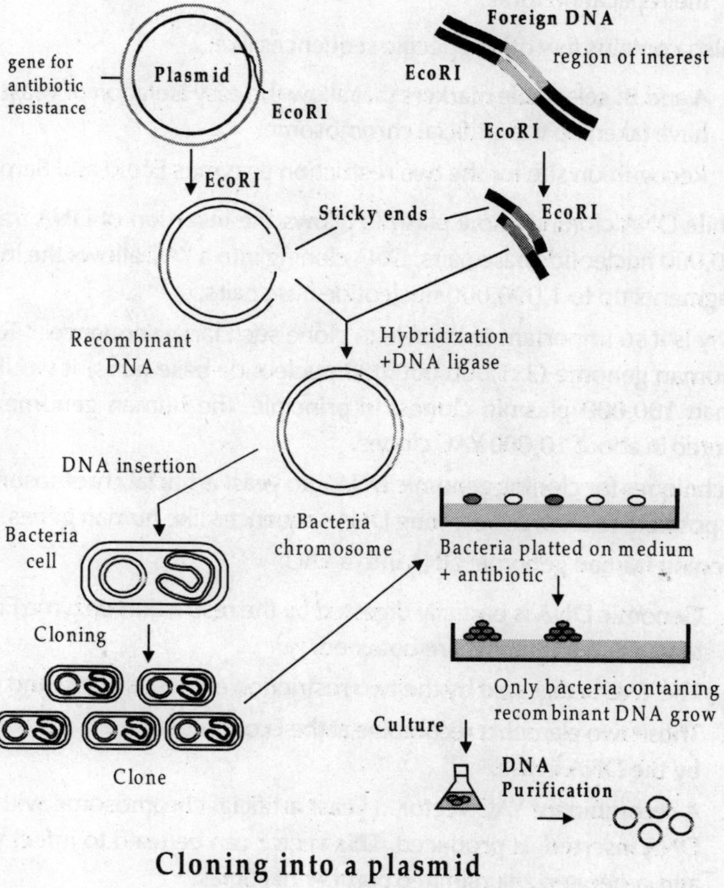

Cloning into a plasmid

▫ cloning into a yac

DNA cloning is a technique that allows the wholesale production of a specific DNA sequence. DNA containing a gene of interest is inserted into the purified DNA genome of a self replicating element, which can be a plasmid, a virus or in this case, a yeast artificial chromosome (YAC).

A YAC can be considered as a functional artificial chromosome (self replicating element), since it includes three specific DNA sequences that enable it to propagate from one cell to its offspring:

- **TEL:** The telomere which is located at each chromosome end, protects the linear DNA from degradation by nucleases.
- **CEN:** The centromere which is the attachment site for mitotic spindle fibers, "pulls" one copy of each duplicated chromosome into each new daughter cell.
- **ORI:** Replication origin sequences which are specific DNA sequences that allow the DNA replication machinery to assemble on the DNA and move at the replication forks.

It also contains few other specific sequences like:

- **A and B:** selectable markers that allow the easy isolation of yeast cells that have taken up the artificial chromosome.
- Recognition site for the two restriction enzymes EcoRI and BamHI.

While DNA cloning into a plasmid allows the insertion of DNA fragment of about 10,000 nucleotide base pairs, DNA cloning into a YAC allows the insertion of DNA fragments up to 1,000,000 nucleotide base pairs.

Why is it so important to be able to clone such large sequences? To map the entire human genome (3x1,000,000,000 nucleotide base pairs) it would require more than 100,000 plasmid clones. In principle, the human genome could be represented in about 10,000 YAC clones.

Techniques for cloning genomic DNA into yeast artificial chromosomes (YAC) make it possible to analyze very long DNA sequences like human genes.

Cloning human genomic DNA into a YAC:

1. Genomic DNA is partially digested by the restriction enzyme EcoRI. Very large DNA fragments are obtained.
2. The YAC is digested by the two restriction enzymes EcoRI and BamHI.
3. Those two elements recombine at the EcoRI sites and are covalently linked by the DNA ligase.
4. A recombinant YAC vector, a yeast artificial chromosome with genomic DNA inserted, is produced. This vector can be used to infect yeast cells and generated an unlimited number of copies.

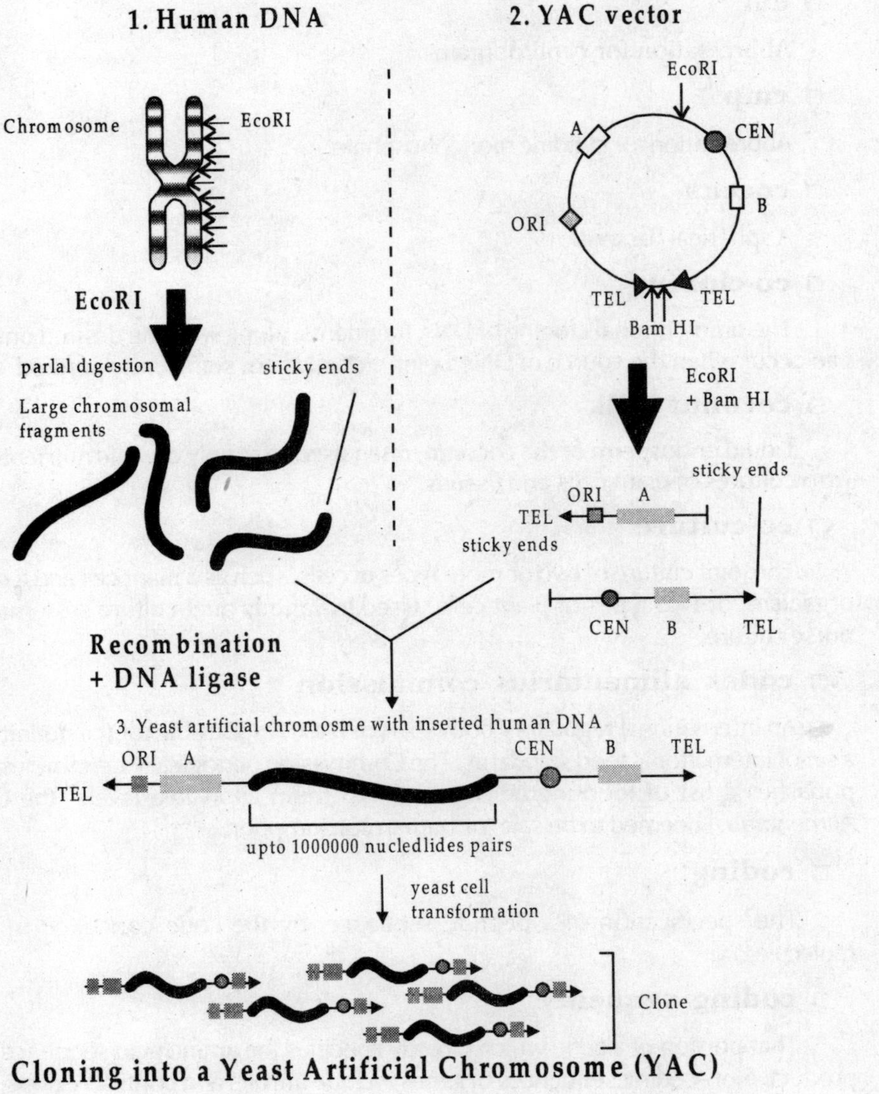

Cloning into a Yeast Artificial Chromosome (YAC)

❑ cloning vector

A small, self-replicating DNA molecule - usually a plasmid or viral DNA chromosome - into which foreign DNA is inserted in the process of cloning genes or other DNA sequences of interest. It can carry inserted DNA and be perpetuated in a host cell.

❑ closed continuous culture

A culture system, in which the inflow of fresh medium is balanced by the outflow of corresponding volumes of spent medium. Cells are separated mechanically from outflowing medium and added back to the culture.

❐ cm

Abbreviation for centiMorgan.

❐ cmp

Abbreviation for cytidine monophosphate.

❐ coccus

A spherical bacterium.

❐ co-cloning

The unintentional cloning of DNA fragments, along with the desired one, that can occur when the source of DNA being cloned is not sufficiently purified.

❐ coconut milk

Liquid endosperm of the coconut, often used to supply organic nutrients to *in vitro* cultures of plant cells and tissues.

❐ co-culture

The joint culture of two or more types of cells, such as a plant cell and a micro-organism, or two types of plant cells. Used in various dual-culture systems or in nurse culture.

❐ codex alimentarius commission

An international regulatory body (part of FAO) responsible for the definition of a set of international food standards. The Commission periodically determines, then publishes a list of food ingredients and maximum allowable levels (the *Codex Alimentarius*) deemed to be safe for human consumption.

❐ coding

The specification of a peptide sequence, by the code contained in DNA molecules.

❐ coding sequence

That portion of a gene which directly specifies the amino acid sequence of its product. Non-coding sequences of genes include introns and control regions, such as promoters, operators, and terminators.

❐ coding strand

The strand of a DNA double helix that contains the same base sequence (after substituting U for T) found in the mRNA molecule resulting from transcription of that segment of DNA. Sometimes called the sense strand. The mRNA molecule is transcribed from the other strand, known as the template or antisense strand. *See*: antisense DNA.

❐ co-dominance

Where both alleles are expressed in the heterozygous state, so that the phenotype reflects a contribution from both alleles. For example, roan coat colour in cattle results from a mixture of red hairs and white hairs, caused by heterozygosity for the red allele and the white allele.

▫ codon

Three bases in a DNA or RNA sequence which specify a single amino acid.

One of the groups of three consecutive nucleotides in mRNA, which represent the unit of genetic coding by specifying a particular amino acid during the synthesis of polypeptides in a cell. Each codon is recognized by a tRNA carrying a specific amino acid, which is incorporated into a polypeptide chain during protein synthesis. In DNA, any informative triplet of bases, including both coding and control sequences.

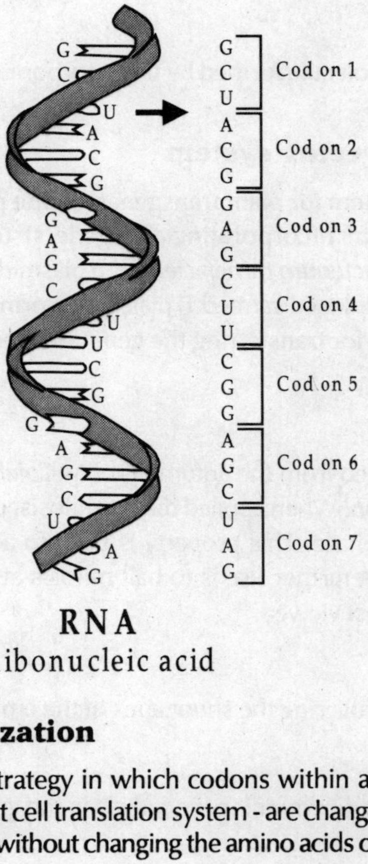

RNA
Ribonucleic acid

▫ codon optimization

An experimental strategy in which codons within a cloned gene - ones not generally used by the host cell translation system - are changed by *in vitro* mutagenesis to the preferred codons, without changing the amino acids of the synthesized protein.

▫ co-enzyme

Synonym for co-factor.

▫ co-evolution

The evolution of complementary adaptations in two species brought about by the selection pressure that each exerts on the other. Common in symbiotic associations, in insect-pollinated plants, etc.

❐ co-factor

An organic molecule or inorganic ion necessary for the normal catalytic activity of an enzyme.

❐ co-fermentation

The simultaneous growth of two micro-organisms in one bioreactor.

❐ coincidence

The ratio of the observed to the expected frequency of double cross-overs, where the expected frequency is calculated by assuming that the two cross-over events occur independently of one another.

❐ co-integrate

A chimeric DNA molecule formed by the incorporation at a single site of two different DNA molecules.

❐ co-integrate vector system

A two plasmid system for plant transgenesis. One plasmid is engineered to carry a T-DNA segment incorporating the gene(s) to be introduced. After introduction into *Agrobacterium tumefaciens*, the plasmid undergoes homologous recombination with a resident disarmed Ti plasmid to form a single plasmid carrying the genetic information for transferring the genetically engineered T-DNA region to plant cells.

❐ colchicine

An alkaloid, obtained from the autumn crocus *Colchicum autumnale*, which inhibits spindle formation. When applied during mitosis, chromosomes are unable to separate during anaphase. This property is used to achieve a doubling of the chromosome number. A further use is to halt mitosis at metaphase, the stage at which karyotypes are best viewed.

❐ coleoptile

Protective sheath covering the shoot apex of the embryo in the grasses.

❐ coleorhiza

A protective sheath surrounding the radicle in the grasses.

❐ co-linearity

1. A general relationship in which the units in one molecule occur in the same sequence as the units in another molecule which they specify; e.g. the nucleotides in a gene are co-linear with the amino acids in its encoded polypeptide.
2. The phenomenon whereby gene order is preserved between distinct species.

▫ collaboration of proteins during replication

The major types of proteins, which must work together during the replication of DNA, are illustrated, showing their positions.

When DNA replicates, many different proteins work together to accomplish the following steps:

1. The two parent strands are unwound with the help of DNA helicases.
2. Single stranded DNA binding proteins attach to the unwound strands, preventing them from winding back together.
3. The strands are held in position, binding easily to DNA polymerase, which catalyzes the elongation of the leading and lagging strands. (DNA polymerase also checks the accuracy of its own work!).
4. While the DNA polymerase on the leading strand can operate in a continuous fashion, RNA primer is needed repeatedly on the lagging strand to facilitate synthesis of Okazaki fragments. DNA primase, which is one of several polypeptides bound together in a group called primosomes, helps to build the primer.
5. Finally, each new Okazaki fragment is attached to the completed portion of the lagging strand in a reaction catalyzed by DNA ligase.

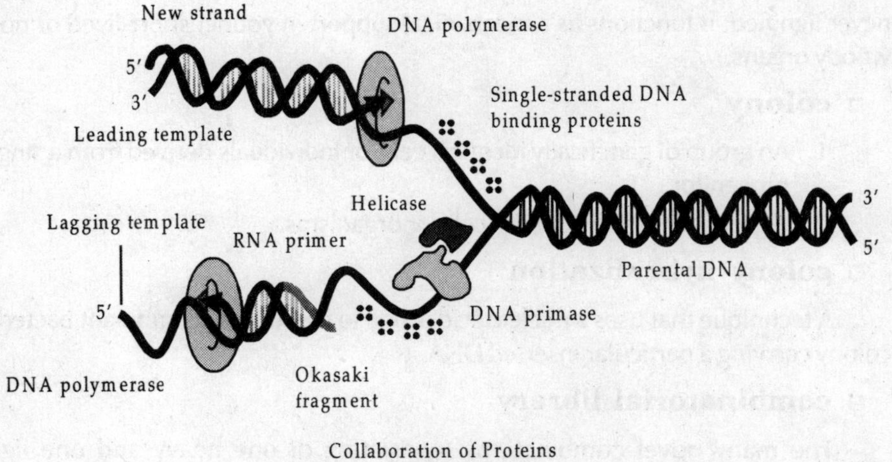

Collaboration of Proteins at the Replication Fork

▫ collagen and elastin

(A) Collagen is a triple helix formed by three extended protein chains that wrap around one another. Many rodlike collagen molecules are cross-linked together in the extracellular space to form unextendable collagen fibrils (top) that have the tensile strength of steel. The striping on the collagen fibril is caused by the regular repeating arrangement of the collagen molecules within the fibril. (B) Elastin polypeptide chains are cross-linked together to form rubberlike, elastic fibers. Each elastin molecule uncoils into a more extended conformation when the fiber is stretched and will recoil spontaneously as soon as the stretching force is relaxed.

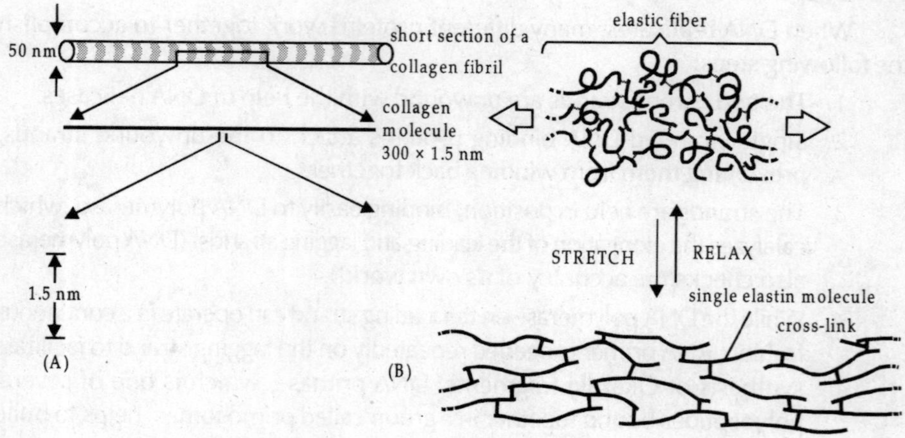

▫ collenchyma

A tissue of living cells, found particularly in midribs and leaf petioles. Characterized by cell walls unevenly thickened with cellulose and hemicellulose, but never lignified; it functions as a mechanical support in young, short-lived or non-woody organs.

▫ colony

1. An group of genetically identical cells or individuals derived from a single progenitor.
2. A group of interdependent cells or organisms.

▫ colony hybridization

A technique that uses a nucleic acid probe to identify a recombinant bacterial colony carrying a particular inserted DNA.

▫ combinatorial library

The many novel combinations (consisting of one heavy and one light immunoglobulin chain coding region) that are generated when a heavy-chain library is combined by random pairing with a light-chain library. These constructs are propagated in a vector, and their gene products screened for novel affinity properties.

▫ commensalism

The interaction of two or more dissimilar organisms where the association is advantageous to one without affecting the other(s).

▫ companion cell

A living cell associated with the sieve cell of phloem tissue in vascular plants.

❐ comparative mapping

The comparison of map locations of genes and markers between species. In comparisons between closely related species, this will usually uncover a high degree of conservation of synteny and co-linearity. In these cases, the likely location of many genes can be predicted from model system data. Comparisons across wider phylogenetic distances reveal increasing loss of synteny.

❐ comparative positional candidate gene

Relates to an indirect means of assigning function to a QTL. Where a QTL has been linked to a marker in one species, and that same marker is linked to a known gene in a model system, inferences may be drawn as to the nature of the QTL.

❐ comparative scale of mapping

The relative mapping of cells, chromosomes and genes illustrated by comparing them to the relative mapping of features on Earth.

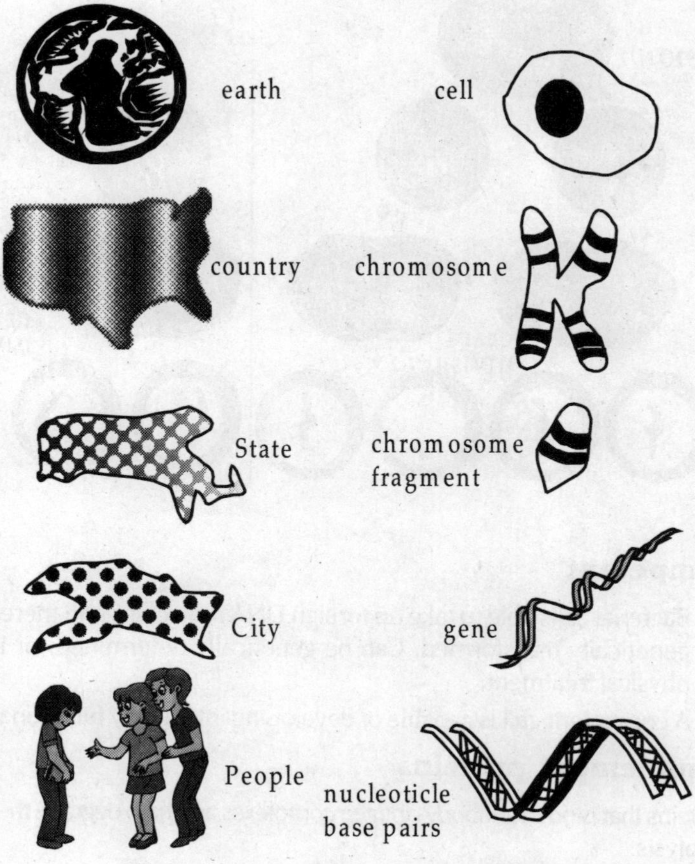

Comparative Scale of Mapping

▫ comparison of meiosis and mitosis

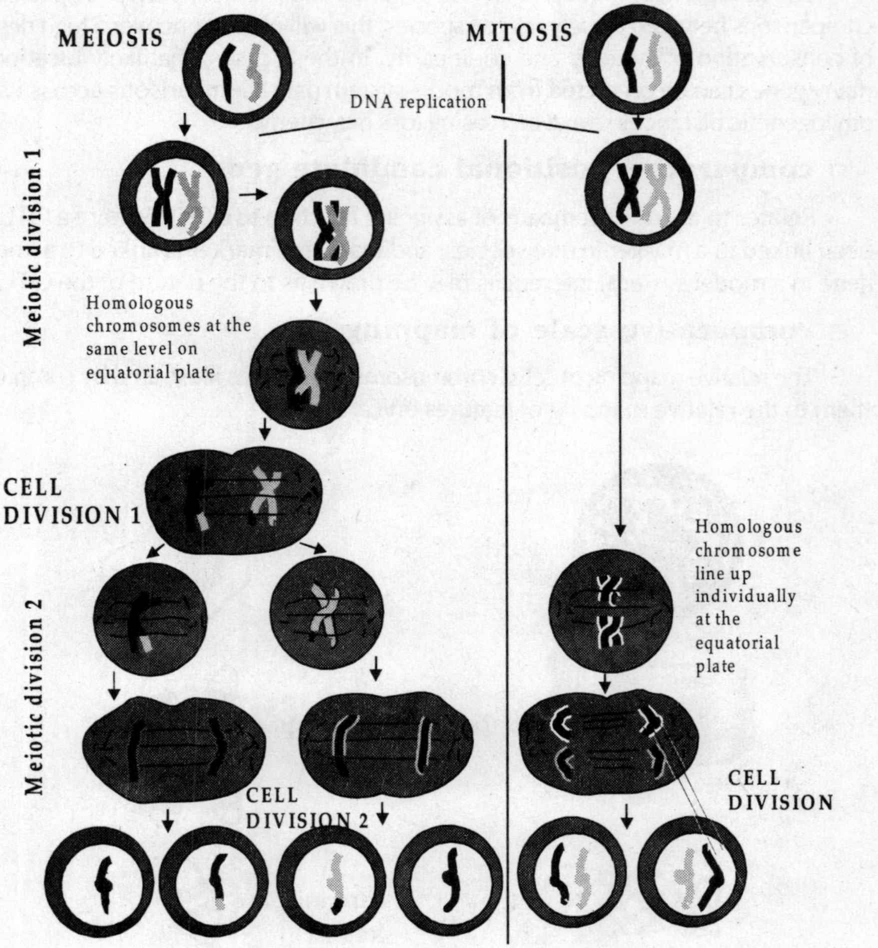

▫ competent

1. Bacterial cells able to take up foreign DNA molecules and thereby become genetically transformed. Can be genetically determined, or induced by physical treatment.
2. A competent cell is capable of developing into a fully functional embryo.

▫ complement proteins

Proteins that bind to antibody-antigen complexes and help degrade the complexes by proteolysis.

▫ complementarity

The similar correspondence between DNA and the mRNA transcribed from it.

complementarity-determining regions

Regions of the light and heavy subunits of the immunoglobulin molecules that interact with the antigen. The primary amino acid sequences of these regions are highly variable between antibodies of the same class. *See*: antibody binding site.

complementary

Two DNA molecules are complementary to one another when each successive base position from the 5' end in the first molecule is matched by the corresponding residue in the second, starting at the 3' end, according to the normal base pair rules (i.e. A for T, C for G). In the appropriate conditions, two complementary single-stranded DNA molecules will renature to form a double-stranded molecule. Complementary nucleotides are members of the pairs adenine-thymine, adenine-uracil, and guanine-cytosine that have the ability to hydrogen bond to one another.

complementary DNA

A DNA strand synthesized *in vitro* from a mature RNA template using reverse transcriptase. DNA polymerase is then used to create a double-stranded molecule. Differs from genomic DNA by the absence of introns. *Synonym*: copy DNA.

complementary entity

1. Synonym of base pair.
2. One of a pair of segments or strands of nucleic acid that will hybridize with one another.

complementary genes

Two or more interdependent genes, such that (in the case of dominant complementarity) the dominant allele from either gene can only produce an effect on the phenotype of an organism if the dominant allele from the other gene is also present; in the case of recessive complementarity, only double homozygous recessive individuals show the effect.

complementary homopolymeric tailing

The process of adding complementary nucleotide extensions to DNA molecules, (e.g. deoxyguanosine) to the 3' end of one DNA molecule and deoxycytidine to the 5' end of another DNA molecule) to facilitate the ligation of the two DNA molecules. *Synonyms*: dA - dT tailing, dG - dC tailing.

complementation test

A genetic method to test whether or not independent mutations are allelic. In a cross between the two mutant individuals, the genotype will be m_1m_2 if the mutations are allelic and m_1 +/+ m_2 if non-allelic. The phenotype of the former will be mutant, but that of the latter will be wild type (normal).

complete digest

The treatment of a DNA preparation with a restriction endonuclease for sufficient time so that all of the potential target sites within that DNA are cleaved.

❐ composite transposon

A transposon formed when two identical or nearly identical transposons insert on either side of a non-transposable segment of DNA.

❐ compound chromosome

A chromosome formed by the fusion of two separate chromosomes, as in attached-X chromosomes or attached-X-Y chromosomes.

❐ concatamer

A DNA segment made up of repeated sequences linked head to tail.

❐ concordance

Identity of matched pairs or groups for a given trait, such as sibs expressing the same trait.

❐ conditional lethal mutation

A mutation that is lethal under one set of environmental conditions (the restrictive conditions, commonly associated with high temperature) but is viable under another set of environmental conditions (the permissive conditions).

❐ conditioning

1. The effects on phenotypic characters of external agents during critical developmental stages.
2. The undefined interaction between tissues and culture medium resulting in the growth of single cells or small aggregates. Conditioning may be accomplished by immersing cells or callus contained within a porous material (such as dialysis tubing) into fresh medium for a period dependent on cell density and a volume related to the amount of fresh medium.

❐ conformation

The various three dimensional shapes that can be adopted by a given molecule. In particular, the different ways in which the primary sequence of a biological polymer may be folded. This is determined by intra-molecular forces, including hydrogen bonding and, in proteins, disulphide bridges. In proteins, conformation is often critical for biological activity, and the functions of some molecules are carried out by switching between two alternative stable conformations. The native conformation found *in vivo* may be changed to typically less ordered, uncharacterized, and usually biologically-inactive forms by denaturing.

❐ conidium

An asexual spore produced by a specialized hypha in certain fungi.

❐ conjugation

1. Union of gametes or unicellular organisms during fertilization.
2. The unidirectional transfer of plasmid DNA from one bacterium cell to another, involving cell-to-cell contact. The plasmid usually encodes the

majority of the functions necessary for its own transfer.

3. Attachment of sugar and other polar molecules to less polar compounds, thus making them more water soluble.

❐ **conjugative functions**

Plasmid-based genes and their products that facilitate the transfer of a plasmid from one bacterium to another via conjugation.

❐ **consanguinity**

Related by descent from a common ancestor.

❐ **consensus sequence**

The part of a gene or signal sequence that is shared over a wide range of members of a gene family, both within a given species, or in comparisons between species.

❐ **conserved sequence**

An identical or highly similar sequence of nucleotides or amino acids which occurs as part, or all of a number of different genes or proteins, in either the same or different species. This conservation can signify which part of the full sequence is responsible for the functionality.

❐ **constant domains**

Regions of antibody chains that have the same amino acid sequence in different members of a particular class of antibody molecules.

❐ **constitutive**

The expression of a gene without any requirement for induction.

❐ **constitutive gene**

A gene that is continually expressed in all cells of an organism.

❐ **constitutive promoter**

An unregulated promoter that allows for continual transcription of its associated gene.

❐ **constitutive synthesis**

Continual synthesis of a gene product by an organism.

❐ **construct**

An engineered chimeric DNA designed to be transferred into a cell or tissue. Typically, the construct comprises the gene or genes of interest, a marker gene and appropriate control sequences as a single package. A repeatedly-used construct may be called a cassette.

❑ construction of a human genomic library

A genomic library comprises a set of bacteria, each carrying a different small fragment of human DNA. For simplicity, cloning of just a few representative fragments (colored) is shown. In reality, all the gray DNA fragments will also be cloned.

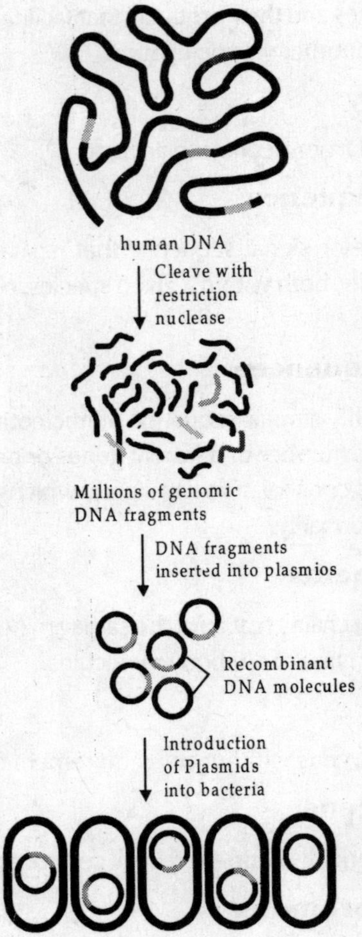

❑ containment

Measures and protocols applied to limit contact of genetically modified organisms or pathogens with the external environment.

❑ contaminant

1. An undesired chemical present in a compound or mixture of compounds.
2. Any micro-organism accidentally introduced into a culture or culture medium. The contaminant may compete with the desired cells and consequently inhibit their growth, or totally replace them.

▫ contig

A set of overlapping cloned DNA fragments that can be assembled to represent a defined region of the chromosome or genome from which they were obtained. Contig definition is a necessary step for assembling whole genome sequences.

▫ continuous culture

A suspension culture continuously supplied with nutrients by the inflow of fresh medium. The culture volume is normally constant.

▫ continuous fermentation

A process in which cells or micro-organisms are maintained in culture in the exponential growth phase by the continuous addition of fresh medium that is exactly balanced by the removal of cell suspension from the bioreactor.

▫ continuous variation

Variation where individuals cannot be classified as belonging to one of a set of discrete classes. Characters showing continuous variation are referred to as quantitative.

▫ control of gene expression

Examples of regulation at each of the steps are known, although for most genes the main site of control is step 1: transcription of a DNA sequence into RNA.

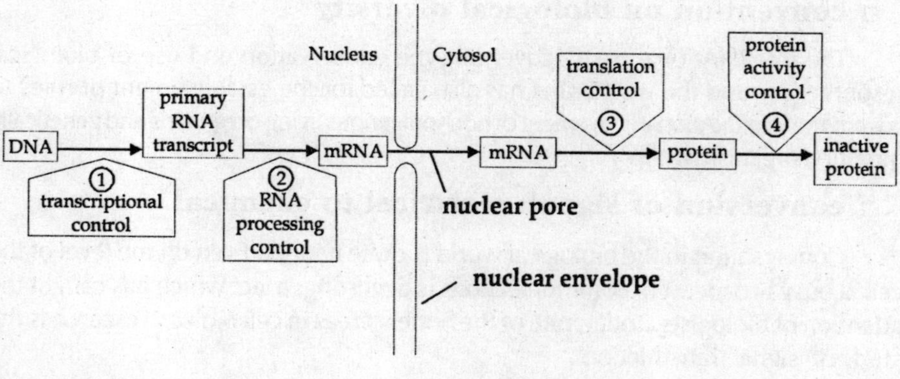

▫ control of the human b-globin gene

The diagram shows some of the gene regulatory proteins thought to control expression of this gene during red blood cell development. Some of the gene regulatory proteins shown, such as CP1, are found in many types of cells, while others, such as GATA-1, are present in only a few types of cells, including red blood cell precursors, and are therefore thought to contribute to the cell-type specificity of beta-globin gene expression. As indicated by the bidirectional arrows, several of the binding sites for GATA-1 overlap those of other gene regulatory proteins; it is thought that occupancy of these sites by GATA-1 excludes binding of other proteins.

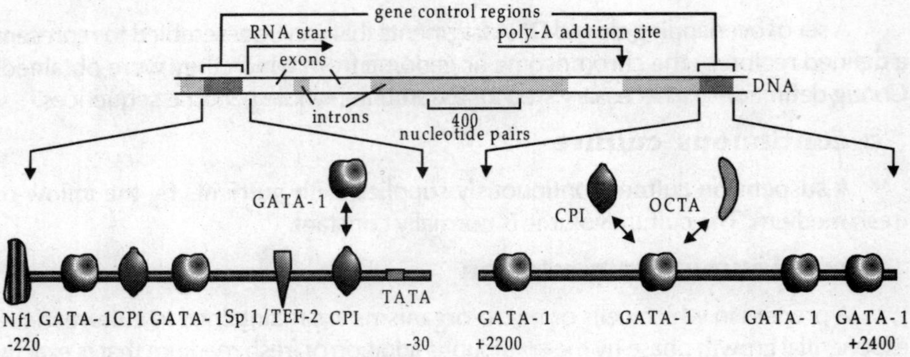

❐ controlled environment

A closed environment in which parameters, such as light, temperature, relative humidity and sometimes the partial gas pressure (and possibly its composition), are fully controlled.

❐ controlling element

In eukaryotes, transposons which affect the activity of known genes. This can occur as a result of the integration within, or close to a gene, thereby disrupting its activity; or following its excision from such a site, thereby restoring activity.

❐ convention on biological diversity

The inteRNAtional treaty governing the conservation and use of biological resources around the world, that has also called for the establishment of rules to govern the inteRNAtional movement of non-indigenous living organisms and genetically modified organisms.

❐ conversion of signal: electrical to chemical

Conversation in the biological world is quite natural. Even on the level of the cell, a busy broadcast of communications is occurring; a fact which has caught the attention of biologists. Today, one of the hottest areas in cell biology research is the study of 'signal transduction'.

Every cell is able to communicate through having evolved the ability to produce, recognize, interpret and respond to signals in its environment. The word 'signals' in this context refers to nothing more than chemical molecules that are floating around. Cells have learned to detect many of these chemicals. Their molecular detection components—produced by the genes they contain— allow the cells to converse in this chemical "language of the cell."

When an action potential reaches a nerve terminal, it opens voltage-gated Ca^{2+} channels in the plasma membrane, allowing Ca^{2+} to flow into the terminal. The increased Ca^{2+} in the nerve terminal stimulates the synaptic vesicles to fuse with the plasma membrane, releasing their neurotransmitter into the synaptic cleft.

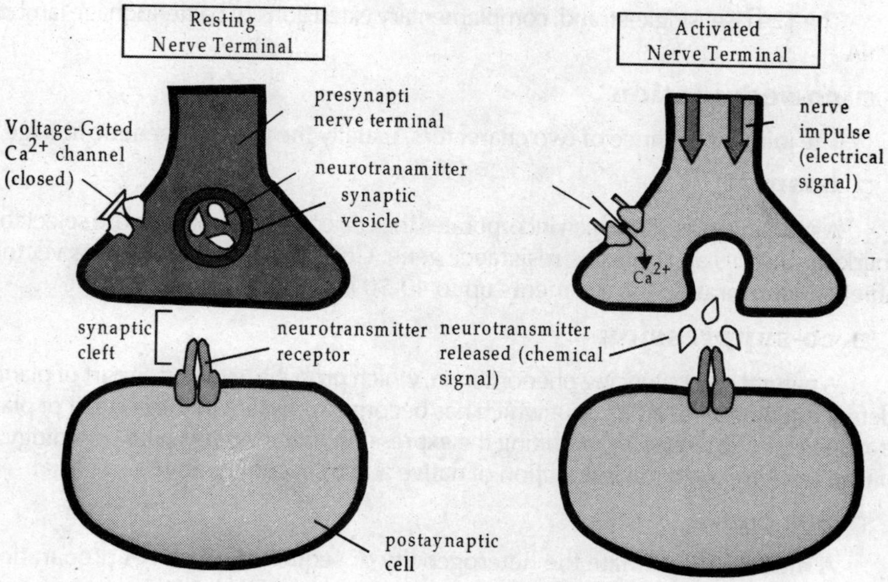

❐ conversion

The development of a somatic embryo into a plant.

❐ coordinate repression

Correlated regulation of a structural gene within an operon by a molecule that interacts with the operator.

❐ copy number

The number of a particular plasmid per bacterium cell, or gene per genome.

❐ co-repressor

An effector molecule that forms a complex with a repressor and turns off the expression of a gene or set of genes.

❐ corpus

A part of the apical meristem below the tunica. In the corpus, cells divide in all directions, and increase in volume.

❐ correlation

A statistical association between variables.

❐ cortex

Primary tissue of a stem or root, bounded externally by the epidermis and internally in the stem by the phloem, and in the root by the pericycle.

❑ cos ends

The 12-base, single-strand, complementary extensions of bacteriophage lambda DNA.

❑ co-segregation

The joint inheritance of two characters, usually the result of genetic linkage.

❑ cosmid

A synthetic plasmid which incorporates the cos ends, and one or more selectable markers such as an antibiotic resistance gene. Cosmids were designed as vectors able to incorporate DNA fragments up to 40-50 kb in size.

❑ co-suppression

A natural gene silencing phenomenon, which probably evolved as part of plants' defence against viral attack, but which has become important in the context of plant transgenesis. Operates by inhibiting the expression of transgenes with homology to native DNA through the interaction of native and transgenic mRNA.

❑ cot curve

A method to estimate the heterogeneity of sequence of a DNA preparation, based on the observation that the more homogenous the DNA, the more easily (and therefore faster) the annealing of single-stranded DNA will occur. The Cot curve plots the extent of annealing from a fully single-stranded preparation over time. The cot (product of initial concentration and time) at which half the DNA has re-natured is the half-cot, a parameter indicating both the degree of heterogeneity in a complex mixture, and of the extent of complementarity in a mixture of two single-stranded DNA molecules.

❑ co-transfection

The procedure by which a baculovirus and a transfer vector are simultaneously introduced into insect cells in culture.

❑ co-transformation

A protocol for producing transgenesis, in which host (plant or animal) cells are transformed simultaneously with two different plasmids, one of which carries a selectable marker, and the other the gene to be transferred. Relies on the observation that given a sufficiently high concentration of both plasmids, transformed cells will have incorporated both plasmids, possibly at different genomic loci. If the transgenes are separable through normal meiotic recombination, transgenic individuals without the selectable marker can be selected in subsequent generations.

❑ cotyledon

Leaf-like structures at the first node of the seedling stem. In some dicotyledons, they represent a food storage organ for the germinating seedling.

❑ coupling

The phase state in which either two dominant or two recessive alleles of two different genes occur on the same chromosome.

❑ Covalent and Ionic Bonds

Atoms can attain a more stable arrangement of electrons in their outermost shell by interacting with one another. An ionic bond is formed when electrons are transferred from one atom to the other. A covalent bond is formed when electrons are shared between atoms. The two cases shown represent extremes; often, covalent bonds form with a partial transfer (unequal sharing of electrons), resulting in a polar covalent bond.

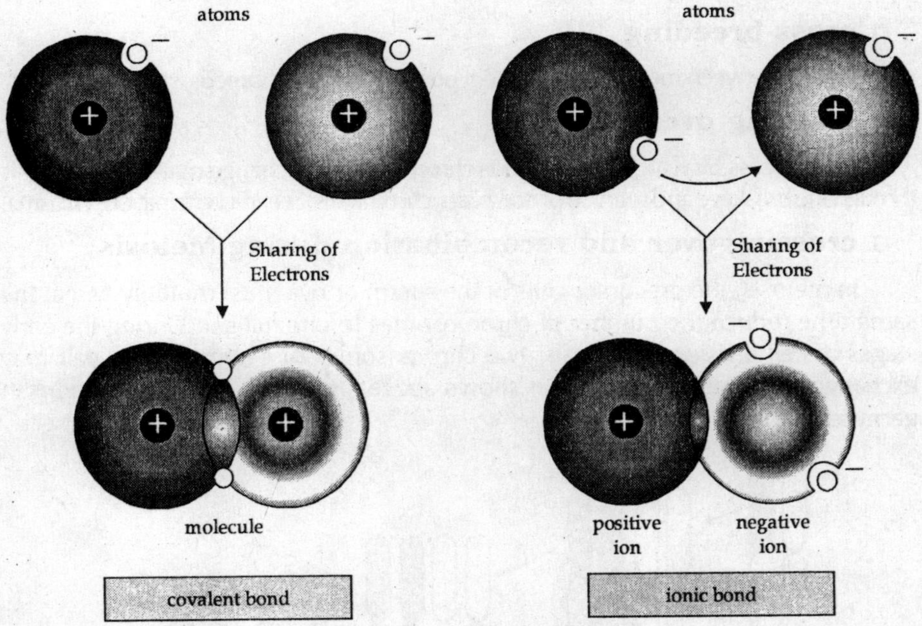

❑ covalently-closed circular DNA

A DNA molecule in which the free ends have ligated to form a circle. The strands remain linked together even after denaturation. Plasmids exist in this form in their *in vivo* state. In its native form, ccc DNA will adopt a supercoiled configuration. *See*: circularization.

❑ cp4 epsps

Abbreviation for CP4 5-enolpyruvyl-shikimate-3-phosphate synthase.

❑ cpdna

Abbreviation for chloroplast DNA.

❑ cross

The mating of two individuals or populations.

❑ cross hybridization

The annealing of a single-stranded DNA sequence to a single-stranded target DNA to which it is only partially complementary. Often, this refers to the use of a

DNA probe to detect homologous sequences in species other than the origin of the probe.

❒ cross pollination

Application of pollen from one plant to another to effect the latter's fertilization.

❒ cross pollination efficiency

The ease with which cross pollination can be achieved. Generally measured by the number of hybrid progeny generated per flower pollinated.1

❒ cross-breeding

Mating between members of different populations (lines, breeds, races or species).

❒ crossing over

The process by which homologous chromosomes exchange material at meiosis through the breakage and reunion of non-sister chromatids. *See:* recombination, chiasma.

❒ crossing-over and recombination during Meiosis

In meiosis, the precursor cells of the sperm or ova must multiply and at the same time reduce the number of chromosomes to one full set. During the early stages of cell division in meiosis, two chromosomes of a homologous pair may exchange segments in the manner shown above, producing genetic variations in germ cells.

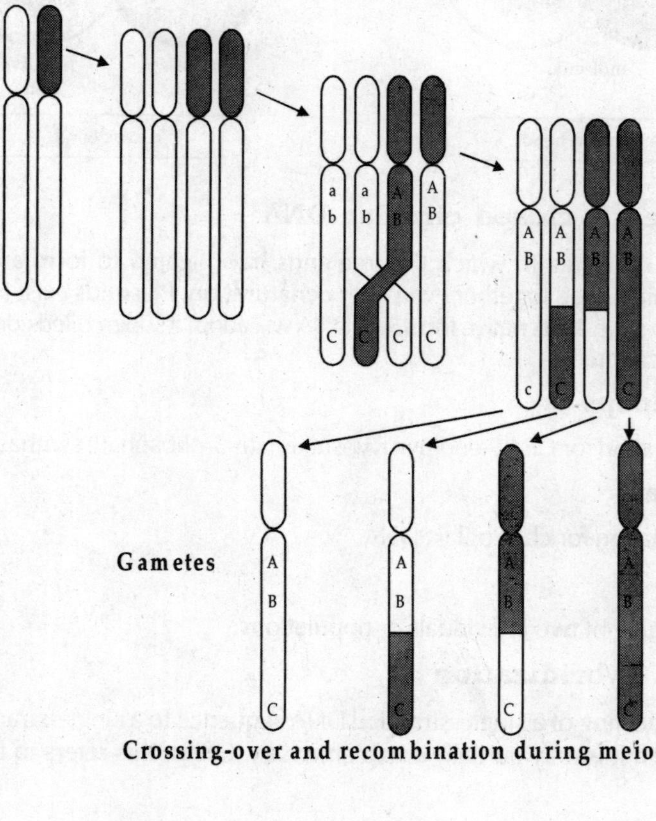

Crossing-over and recombination during meiosis

❒ crown

The base of the stem of cereals and forage species from which tillers or branches arise. In woody plants, the root-stem junction. In forestry, the top portion of the tree.

❒ crown gall

A tumorous growth at the base of certain plants characteristic of infection by *Agrobacterium tumefaciens*. The gall is induced by the transformation of the plant cell by portions of the Ti plasmid.

❒ crp

Abbreviation for catabolite regulator protein.

❒ cry proteins

A class of crystalline proteins produced by strains of *Bacillus thuringiensis*, and engineered into crop plants to give resistance against insect pests. These proteins are toxic to certain categories of insects (e.g. corn borers, corn rootworms, mosquitoes, black flies, armyworms, tobacco hornworms, some types of beetles, etc.), but are harmless to mammals and most beneficial insects.

❒ cryobiological preservation

The preservation of germplasm resources in a dormant state by storage at ultra-low temperatures, often in liquid nitrogen. Currently applied to storage of plant seeds and pollen, micro-organisms, animal sperm, and tissue culture cell lines.

❒ cryogenic

At very low temperature.

❒ cryoprotectant

Compound preventing cell damage during successive freezing and thawing processes. Cryoprotectants are agents with high water solubility and low toxicity. Two types commonly used: permeating (glycerol and DMSO) and non-permeating (sugars, dextran, ethylene glycol, polyvinylpyrrolidone and hydroxyethyl starch).

❒ cryptic

Anything hidden.

1. Structurally heterozygous individuals that are not identifiable as they do not show abnormal meiotic chromosome pairing configurations ('cryptic structural hybrids').
2. A form of polymorphism controlled by recessive genes ('cryptic polymorphism').
3. Any mutation which is exposed by a sensitizing mutation and otherwise poorly detected (such mutations probably escape detection because of the plasticity of composition of the corresponding polypeptide).

4. Phenotypically very similar species (cryptic species) which do not hybridize under normal conditions.
5. Cryptic genetic variation refers to the existence of, for example, alleles conferring high performance for a trait, in a breed that has low performance for that trait.

❐ ctp

Abbreviation for:

1. cytidine 5'-triphosphate, which is required for RNA synthesis since it is a direct precursor molecule;
2. Chloroplast transit peptide.

❐ cultigen

A cultivated plant species with no known wild progenitor.

❐ cultivar

An inteRNAtionally accepted term denoting a variety of a cultivated plant. Must be distinguishable from other varieties by stated characteristics and must retain their distinguishing characters when reproduced under specific conditions.

❐ culture

A population of plant or animal cells or micro-organisms grown under controlled conditions.

❐ culture alteration

A term used to indicate a persistent change in the properties of a culture's behaviour (e.g. altered morphology, chromosome constitution, virus susceptibility, nutritional requirements, proliferative capacity, etc.). The term should always be qualified by a precise description of the change which has occurred in the culture.

❐ culture medium

Any nutrient system for the cultivation of cells, bacteria or other organisms; usually a complex mixture of organic and inorganic nutrients.

❐ culture room

A dedicated room for maintaining cultures, often in a controlled environment.

❐ curing

The elimination of a plasmid from its host cell. Many agents which interfere with DNA replication, e.g. ethidium bromide, can cure plasmids from either bacterial or eukaryotic cells.

❐ cuticle

Layer of cutin or wax, formed on the outer surface of leaves and fruits, thought to have evolved to reduce evaporative water loss.

❑ cutting

A detached plant part that, with appropriate treatment, can regenerate into a complete plant.

❑ cybrid

A hybrid, originating from the fusion of a cytoplast (the cytoplasm without nucleus) with a whole cell derived from a different species.

❑ cyclic adenosine monophosphate

A "messenger" molecule that regulates many intracellular reactions by transducing signals from extracellular growth factors to cellular metabolic pathways.

❑ cyclic amp

Abbreviation for cyclic adenosine monophosphate.

❑ cyclodextrin

Cyclic oligomer of glucose.

❑ cycloheximide

A molecule that inhibits protein synthesis in eukaryotes, but not in prokaryotes. It blocks peptide bond formation by binding to the large ribosomal subunits. *Synonym*: actidione.

❑ cytidine

The (ribo)nucleoside resulting from the combination of the base cytosine (C) and the sugar D-ribose. The corresponding deoxyribonucleoside is called deoxycytidine.

❑ cytidylic acid

Synonym for cytidine monophosphate (abbreviation: CMP), a (ribo)nucleotide containing the nucleoside cytidine. The corresponding deoxyribonucleotide is called deoxycytidine 5'-monophosphate or deoxycytidylic acid.

❑ cytochrome

A class of pigments in plant and animal cells, usually in the mitochondria. They function as electron carriers in respiration.

❑ cytochrome p450

A highly diversified set (more than 1500 known sequences) of heme-containing proteins. Frequently called hydroxylases, although P450 proteins can perform a wide spectrum of other reactions. In bacteria they are soluble and approximately 400 amino acids long; eukaryotic P450s are larger - about 500 amino acids. In mammals they are critical for drug metabolism, haemostasis, cholesterol biosynthesis and steroidogenesis; in plants they are involved in plant hormone synthesis, phytoalexin synthesis, flower petal pigment biosynthesis and many unknown functions. In fungi they make ergosterol and they are involved in pathogenesis. Bacterial P450s are key elements in antibiotic synthesis.

▫ cytogenetic map

A type of genetic map relating gene positions to chromosomal banding patterns. The maps are built from relating the positions of genes to cytogenetic markers or by in situ hybridization.

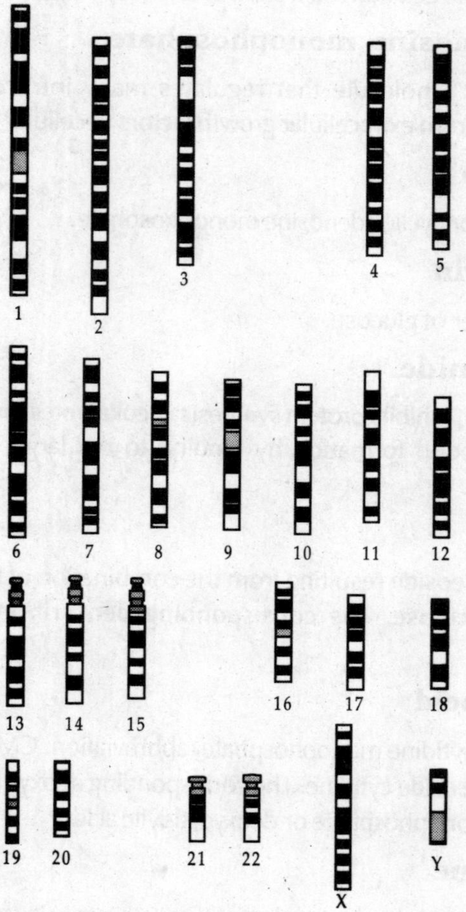

▫ cytogenetics

The biology of chromosomes and their relation to the transmission and recombination of genes.

▫ cytokine

A generic name for a diverse group of soluble proteins and peptides which act as humoral regulators at extremely small concentrations and which, either under normal or pathological conditions, modulate the functional activities of individual cells and tissues.

❐ cytokinesis

Cytoplasmic division and other changes exclusive of nuclear division that are a part of mitosis or meiosis.

❐ cytokinin

Plant growth regulators characterized as substances that induce cell division and cell differentiation. In tissue culture, these substances are associated with enhanced callus and shoot development. The compounds are derivatives of adenine.

❐ cytology

The study of the structure and function of cells.

❐ cytolysis

Cell disintegration.

❐ cytoplasm

The living material of the cell, exclusive of the nucleus, consisting of a complex protein matrix or gel, and where essential membranes and cellular organelles (mitochondria, plastids, etc.) reside.

❐ cytoplasmic genes

Genes located on DNA outside the nucleus, i.e. on plastids.

❐ cytoplasmic inheritance

Hereditary transmission dependent on cytoplasmic genes.

❐ cytoplasmic male sterility

Genetic defect due to faulty functioning of mitochondria in pollen development, preventing the formation of viable pollen. Commonly found or inducible in many plant species and exploited for some F_1 hybrid seed programmes.

❐ cytoplasmic organelles

Discrete sub-cellular structures located in the cytoplasm of cells - mitochondria, plastids and lysosomes.

❐ cytosine

One of the bases found in DNA and RNA.

❐ cytoskeleton filaments

The cytoskeleton is the internal scaffolding within a biological cell that helps it to maintain or alter its shape. It is composed of actin, microtubules, and other proteins. In addition to holding the cell together, the cytoskeleton is crucial in cell motion, especially amoeba-like motion.

- The cytoskeleton functions analogously to the human skeleton.
- It defines the *shape* of the cell and determines
- its rigidity facilitates intra-cellular transport and is responsible for *cell motility*.

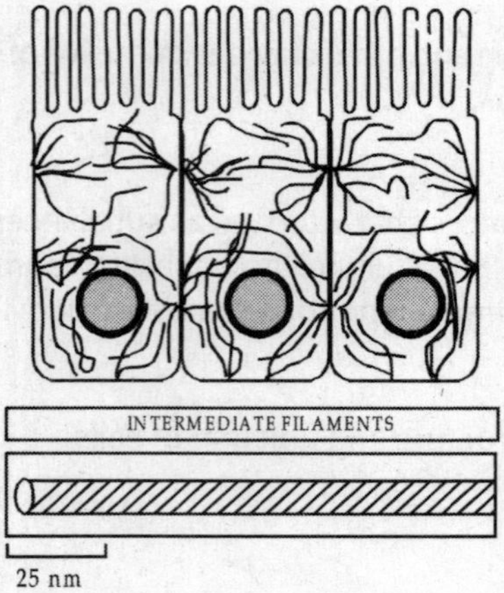

Intermediate filaments are ropelike fibers with a diameter of about 10nm; they are made of intermediate filament proteins, which constitute a large and heterogeneous family. One type of inter mediate filament forms a meshwork called the nuclear lamina just beneath the inner nuclear membrane. Other types extend across the cytoplasm, giving cells mechanical strength and carrying the mechanical stresses in an epithelial tissue by spanning the cytoplasm from one cell-cel junction to another.

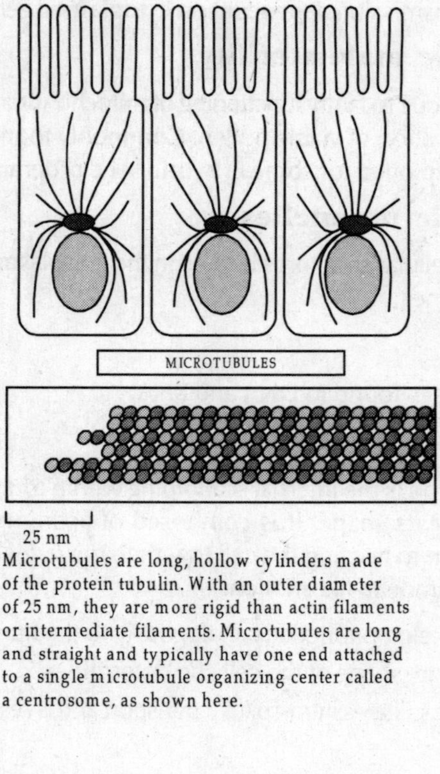

Microtubules are long, hollow cylinders made of the protein tubulin. With an outer diameter of 25 nm, they are more rigid than actin filaments or intermediate filaments. Microtubules are long and straight and typically have one end attached to a single microtubule organizing center called a centrosome, as shown here.

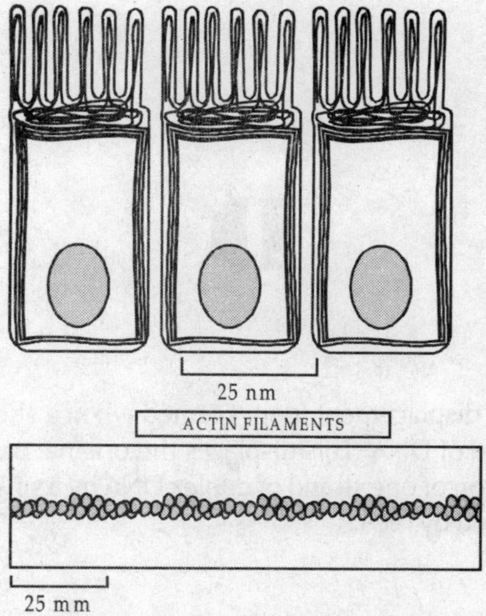

Actin filaments (also known as microfilaments) are helical polymers of the protein actin. They appear as flexible structures, with a diameter of about 7 nm, that are organized into a variety of linear bundies, two dimensional networks, and three-dimensional gels. Although actin filaments are dispersed throughout the cell, they are most highly concentrated in the cortex, just beneath the plasma membrane.

❐ cytosol

The fluid portion of the cytoplasm, i.e. the cytoplasm minus its organelles.

❐ cytotoxicity

Poisoning of the cell.

❐ cytotype

A maternally inherited cellular condition in *Drosophila* that regulates the activity of transposable P elements.

D

▫ d loop

Abbreviation for displacement loop. Formed when a short stretch of RNA is paired with one strand of DNA. This displaces the original partner DNA; also the displacement of a region of one strand of duplex DNA by a single-stranded invader in the reaction catalyzed by recA.

▫ dab

Antibodies with only one (instead of two) protein chain derived from only one of the two domains of the normal antibody structure. Exploits the finding that for some antibodies, half of the molecule binds to its target antigen almost as well as the whole molecule. The major advantage of dAbs over other antibodies is that they can be cloned and expressed into bacteria, so that large numbers of antibodies can be generated and screened in parallel.

▫ dalton

A unit of atomic mass roughly equivalent to the mass of a hydrogen atom. Used as to express molecular weight, which for biological macromolecules is usually in the range kilo- (kDa) to megaDaltons (MDa).

▫ darwinian cloning

Selection of a clone from a large number of essentially random starting points, rather than isolating a natural gene or making a carefully designed artificial one. Molecules which are more similar to those needed are selected, mutated to generate new variants, and re-selected. The cycle proceeds until the required molecule is found. The advantage of the system is that the selection is from a vast number of possibilities.

▫ datp

Abbreviation for deoxyadenosine 5'-triphosphate. dATP is required for DNA synthesis since it is a direct precursor molecule.

▫ dctp

Abbreviation for deoxycytidine 5'-triphosphate. dCTP is required for DNA synthesis since it is a direct precursor molecule.

❐ ddntp

Abbreviation for di-deoxynucleotide.

❐ death phase

The final growth phase of cell culture, during which nutrients have been depleted and cell number decreases.

❐ deceleration phase

The phase of declining growth rate, following the linear phase and preceding the stationary phase in most batch-suspension cultures.

❐ de-differentiation

The process, in response to wounding and in tissue cultures, by which plant cells can become unspecialized and start to proliferate by cell division to form a mass of undifferentiated cells (or callus) which, in response to appropriate stimuli, may later differentiate again to form either the same cell type or a different one.

❐ defective virus

A virus that, by itself, is unable to reproduce when infecting its host cell, but that can grow in the presence of another virus. This other virus provides the necessary molecular machinery that the first virus lacks.

❐ deficiency

Lack of adequate supply of nutritional, enzymatic, or environmental requirements, so that development, growth or physiological functions are affected.

❐ defined

1. Fixed conditions of medium, environment and protocol for growth.
2. Precisely known and stated elements of a tissue culture medium.

❐ degeneracy

The specification of one amino acid by more than one codon. It arises from the inevitable redundancy resulting from the 64 possible codons encoding only 20 amino acids.

❐ degeneration

1. Changes in cells, tissues or organs due to disease.
2. The reduction in size or complete loss of organs during evolution.

❐ dehalogenation

The removal of halogen atoms (fluorine, chlorine, bromine, iodine) from molecules, for example during biodegradation.

❐ dehiscence

The spontaneous and often violent opening of a fruit, seed pod or anther to release and disperse the seeds or pollen.

❑ dehydrogenase

An enzyme that catalyses the removal of hydrogen atoms in biological reactions.

❑ dehydrogenation

A chemical reaction in which hydrogen is removed from a compound.

❑ de-ionized water

Water from which most salts have been removed - with varying degrees of efficiency - by ion exchange.

❑ deletion

A type of mutation caused by loss of one or more nucleotides from a DNA segment. Deletions can be very large, encompassing many genes and megabases of DNA, to the point of producing a visible cytological abnormality in a chromosome. Small deletions within a gene can alter the reading frame, and thus the amino acid sequence of the encoded protein.

A mutation involving the removal of one or more base pairs in a DNA sequence. Large deletions are sometimes microscopically visible in karyotype analyses.

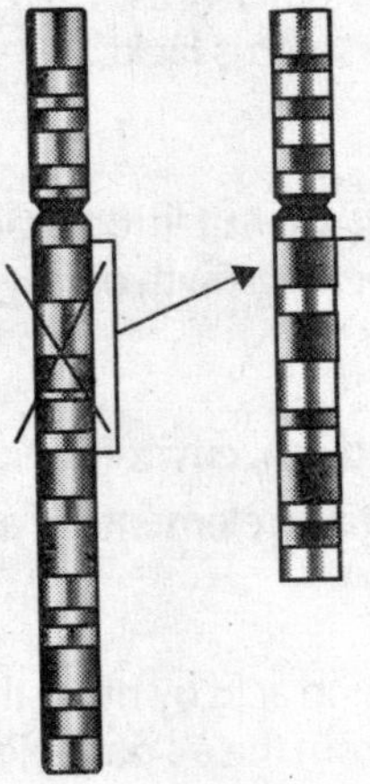

❑ deliberate release

In a biotechnology context, the intentional release of genetically modified organisms.

❑ deme

A group of organisms in the same taxon.

❑ demineralize

To remove the mineral content (salts, ions) from a substance, especially water. Removal methods include distillation, electrodialysis and ion exchange. *See*: de-ionized water.

▫ denature

To disrupt the normal *in vivo* conformation of a nucleic acid or (more usually) a protein by physical or chemical means, usually accompanied by the loss of activity.

▫ denatured DNA

Double-stranded DNA that has been converted to single strands by breaking the hydrogen bonds linking complementary nucleotide pairs. Often reversible. Usually achieved by heating.

▫ denatured protein

Altering the *in vivo* conformation of a protein by heat or salt treatment, thereby destroying its biological activity. Unlike denatured DNA, denatured proteins are seldom able to be renatured.

▫ denaturing gradient gel electrophoresis

An electrophoresis method for separating similar sized DNA fragments on the basis of their sequence, by applying across the gel a gradient of increasingly denaturing conditions (usually by increasing the concentration of a denaturing chemical, such as formamide or urea). As the double-stranded molecules denature into a partially and eventually a fully single-stranded state, their electrophoretic mobility changes.

▫ dendrimer

A polymer that repeatedly branches until stopped by the physical constraint of having formed a complete, hollow sphere. These structures possess sites on their exterior surface to which DNA fragments can be attached, and are thus useful as carriers of DNA for transgenesis.

▫ denitrification

A chemical process in which nitrates in the soil are reduced to molecular nitrogen, which is released to the atmosphere.

▫ density gradient centrifugation

High-speed centrifugation in which molecules are separated on the basis of their different densities using a concentration gradient of caesium chloride or sucrose. The density gradient may either be formed before centrifugation by mixing two solutions of different density (as in sucrose density gradients) or it can be formed by the process of centrifugation itself (as in CsCl and Cs_2SO_4 density gradients).

▫ deoxythymidine

Strictly correct but rarely used synonym for thymidine.

▫ derepression

The process of "turning on" the expression of a gene or set of genes whose expression has been repressed (turned off), usually by the displacement of a repressor from a promoter, since, when attached to the DNA, the repressor prevents transcription.

derivative

1. Resulting from or derived from.
2. Term used to identify a variant during meristematic cell division.

desiccant

Any compound used to remove moisture or water.

desoxyribonucleic acid

Obsolete spelling of deoxyribonucleic acid.

detergent

Substance which lowers the surface tension of a solution, improving its cleaning properties.

determinate growth

Growth determined and limited in time, with a bud or flower terminating the growth of the main axis. Once established, it is usually irreversible.

determination

Process by which undifferentiated cells in an embryo become committed to develop into specific cell types, such as neurons, fibroblasts or muscle cells.

determined

Describing embryonic tissue at a stage when it can develop only as a certain kind of tissue.

development

The sum total of events that contribute to the progressive elaboration of an organism. The two major aspects of development are growth and differentiation.

deviation

1. An alteration from the typical form, function or behaviour. Mutation or stress are the common reasons behind deviation.
2. A statistical term describing the difference between an actual observation and the mean of all observations.

dextrin

An intermediate polysaccharide compound resulting from the hydrolysis of starch to maltose by amylase enzymes.

dgtp

Abbreviation for deoxyguanosine 5'-triphosphate. dGTP is required for DNA synthesis since it is a direct precursor molecule.

diagnostic procedure

A test or assay used to determine the presence of a specific substance, organism or nucleic acid sequence alteration, etc.

▢ diagram of a retrovirus

Retroviruses are infectious particles consisting of an RNA genome packaged in a protein capsid, surrounded by a lipid envelope. This lipid envelope contains polypeptide chains including receptor binding proteins which link to the membrane receptors of the host cell, initiating the process of infection.

Retroviruses contain RNA as the hereditary material in place of the more common DNA. In addition to RNA, retrovirus particles also contain the enzyme reverse transcriptase (or RTase), which causes synthesis of a complementary DNA molecule (cDNA) using virus RNA as a template.

When a retrovirus infects a cell, it injects its RNA into the cytoplasm of that cell along with the reverse transcriptase enzyme. The cDNA produced from the RNA template contains the virally derived genetic instructions and allows infection of the host cell to proceed.

The virus that causes AIDS (*acquired immune deficiency syndrome*) is a retrovirus. It is called HIV for *human immunodeficiency virus*.

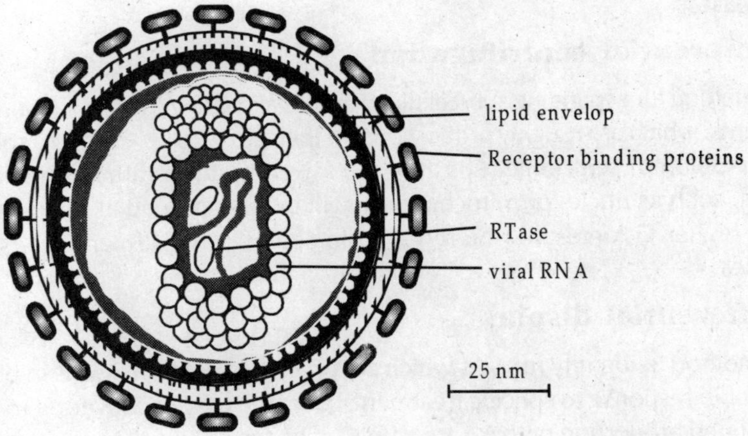

▢ diakinesis

A stage of meiosis at the end of prophase I, in which the contraction of the chromosomes is almost at a maximum, pairing configurations are well defined, the nucleolus normally disappears and the nuclear envelope is disrupted.

▢ dialysis

A biochemical technique by which large molecules such as proteins in solution are separated from smaller species such as salts. The technique is based on the properties of certain membrane structures, which selectively only allow the passage of the smaller molecules. A frequently used method for the purification of proteins.

▢ diazotroph

An organism that can fix atmospheric nitrogen.

❐ dicentric chromosome

A chromosome having two active centromeres.

❐ dichogamy

The condition in which the male and the female reproductive organs of a flower (or certain hermaphroditic animals) mature at different times, thereby making self-fertilization improbable or impossible.

❐ dicotyledon

A plant with two cotyledons. One of the two major classes of flowering plants (along with the monocotyledons). Examples include many crop plants (potato, pea, beans), ornamentals (rose, ivy) and timber trees (oak, beech, lime).

❐ di-deoxynucleotide

A synthetic deoxynucleotide that lacks a 3'-hydroxyl group, and is thus unable to form the 3'5' phosphodiester bond necessary for chain elongation. Used as strand terminators in the Sanger DNA sequencing reaction and in the treatment of some viral diseases.

❐ differential centrifugation

A method for separating sub-cellular particles according to their sedimentation coefficients, which are roughly proportional to their size. Cell extracts are subjected to a succession of centrifuge runs at progressively faster rotation speeds. Large particles, such as nuclei or mitochondria, will be precipitated at relatively slow speeds; higher G forces will be required to sediment small particles, such as ribosomes.

❐ differential display

A method to identify mRNAs which are present at different levels in different tissues, or in response to specific treatments. The mRNAs are converted to cDNA, and a defined proportion of these are amplified by the polymerase chain reaction, and separated by electrophoresis.

❐ differentially permeable

Referring to a membrane, through which different substances diffuse at different rates. Some substances may be unable to diffuse through such a membrane, usually because they are too large to fit through the pores of the membrane.

❐ differentiation

A process as a result of which unspecialized cells develop structures and functions characteristic of a particular type of cell, typically during the process of development from one cell to many cells, accompanied by a modification of the new cells for the performance of particular functions. The process is generally irreversible *in vivo* in higher organisms. In tissue culture, the term is used to describe the formation of different cell types.

❐ diffusion

The spontaneous movement of molecules from a region of higher concentration to a region of lower concentration.

❐ digest

To treat DNA molecules with one or more restriction endonucleases in order to cleave them into smaller fragments.

❐ dihaploid

An individual which arises from a doubled haploid.

❐ dihybrid

An individual that is heterozygous for two pairs of alleles; the progeny of a cross between homozygous parents differing at two loci.

❐ dimer

1. A molecule formed by the covalent combination of two monomers, generally accompanied by elimination of water.
2. The reversible association of two similar (or nearly similar) molecules. The active form of many enzymes is as a dimer between two non-active monomeric subunits.

❐ dimethyl sulphoxide

A highly hygroscopic liquid and powerful solvent with little odour, colour or toxicity when pure. It is employed in small quantities to dissolve organic substances in tissue culture media preparation and has uses as a cryoprotectant and in promoting the passage of chemicals through skin.

❐ dimorphism

The existence of two distinctly different types of individuals within a species. An obvious example is sexual dimorphism in mammals.

❐ dinucleotide

A nucleotide dimer.

❐ dioecious

A plant species in which male and female flowers form on different plants.

❐ diploid

The status of having two complete sets of chromosomes, most commonly one set of paternal origin and the other of maternal origin. Somatic tissues of higher plants and animals are ordinarily diploid in chromosome constitution, in contrast with the haploid gametes.

❐ diplonema

Stage in prophase I of meiosis following the pachytene stage, but preceding diakinesis, in which one pair of sister chromatids begin to separate from the other pair.

□ direct embryogenesis

The formation in culture, on the surface of zygotic or somatic embryos or on explant tissues (leaf section, root tip, etc.), of embryoids without an intervening callus phase.

□ direct organogenesis

Formation of organs directly on the surface of cultured intact explants. The process does not involve callus formation.

□ direct repeat

Two or more stretches of DNA within a single molecule which have the same nucleotide sequence in the same orientation. Direct repeats may be either adjacent to one another or far apart on the same molecule.

□ directed amplification of minisatellite dna

A polymerase chain reaction technique used for obtaining molecular markers in the region of minisatellites. To target these regions, one of the primers is directed to a VNTR core sequence.

□ directed mutagenesis

The generation of changes in the nucleotide sequence of a cloned gene by one of several procedures. Undertaken to explore the relationship between nucleotide sequence and gene function, and to modify gene products.

□ directional cloning

The technique by which a vector and a DNA insert are both digested with two different restriction endonucleases to create non-complementary sticky ends at either end of both molecules, so favouring the insert to be ligated into the vector in a specific orientation, while also preventing the vector from re-circularizing.

□ disaccharide

A dimer consisting of two covalently linked monosaccharides.

□ disarm

The deletion from a plasmid or virus of genes that are pathogenic.

□ discontinuous variation

Variation where individuals can be classified as belonging to one of a set of discrete, non-overlapping classes. Generated by simple genetic control of a trait (one or a small number of genes, each of large effect) and involving minimal non-genetic effect. Characters showing discontinuous variation are referred to as qualitative.

□ discordant

Members of a pair showing different, rather than similar, characteristics.

disease resistance

The genetically determined ability to prevent the reproduction of a pathogen, thereby remaining healthy. Some resistances operate by pathogen exclusion, some by preventing pathogen spread, and some by tolerating pathogen toxin.

disease-free

A plant or animal certified through specific tests as being free of specified pathogens. Should be interpreted to mean "free from any *known* disease" as "new" diseases may yet be discovered to be present.

disease-indexing

Disease-indexed organisms have been assayed for the presence of known diseases according to standard testing procedures.

disinfection

Attempted elimination by chemical means of internal micro-organisms (particularly pathogens) from a culture or sample; rarely attained.

disinfestation

The elimination or inhibition of the activity of surface-adhering micro-organisms and removal of insects.

disjunction

Separation of homologous chromosomes during anaphase I of meiosis, or of sister chromatids during anaphase of mitosis and anaphase II of meiosis.

disomy

The presence of a pair of a specific homologous chromosomes. This is the norm for diploids.

dispense

The transfer of a measured volume of a solution.

disrupter gene

Used to enforce the sterility of seed saved from a genetically engineered crop.

dissecting microscope

A microscope with a magnifying power of about 50x, used as an aid in the manipulation of small objects, e.g. excision of embryos from young zygotes.

dissection

Separation of a tissue by cutting into components, for analysis or observation.

distillation

The process of heating a mixture to separate the more volatile from the less volatile parts, and then condensing fractions of the resulting vapour so as to produce a more nearly pure or refined substance.

❒ disulphide bridge

A chemical bond between pairs of sulphur atoms that stabilizes the three-dimensional structure of proteins, and hence the protein's normal function. These form particularly readily between cysteine residues in the same or different peptide molecules.

❒ ditype

In fungi, a tetrad that contains two kinds of meiotic products (spores), e.g. 2AB and 2ab.

❒ diurnal

An event that occurs repetitively on a daily basis, generally during daylight hours.

❒ dizygotic twins

Two-egg twins, i.e. a pair of individuals that shared the same uterus at the same time, but which arose from separate and independent fertilization of two ova.

❒ dna

Abbreviation for deoxyribonucleic acid, former spelling desoxyribonucleic acid. A long chain polymer of deoxyribonucleotides. DNA constitutes the genetic material of most known organisms and organelles, and usually is in the form of a double helix, although some viral genomes consist of a single strand of DNA, and others of a single- or a double-stranded RNA.

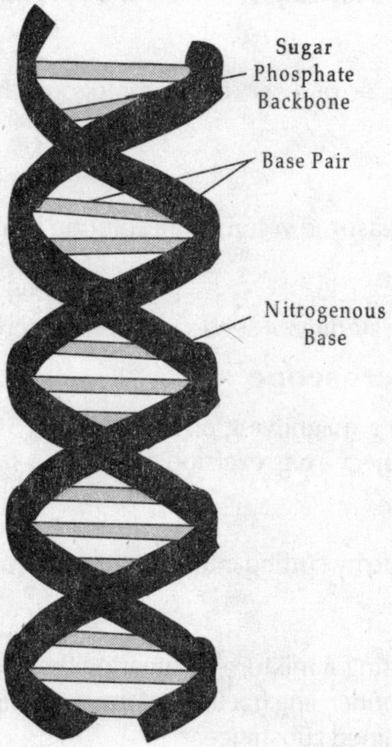

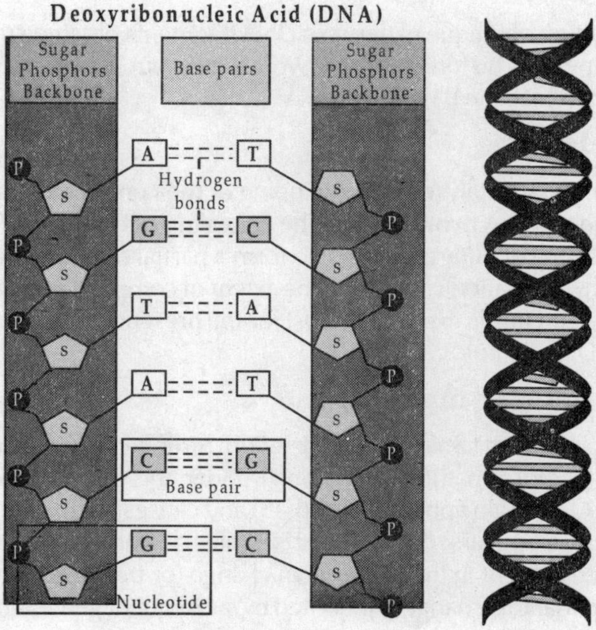

❐ dna amplification fingerprinting

A arbitrarily primed polymerase chain reaction technique for obtaining molecular markers using very short (5-8 bp) primers.

❐ dna amplification

Many-fold multiplication of a particular DNA sequence either *in vivo* in a plasmid, phage or other vector; or *in vitro* using, most commonly, the polymerase chain reaction.

❐ dna construct

A chimeric DNA molecule, carrying all the genetic information necessary for its transgenic expression in a host cell.

❐ dna delivery system

A generic term for any procedure that transports DNA into a recipient cell.

❐ dna diagnostics

The use of DNA polymorphisms to detect the presence of a specific sequence, which could indicate the presence of a contaminant, of a pathogen, or of a particular allele at a target gene. Most commonly utilises the polymerase chain reaction.

❐ dna fingerprint

A description of the genotype of an individual from the pattern of DNA fragments obtained from DNA fingerprinting.

❒ dna fingerprinting

The derivation of unique patterns of DNA fragments obtained using a number of marker techniques; historically these were RFLPs, but latterly they are generally polymerase chain reaction based.

❒ dna helicase

An enzyme that catalyses the unwinding of the complementary strands of a DNA double helix. DNA hybridization The annealing of two single-stranded DNA molecules, possibly of different origin, to form a partial or complete double helix. The degree of hybridization varies with the extent of complementarity between the two molecules, and this is exploited to test for the presence of a specific nucleotide sequence in a DNA sample.

❒ dna is genetic material

Two closely related strains of the bacterium Streptococcus pneumoniae differ from each other both in their appearance under the microscope and in their pathogenicity. One strain appears smooth (S) and causes death when injected into mice, and the other appears rough (R) and is nonlethal. The experiment in (A) shows that a substance present in the S strain can change (or transform) the R strain into the S strain and that this change is inherited by subsequent generations of bacteria. The experiment in (B) identifies the substance as DNA.

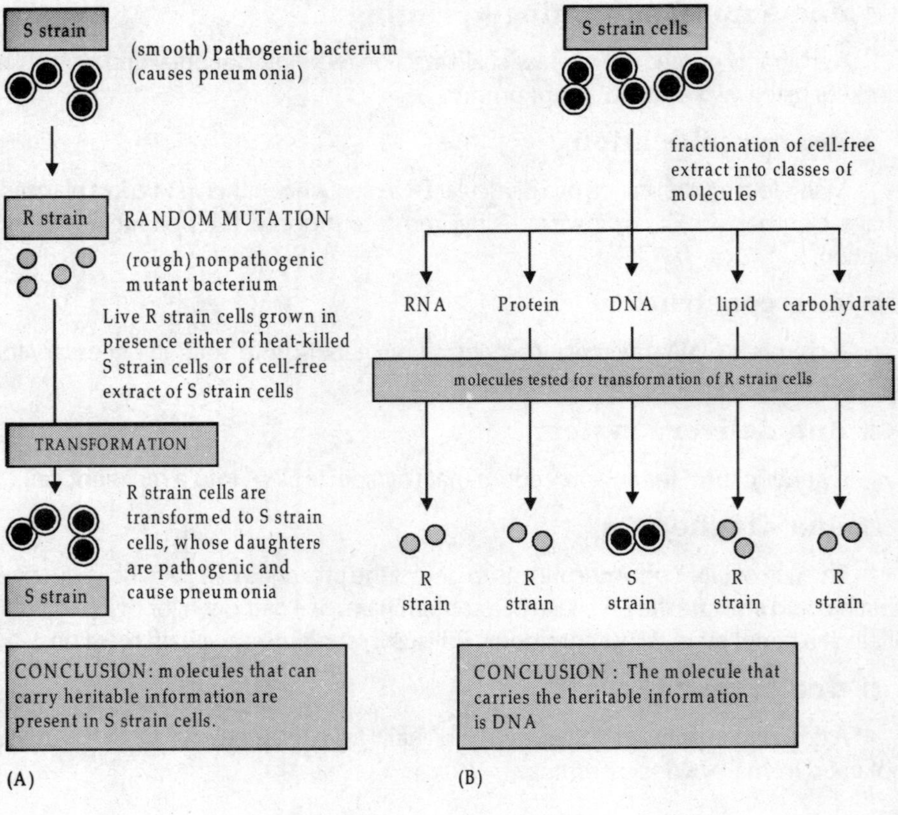

❐ dna ligase

An enzyme that catalyses a reaction to link two separate DNA molecules via the formation of a phosphodiester bond between the 3'-hydroxyl end of one and the 5'-phosphate of the other. Its natural role lies in DNA repair and replication. An essential tool in recombinant DNA technology, as it enables the incorporation of foreign DNA into vectors.

❐ dna molecule - two views

The double helix of the DNA is shown along with details of how the bases, sugars and phosphates connect to form the structure of the molecule.

DNA is a double-stranded molecule twisted into a helix (think of a spiral staircase). Each spiraling strand, comprised of a sugar-phosphate backbone and attached bases, is connected to a complementary strand by non-covalent hydrogen bonding between paired bases. The bases are adenine (A), thymine (T), cytosine (C) and guanine (G).

A and T are connected by two hydrogen bonds. G and C are connected by three hydrogen bonds.

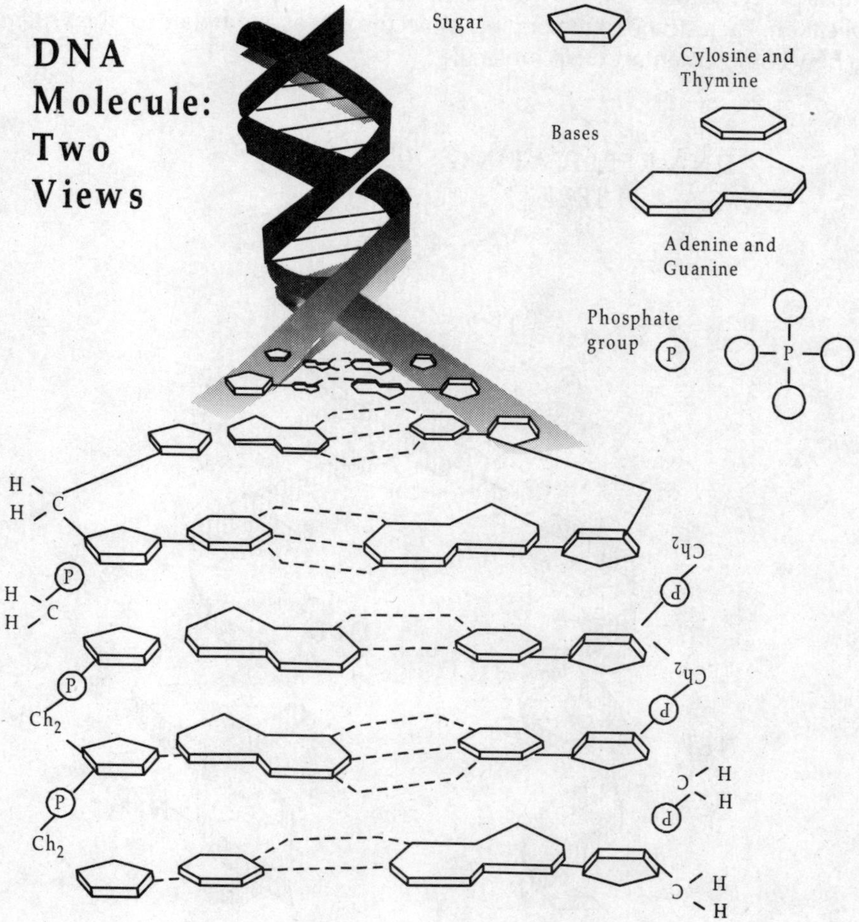

▢ dna polymorphism

The existence of two or more alternative alleles at a DNA-based marker locus.

▢ dna primase

An enzyme that catalyses the synthesis of the short strands of RNA that initiate the synthesis of DNA strands.

▢ dna repair

A variety of mechanisms that repair errors (e.g. the incorporation of a non-complementary nucleotide) that occur naturally during DNA replication.

▢ dna replicating itself

A simplified representation of a DNA molecule separating to form two new molecules.

To reproduce, a cell must copy and transmit its genetic information (DNA) to all of its progeny. To do so, DNA replicates, following the process of semiconservative replication. Each strand of the original molecule acts as a template for the synthesis of a new complementary DNA molecule.

DNA REPLICATING ITSELF

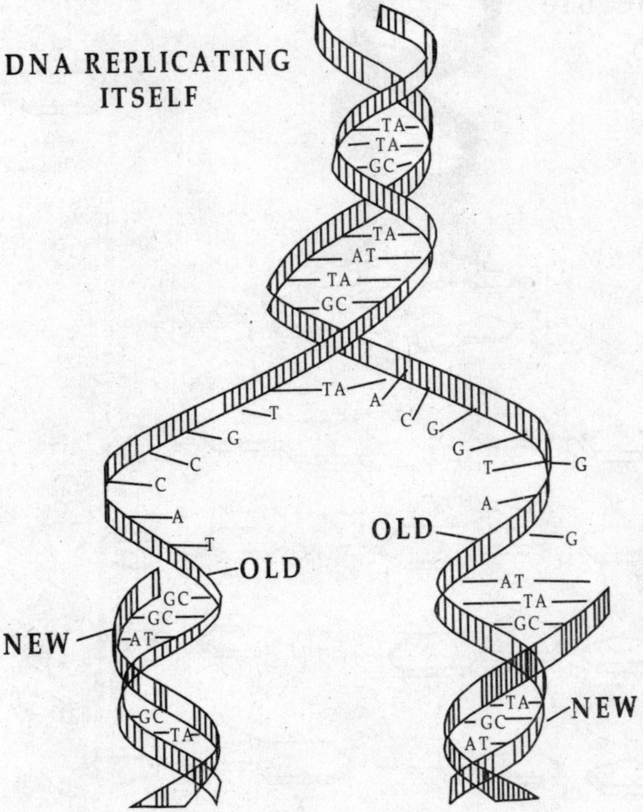

The two strands of the double helix are first separated by enzymes. With the assistance of other enzymes, spare parts available inside the cell are bound to the individual strands following the rules of complementary base pairing: adenine (A) to thymine (T) and guanine (G) to cytosine (C).

Two strands of DNA are obtained from one, having produced two daughter molecules which are identical to one another and to the parent molecule.

▫ dna replication

The process whereby DNA copies itself, under the action of and control of DNA polymerase.

▫ dna sequencing

Procedures for determining the nucleotide sequence of a DNA fragment. Two common methods available:

1. The Maxam Gilbert technique, which uses chemicals to cleave DNA into fragments at specific bases; or, most commonly,
2. The Sanger technique (also called the di-deoxy or chain-terminating method) which uses DNA polymerase to make new DNA chains, in the presence of di- eoxynucleotides (chain terminators) to stop the chain randomly as it grows. In both cases, the DNA fragments are separated according to length by polyacrylamide gel electrophoresis, enabling the sequence to be read directly from the gel. The procedure has become increasingly automated and large-scale in recent years.

▫ dna topo- somerase

An enzyme that catalyses the introduction or removal of supercoils in DNA.

▫ dna vaccine

A vaccine generated by the injection of specific DNA fragments to stimulate an immune response.

▫ dnase

Abbreviation for deoxyribonuclease. Any enzyme that catalyses the cleavage of DNA phosphodiester bonds. DNAse I is a digestive endonuclease secreted by the pancreas, that degrades DNA into shorter fragments. Many other endonucleases and exonucleases are involved in DNA repair and replication.

▫ dolly

The first mammal (a sheep) to be created (via nuclear transfer) by the cloning of an adult cell (from the mammary tissue of a ewe). This showed that the process of differentiation into adult tissue is not, as previously thought, irreversible.

▫ domain

A portion of a protein or DNA molecule that has a discrete function or conformation. At the protein level, can be as small as a few amino acid residues or as large as half of the entire protein.

❐ dominance

The gene action exhibited by a dominant allele.

❐ dominant

1. Of alleles, one whose effect with respect to a particular trait is the same in heterozygotes as in homozygotes. The opposite is recessive.
2. Of an individual animal, one that is allowed priority in access to food, mates, etc., by others of its species because of its success in previous aggressive encounters.
3. Of an animal or plant species, the most conspicuously abundant and characteristic in a particular location or environment.

❐ dominant (-acting) oncogene

A gene that stimulates cell proliferation and contributes to oncogenesis when present in a single copy.

❐ dominant marker selection

Selection of cells via a gene encoding a product that enables only the cells that carry the gene to grow under particular conditions. For example, plant and animal cells that express the introduced *neo*^{r} gene are resistant to neomycin and analogous antibiotics, while cells that do not carry *neo*^{r} are killed.

❐ dominant selectable marker

A gene that allows the host cell to survive under conditions where it would otherwise die.

❐ donor junction site

The junction between the 5' end of an exon and the 3' end of an intron.

❐ dormancy

A period in the life of an animal (hibeRNAtion and aestivation) or plant during which growth slows or completely ceases. Evolved to allow survival of adverse environmental conditions. Annual plants survive the winter as dormant seeds, while many perennial plants survive as dormant tubers, rhizomes, or bulbs. Premature breaking of seed dormancy post harvest can be a major problem for maintaining nutritional and/or functional quality, while difficulties in breaking dormancy will lead to poor germination of the crop.

❐ dosage compensation

A regulatory mechanism for sex-linked genes, to allow equivalent levels of gene expression from (in mammals) XY or XX genotypes, even though the gene copy number in XX is double that in XY.

❐ double crossing-over

The formation of two chiasmata within a chromosome arm, leading to the generation of a double recombinant gamete with respect to genes located within the segment defined by the two genes concerned.

double fertilization

A process, unique to flowering plants, in which two male nuclei, which have travelled down the pollen tube, separately fuse with different female nuclei in the embryo sac. The first male nucleus fuses with the egg cell to form the zygote; the second male nucleus fuses with the two polar nuclei to form a triploid nucleus that develops into the endosperm.

double helix

Describes the coiling of the two strands of the double-stranded DNA molecule, resembling a spiral staircase in which the base pairs form the steps and the sugar-phosphate backbones form the rails on each side. One strand runs 3'?5', while the complementary one runs 5'?3'

double recessive

An organism homozygous for a recessive allele at each of two loci.

double-stranded complementary DNA

A double-stranded DNA molecule created from a cDNA template.

double-stranded DNA

Two complementary strands of DNA annealed in the form of a double helix.

down promoter mutation

A mutation that decreases the frequency of initiation of transcription. This leads to a fall in the level of mRNA compared to the wild type state.

down-regulate

To induce genetically a reduction in the level of a gene's expression.

downstream

1. With respect to DNA, the nucleotides that lie in the 3' direction from the point of reference, which is frequently the site at which transcription is initiated. This is generally designated +1, with downstream nucleotides numbered +2, +10 etc.
2. In chemical engineering, those phases of a manufacturing process that follow the biotransformation stage. Usually refers to the recovery and purification of the product of a fermentation process. *See*: downstream processing.

downstream processing

A general term for biotechnological processes which follow the biology, i.e. fermentation of a micro-organism or growth of a plant. Particularly relevant to fermentation processes, which produce a large quantity of a dilute mixture of substances, products and micro-organisms. These must be separated, and the product concentrated, purified and converted into a useful form.

drosophila melanogaster

The fruit fly, used for many years as a model for eukaryotic genetics. Of the nearly 300 disease-causing genes in the human genome, more than half have an analogous gene in the Drosophila genome.

◻ drug delivery

Method by which a drug is delivered to its site of action. For traditional drugs this is another name for *formulation*. However, biotechnology has allowed the development of a range of new therapeutic-agent delivery systems, such as liposomes and other encapsulation techniques, and a range of mechanisms that target a therapeutic agent to a particular cell or tissue.

◻ dry weight

The weight of tissue obtained following sufficiently prolonged oven-drying at high temperature to remove all water. Freeze-drying may also be employed but generates a slightly different result because bound water is not removed.

◻ dttp

Rarely used but strictly correct abbreviation for deoxythymidine 5'-triphosphate. Required for DNA synthesis since it is a direct precursor molecule.

◻ dual culture

A culture made of a plant tissue and one organism (such as a nematode) or an obligate parasite/micro-organism (such as a fungus). Dual culture techniques are used for a variety of purposes, including assessing host-parasite interactions and the production of axenic cultures.

◻ duplication

A particular kind of mutation: production of one or more copies of any piece of DNA, including a gene or even an entire chromosome.

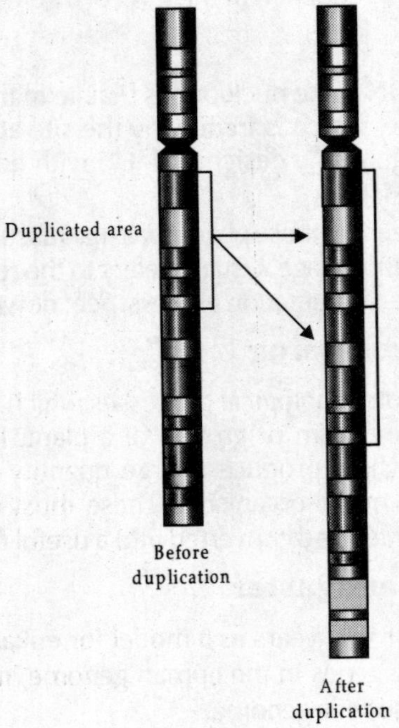

E

❐ ecdysone

A steroid hormone in insects stimulating the synthesis of proteins involved in moulting and metamorphosis.

❐ eclosion

1. Emergence of an adult insect from the pupal stage.
2. Initial phase of germination of fungal spores.

❐ economic trait locus

A locus influencing a trait that contributes to producer's income.

❐ ecosystem

The complex of a living community and its environment, functioning as an ecological unit in nature.

❐ ecotype

A population or a strain of an organism that is adapted to a particular habitat.

❐ ectopic

Anomalous situation or relation, particularly with respect to pregnancy, where the foetus is implanted outside the uterus.

❐ edible vaccine

Edible antigen-containing material, that activates the immune system via gut-associated lymphoid tissues. A preferred route for vaccine administration, particularly in areas where the technological infrastructure needed for maintenance of vaccines is absent. The vaccine is synthesized *in vivo* in the edible parts of transgenic plants (e.g. grains, tubers, fruits, etc.) or eggs.

❐ edv

Abbreviation for essential derivation of varieties.

❐ effector cells

Cells of the immune system that are responsible for the production of cell-mediated cytotoxicity.

▢ effector molecule

A molecule that influences the behaviour of a regulatory molecule, such as a repressor protein, thereby influencing gene expression.

▢ egg

1. The fertilized zygote in egg-laying animals.
2. The mature female reproductive cell in animals and plants.

▢ elastin

A fibrous protein that is the major constituent of the yellow elastic fibres of animal connective tissue.

▢ electro-blotting

The electrophoretic transfer of DNA, RNA or protein from a gel, in which they have been separated, to a support matrix, such as nitrocellulose. A transfer technique employed in Southern and northern blotting.

▢ electrochemical sensor

Biosensors, such as an enzyme electrode, in which a biological process is harnessed to an electrical sensor system. Other types couple a biological event to an electrical one via a range of mechanisms, including the reduction of oxygen or pH change.

▢ electron microscope

A microscope that uses an electron beam focussed by magnetic 'lenses'. electrophoresis A ubiquitous molecular biology technique, with many variants, used to resolve complex mixtures of macromolecules into their components. Its principle is to subject samples to an electric field applied across a porous matrix. Molecules will migrate under these conditions at a rate dependent on their net electric charge and/or their molecular weight. *See*: agarose gel electrophoresis, polyacrylamide gel electrophoresis, denaturing gradient gel electrophoresis, capillary electrophoresis, sodium dodecyl sulphate polyacrylamide gel electrophoresis, thermal gel gradient electrophoresis pulsed-field gel electrophoresis, and iso-electric focusing gel.

▢ electroporation

The induction of transient pores in bacterial cells or protoplasts by the application of a pulse of electricity. These pores allow the entry of exogenous DNA into the cell. Widely used for the transformation of bacteria.

▢ elisa

Abbreviation for enzyme-linked immunosorbent assay. An immunoassay, i.e. an antibody-based technique for the diagnosis of the presence and quantity of specific molecules in a mixed sample. It combines the specificity of an immunoglobulin with the detectability of an enzyme-generated coloured product. In one form, the primary antibody (specific to the test protein) is adsorbed onto a solid substrate, and a known amount of the sample is added; all the antigen in the sample is bound by the antibody. A second antibody (conjugated with an enzyme) specific for a second site on the test protein is added; and the enzyme generates a colour change in the presence of a substrate reagent.

❒ elite tree

A phenotypically superior tree in a tree breeding programme.

❒ elongation factors

Soluble proteins required for the elongation of polypeptide chains on ribosomes.

❒ embryo

An immature organism in the early stages of development. In mammals, develops in the first months in the uterus. In plants, it is the structure that develops in the megagametophyte, as result of the fertilization of an egg cell, or occasionally without fertilization. Somatic embryos can often be induced in *in vitro* plant cell cultures.

❒ embryo cloning

The creation of identical copies of an embryo by embryo splitting or by nuclear transfer from undifferentiated embryonic cells.

❒ embryo culture

The culture of embryos on nutrient media.

❒ embryo multiplication and transfer

The cloning of animal embryos and their subsequent transfer to recipients via artificial inembryonation. The cloned embryos can be derived from embryonic or adult tissue.

❒ embryo rescue

A sequence of tissue culture techniques utilized to enable a fertilized immature embryo resulting from an interspecific cross to continue growth and development, until it can be regenerated into an adult plant.

❒ embryo sac

The mature female gametophyte in angiosperms. Generally a seven-celled structure - two synergids, one egg cell, three antipodal cells (each with a single haploid nucleus) and one endosperm mother cell with two haploid nuclei.

❒ embryo sexing

The determination of the sex of an embryo prior to birth. Typically achieved by the polymerase chain reaction-mediated amplification of DNA extracted from a sample of embryonic tissue. Dependent on the availability of reliable markers for the differential sex chromosome.

❒ embryo splitting

The splitting of young embryos into several sections, each of which develops into an animal. A form of animal cloning, i.e. of producing animals that are genetically identical. In practice, the number of animals that can be produced from a single embryo is less than 10.

❑ embryo storage

Cryogenic preservation of animal embryos, allowing embryonation or other manipulations long after embryo formation.

❑ embryo technology

Generic name for any modification of mammalian embryos. It encompasses embryo cloning, embryo splitting, embryo storage, and *in vitro* fertilization.

❑ embryogenesis

1. (General) Development of an embryo.
2. (In plants) *In vitro* formation of plants from plant tissues, through a pathway closely resembling normal embryogeny from the zygote. Somatic cell embryogenesis is an alteRNAtive technique. The generation of embryos has two stages: initiation and maturation. Initiation needs a high level of the group of plant hormones called auxins; maturation needs a lower level.

❑ embryoid

Plant biotechnology term no longer commonly used. An embryo-like body developing *in vitro*, forming a complete, self-contained plantlet with no vascular connection with the callus.

❑ embryonic stem cells

Cells of the early embryo that can give rise to all differentiated cells, including germ line cells.

❑ emission wavelength

The specific wavelength of light emitted by a fluorescent molecule, such as a labelled probe, upon absorption of light at the (higher) excitation wavelength.

❑ encapsidation

The process by which the nucleic acid of a virus is enclosed in a capsid.

❑ encapsulating agents

Anything which forms a shell around an enzyme or bacterium, common agents being polysaccharides such as alginate or agar. The agents are inert and allow nutrients and oxygen to diffuse readily into and out of the sphere, and are easy to convert from gel (solid) to sol (liquid) or solution form by altering the temperature or the concentration of ions.

❑ encapsulation

Any method packaging an enzyme or bacterium and maintaining its normal functions. Used to immobilize cells in a bioreactor.

❑ encode

The gene product specified by a particular nucleic acid sequence. *See*: genetic code.

❑ endangered species

A plant or animal species in immediate danger of extinction because its population number has reached a critical level, or its habitat has been drastically reduced.

❑ endemic

Describing an organism, often a disease or pest, that is always present in a stated area.

❑ end-labelling

The introduction of a readily-visualized tag at the end of a DNA or RNA molecule. A commonly used method is to introduce a ^{32}P atom onto the end of a DNA molecule by means of the enzyme T4 polynucleotide kinase.

❑ endocrine gland

Any gland in an animal that manufactures hormones and secretes them directly into the bloodstream to act at distant sites in the body, known as target organs or cells.

❑ endocrine interference

Interference with the normal balance of hormones.

❑ endocytosis

The process by which materials enter a cell without passing through the cell membrane. The membrane folds around material outside the cell, resulting in the formation of a sac-like vesicle inside which the material is entrapped. This vesicle is then pinched off from the cell surface so that it lies within the cell.

❑ endoderm

The internal layer of cells of the gastrula, which develops into the alimentary canal (gut) and digestive glands of the adult.

❑ endodermis

The layer of living cells, with various characteristically thickened walls and no intercellular spaces, which surrounds the vascular tissue of certain plants and occurs in nearly all roots and certain stems and leaves. It separates the cortical cells from cells of the pericycle.

❑ endogenous

Derived from within; from the same cell type or organism.

❑ endomitosis

Duplication of chromosomes without division of the nucleus, resulting in a doubling (or more) in the chromosome number within a cell.

❐ endonuclease

An enzyme that cleaves a phosphodiester bond within a DNA strand, forming two smaller strands.

❐ endophyte

An organism that lives inside a plant.

❐ endoplasmic reticulum

A cytoplasmic net of membranes, adjacent to the nucleus, visible under the electron microscope. The sites of protein synthesis.

❐ endopolyploidy

The net result of endomitoses. The somatic chromosome number has doubled (or more), forming a polyploid cell line. If these differentiate into a germ line, then the gametic number will have also increased proportionately, giving rise to homogeneously polyploid individuals, termed endopolyploids.

❐ endoprotease

An enzyme that cleaves internal peptide bonds within a polypeptide molecule. Site of cleavage is usually specific to certain amino acid residues.

❐ endoreduplication

Chromosome reproduction during interphase. Four-chromatid chromosomes (diplochromosomes) are seen during this phase.

❐ endosperm

The nutritive tissue that develops in the seed of most angiosperms, containing varying proportions of carbohydrate (usually starch), protein and lipid. In most diploid plants, the endosperm is triploid.

❐ endosperm mother cell

One of the seven cells of the mature plant embryo sac, containing the two polar nuclei and, which, following fertilization, gives rise to the primary endosperm cell from which the endosperm develops.

❐ endotoxin

A component of the cell wall of gram-negative bacteria that elicits, in mammals, an inflammatory response and fever.

❐ end-product inhibition

The inhibition of an enzyme by a metabolite. Typically, the enzyme is the first enzyme in a biosynthetic pathway, and the metabolite the product of the last step in the pathway. *See*: feedback inhibition.

❐ enhancer

1. A substance or object that increases a chemical activity or a physiological process.

2. A eukaryotic DNA sequence (also found in some eukaryotic viruses) which increases the transcription of a gene. Located up to several kbp, usually (but not exclusively) upstream of the gene in question. In some cases can activate transcription of a gene with no (known) promoter.
3. A major or modifier gene that increases the rate of a physiological process.

▫ enolpyruvyl-shikimate-3-phosphate synthase

An enzyme produced by virtually all plants, which is essential for normal metabolism, and for the biosynthesis of aromatic amino acids. Glyphosate- and sulfosate-containing herbicides act by inhibiting EPSP synthase activity, but because strain CP4 of *Agrobacterium* sp. is unaffected by glyphosate, the introduction of the *CP4 EPSPS* gene into crop plants generates a tolerance of glyphosate-containing herbicides.

▫ enterotoxin

A bacterial protein that, following release into the intestine, causes cramps, diarrhoea and nausea.

▫ enucleated ovum

Egg cell from which the nucleus has been removed, usually as a preparatory step for nuclear transfer.

▫ enzyme

A protein which, even in very low concentration, catalyses specific chemical reactions but is not used up in the reaction. Enzymes are classified into six major groups (1-6), according to the type of reaction they catalyse:

1. oxidoreductases;
2. transferases;
3. hydrolases;
4. lyases;
5. isomerases;
6. ligases.

Generally enzymes are named by the addition of the suffix -ase to the name of their substrate, and are classified by a standard numerical system: the Enzyme Commission (EC) number.

Mechanism of enzyme activity

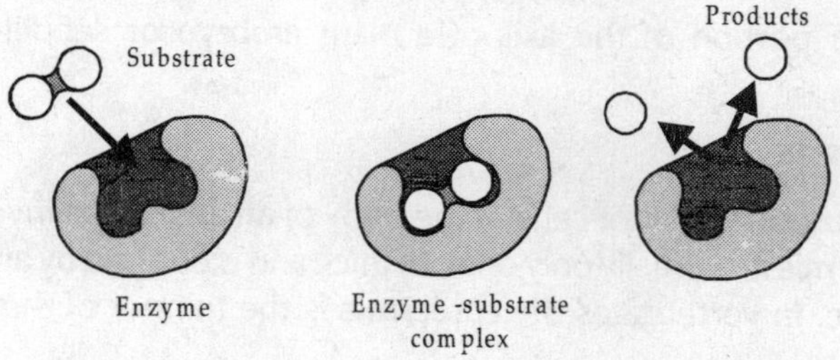

▫ enzyme bioreactor

A reactor in which a chemical conversion reaction is catalysed by an enzyme.

▫ enzyme Commission number

Systematic label which identifies an enzyme in the technical literature. Consists of four numbers separated by dots: the first classifies the enzyme into one of the six broad enzyme groups (*see*: enzymes); each group is subdivided into sub-groups, each sub-group into sub-sub-groups, and the last number is specific for the enzyme, e.g. EC 3.1.21.1 is deoxyribonuclease I.

▫ enzyme electrode

A type of biosensor, in which an enzyme is immobilized onto the surface of an electrode. When the enzyme catalyses its reaction, electrons are transferred from the reactant to the electrode, and so a current is generated. There are two types of enzyme electrodes:

1. Ampometric (measuring current flow) where the electrode is kept as near zero voltage as possible. When the enzyme catalyses its reaction, electrons move into the electrode, and so a current flows;
2. Potentiometric (measuring changes in electrical potential) when the electrode is held at a voltage which counteracts the voltage determined by the enzyme's tendency to push electrons into it. Usually enzymes transfer their electrons inefficiently to the electrode, so a mediator compound is coated onto the electrode to help the transfer.

▫ enzyme immunoassay

A range of immunoassay techniques employing enzymes, which includes ELISA.

▫ enzyme kinetics

The quantitative characteristics of enzyme reactions.

▫ enzyme stabilization

Maintaining the active conformation of an enzyme. This can be achieved *in vitro* by providing the appropriate chemical environment and cofactors. In some cases the criticality of these factors can be reduced by binding an antibody to the enzyme, in such a way that the active site of the enzyme is left unblocked.

▫ epicotyl

The upper portion of the axis of a plant embryo or seedling, above the cotyledons.

▫ epidermis

1. The outmost layer of cells of the body of an animal. In invertebrates the epidermis is normally only one cell thick and is covered by an impermeable cuticle. In vertebrates the epidermis is the thinner of the two layers of skin.

2. The outermost layer of cells covering a plant. It is overlaid by a cuticle and its functions are principally to protect the plant from injury and to reduce water loss. Some epidermal cells are modified to form guard cells or hairs of various types. In woody plants the functions of the shoot epidermis are taken over by the periderm tissues and in mature roots the epidermis is sloughed off and replaced by the hypodermis.

❒ epigenesis

Describes the developmental process whereby each successive stage of normal development is built up on the foundations created by the preceding stages of development; an embryo is built up from a zygote, a seedling from an embryo, and so on.

❒ epigenetic variation

Non-hereditary and reversible variation; often the result of a change in gene expression due to methylation of DNA.

❒ epinasty

A process by which the growth of branches or petioles is abnormally pointing downward. This phenomenon is caused by the more rapid growth of the upper side. Epinasty may result from either nutritional deficiencies or irregularities at the plant growth regulator level. Not to be confused with wilting, as epinastic tissues are turgid.

❒ epiphyte

A plant that grows upon another plant, but is neither parasitic on it nor rooted in the ground.

❒ episome

A genetic extrachromosomal element (e.g. the F factor in *Escherichia coli*) which replicates within a cell independently of the chromosome and is able to integrate into the host chromosome. The step of integration may be governed by a variety of factors and so the term episome has lost favour and been superseded by the wider term plasmid.

❒ epistasis

Interaction between genes at different loci, e.g. one gene suppresses the effect of another gene that is situated at a different locus. Dominance is associated with members of allelic pairs, whereas epistasis describes an interaction among products of non-alleles.

❒ epitope

Synonym for antigenic determinant.

❒ epizootic

A disease simultaneously affecting a large number of animals.

❒ epsps

Abbreviation for enolpyruvyl-shikimate 3-phosphate synthase.

❒ epsp synthase

Abbreviation for enolpyruvyl-shikimate 3-phosphate synthase.

❒ equational division

A chromosome division in which the two chromatids of each duplicated chromosome separate longitudinally, prior to being incorporated into two daughter nuclei. Seen at the mitotic-type second division of meiosis; also in somatic mitosis and the non-reductional division of meiosis. The number of chromosomes is the same at the end of the division as at the beginning.

❒ equilibrium density gradient centrifugation

A procedure used to separate macromolecules based on their density (mass per unit volume).

❒ erlenmeyer flask

A conical flat-bottomed laboratory flask with a narrow neck, widely used for culturing micro-organisms.

❒ escherichia coli

A commensal bacterium inhabiting the colon of many animal species, including human. *E. coli* is widely used as a model of cell biochemical function, and as a host for cloning DNA. In environmental studies, its presence is a key indicator of water pollution due to human sewage effluent. Some strains, notably *E. coli* 0157:H7, are significant pathogens.

❒ essential amino acid

An amino acid required for normal metabolism, but which cannot be synthesized by an organism. It therefore has to be supplied via feed or food.

❒ essential derivation of varieties

Genotypes very similar to an originating cultivar, obtained, for example, by the selection of a mutant or a variant individual from plants of the initial variety, or by backcrossing or transformation.

❒ essential element

Any of a number of elements required by living organisms to ensure normal growth, development and maintenance.

❒ essential nutrient

Any substance required by living organisms to ensure normal growth, development and maintenance.

❒ essential requirement

In plant cell tissue culture, comprises inorganic salts, including all of the elements necessary for plant metabolism; organic factors (amino acids, vitamins); usually also endogenous plant growth regulators (auxins, cytokinins and often gibberellins); as well as a carbon source (sucrose or glucose).

❐ established culture

1. An aseptic viable explant (*See*: micropropagation).
2. A suspension culture subjected to several passages with a constant cell number per unit time.

❐ estimated breeding value

Twice the expected progeny difference. The difference is doubled because breeding value is a reflection of all the genes of an individual, in contrast to progeny difference, which is a reflection of a sample half of an individual's genes. The predicted performance of the offspring of the mating between any two parents is the average of their EBVs (averaged because each parent makes an equal contribution to each offspring).

❐ et

Abbreviation for embryo transfer.

❐ ethanol

Commonly used to disinfect plant tissues, glassware utensils and working surfaces in tissue culture manipulations; to precipitate aqueous solutions of nucleic acids; and to dissolve water-insoluble components of culture media.

❐ ethephon

A synthetic compound commonly used as a source of ethylene, a gaseous plant growth regulator.

❐ ethidium bromide

A fluorescent dye which can intercalate between base pairs of double-stranded DNA, and hence is much used to stain DNA in gels. The dye fluoresces when exposed to UV light. It is a known to be a strong mutagen and is also possibly both a carcinogen and a teratogen.

❐ ethylene

A gaseous plant growth regulator acting on various aspects of vegetative growth, fruit ripening and abscission of plant parts.

❐ ethylenediamine tetraacetic acid

A chelating compound used to keep nutrients, such as iron, bound in a soluble form that leaves them still available to the plant cells *in vitro*. Also a potent inhibitor of DNase activity and therefore used as an additive for long-term storage of dissolved DNA.

❐ etiolation

An abnormal increase in stem elongation, accompanied by poor (if any) leaf development. Physiological etiolation is caused by a lack of chlorophyll, and is typical of plants growing under low light intensity or in complete darkness. It can also be induced by some fungal pathogens.

❒ euchromatin

Chromosomal material that is stained less intensely by certain dyes. Thought to be the chromosomal domains which are gene-rich, since the DNA in these regions remains less contracted than those rich in repetitive DNA - the heterochromatin.

❒ eugenics

The application of the principles of genetics to the 'improvement' of humankind. Wholly discredited as a scientific approach since the Nazi period.

❒ eukaryote

One of the two major evolutionary clades, characterized by having the nucleus enclosed by a membrane, and possessing chromosomes that undergo mitosis and meiosis. Eukaryotic organisms include animals, plants, fungi and some algae.

❒ euploid

An organism or cell having a chromosome number that is an exact multiple of the haploid number. Terms used to identify different levels in an euploid series are diploid (2x), triploid (3x), tetraploid (4x) etc.

❒ evapotranspiration

The net water loss (in vapour form) per unit area of land, both directly from the land surface, and indirectly through transpiring leaves.

❒ evolution

The process by which the present diversity of plant and animal life has arisen, and which continues to drive changes in form and mode of existence of all living organisms.

❒ ex vitro

Organisms removed from tissue culture and transplanted; generally plants to soil or potting mixture.

❒ *ex vivo* gene therapy

The delivery of a gene or genes to the isolated cells of an individual, with the intention of alleviating a genetic disorder. After culturing, the transformed cells are re-introduced into the individual by transfusion, infusion or injection.

❒ examples of viruses

Illustration of three viruses: the tobacco mosaic virus (TMV), the T4 bacteriophage, and the Human Immunodeficiency Virus (HIV).

1. **Tobacco mosaic virus(TMV):** consists of a single long RNA molecule (6000 nucleotides) enclosed in a cylindrical protein coat (2130 copies of the same coat protein). It is approximately 300 nanometers long and 15 nanometers wide.

2. **T4 bacteriophage:** consists of a DNA molecule tightly packaged in a protein head. The virus attaches to the surface of a bacteria cell and injects its DNA into the cell through its tail.
3. **Human Immunodeficiency Virus (HIV):** consists of an outer lipid envelope derived from the host cell. It is a retrovirus, i.e., an RNA-containing virus that replicates by making a DNA intermediate.

TOBACCO MOSAIC VIRUS

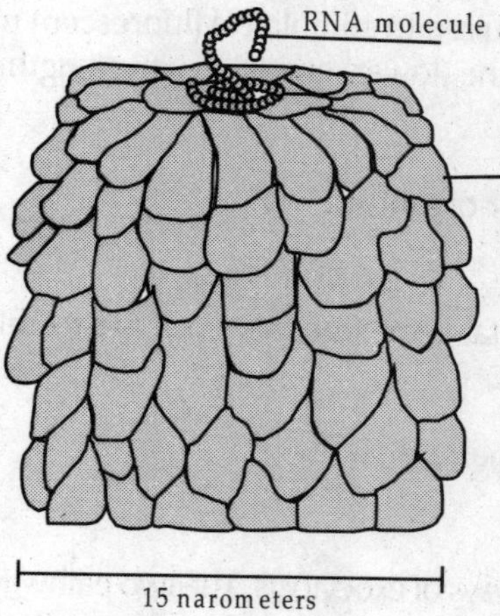

T4 BACTERIOPHAGE

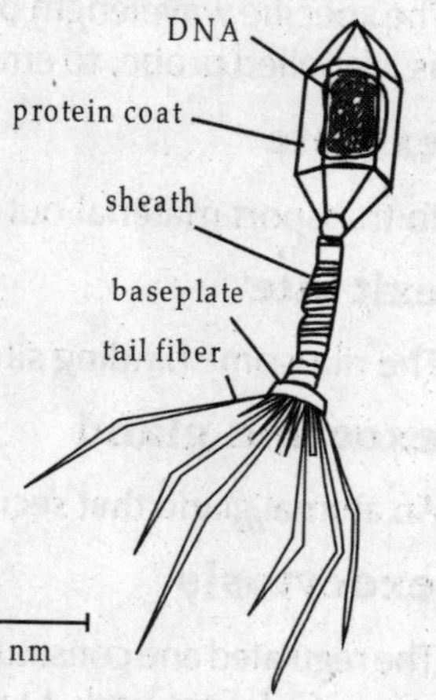

HIV VIRUS WITH ITS LIPID ENVELOPE

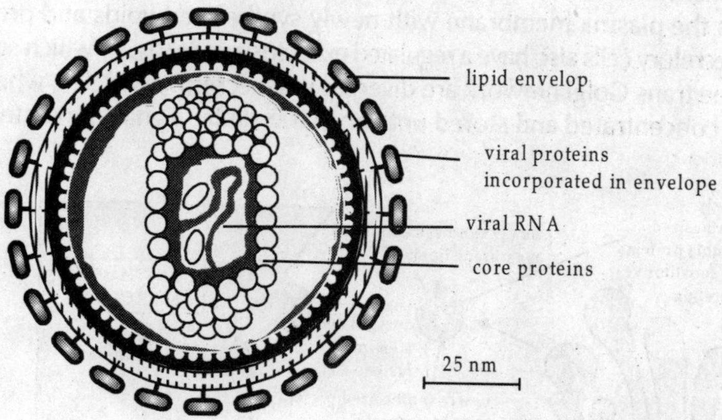

❐ excinuclease

The endonuclease-containing protein complex that excises a segment of damaged DNA during excision repair.

❐ excision

1. The natural or *in vitro* enzymatic removal of a DNA segment from a chromosome or cloning vector.

2. The cutting out and preparation of a tissue, organ, etc., for culture.
3. The removal of adventitious shoots from callus tissue.

❒ excision repair

DNA repair processes that involve the removal of a damaged or incorrect segment of one strand of double-stranded DNA and its replacement by the synthesis of a new segment using the complementary strand of DNA as template.

❒ excitation wavelength

The specific wavelength of light required to stimulate a fluorescent molecule, such as a labelled probe, to emit light at the (lower) emission wavelength.

❒ excrete

To transport material out of a cell or organism.

❒ exit site

The ribosome binding site that contains the free tRNA prior to its release.

❒ exocrine gland

An animal gland that secretes through a duct.

❒ exocytosis

The regulated and constitutive pathways of exocytosis. The two pathways diverge in the trans Golgi network. Many soluble proteins are continually secreted from the cell by the constitutive secretory pathway, which operates in all cells. This pathway also supplies the plasma membrane with newly synthesized lipids and proteins. Specialized secretory cells also have a regulated exocytosis pathway, by which selected proteins in the trans Golgi network are diverted into secretory vesicles, where the proteins are concentrated and stored until an extracellular signal stimulates their secretion.

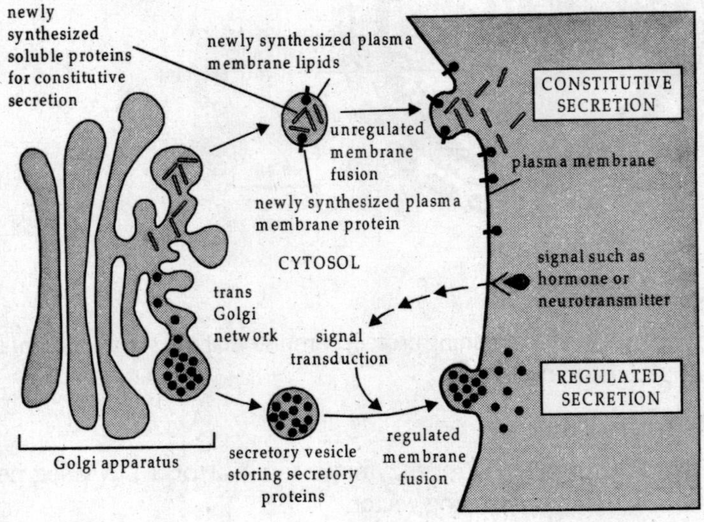

❐ exogenous DNA

DNA that has been derived from one organism, and is to be introduced into a cell a different species. Also referred to as foreign DNA or heterologous DNA.

❐ exogenous

Produced outside of; originating from, or due to, external causes.

❐ exon

The region of a gene that contains the code for producing the gene's protein. Each exon codes for a specific portion of the complete protein. In some species (including humans), a gene's exons are separated by long regions of DNA (called introns or sometimes "junk DNA") that have no apparent function.

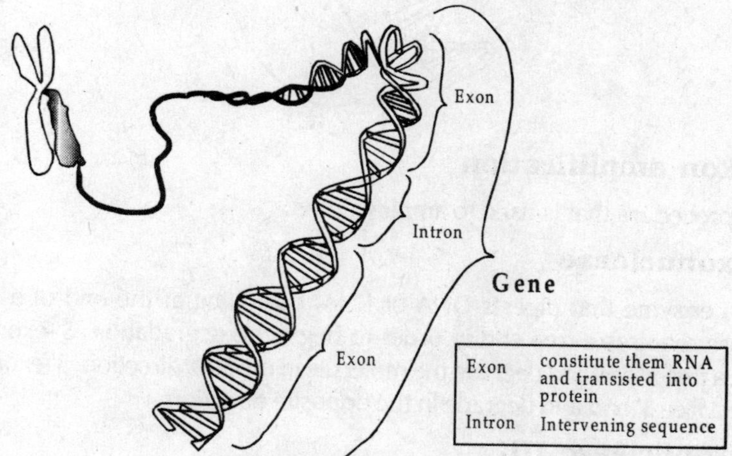

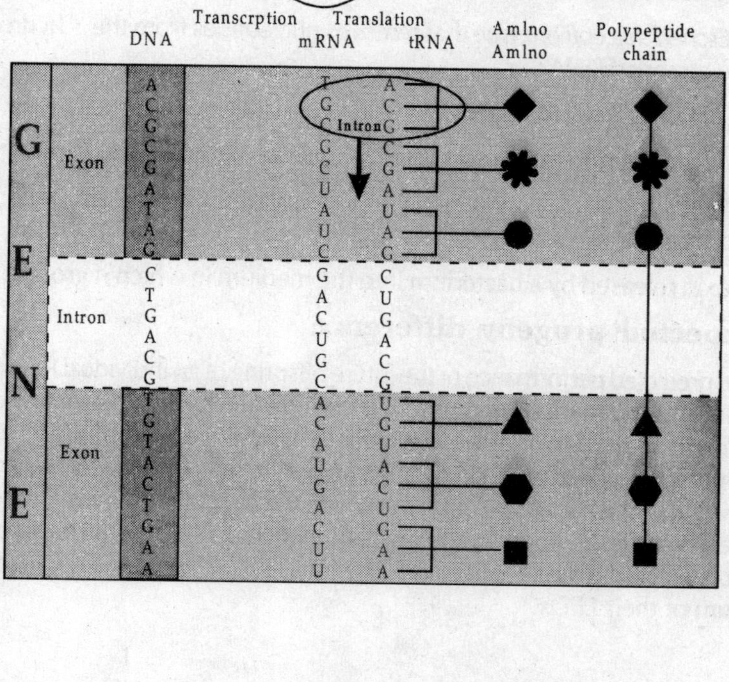

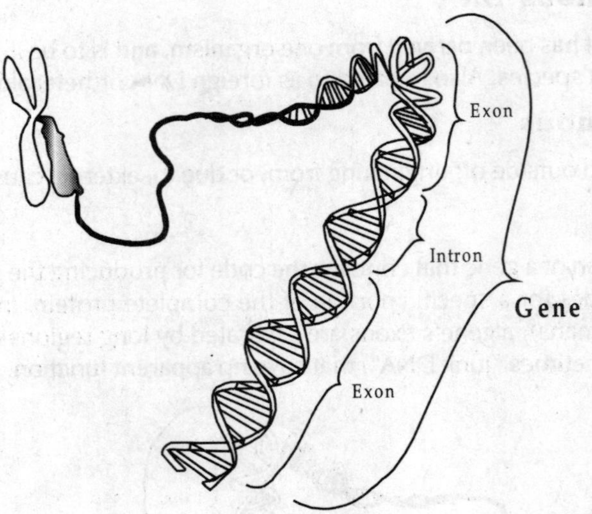

❐ exon amplification

A procedure that is used to amplify exons.

❐ exonuclease

An enzyme that digests DNA or RNA, beginning at the end of a strand. It therefore requires a free end in order to begin the degradation. 5'-exonucleases require a free 5' end and degrade the molecule in the 5'?3' direction. 3'-exonucleases require a free 3' end and degrade in the opposite direction.

❐ exonuclease III

An *Escherichia coli* enzyme that removes nucleotides from the 3' hydroxyl ends of double-stranded DNA.

❐ exopolysaccharide

A polysaccharide that is secreted by a micro-organism into the surrounding environment.

❐ exotoxin

A toxin released by a bacterium into the medium in which it grows.

❐ expected progeny difference

The predicted performance of the future offspring of an individual for a particular trait, calculated from measurement(s) of the individual's own performance and/or the performance of one or more of its relatives, for the trait in question and/or for one or more correlated traits. Typically, the prediction is expressed as a deviation from a well-defined base population, assuming the individual in question is mated to a sample of individuals whose average genetic merit equals that of the base population. The predicted performance of the offspring of the mating between any two individuals is the sum of their EPDs.

❐ explant

A portion of a plant aseptically excised and prepared for culture in a nutrient medium.

❐ explant donor

The plant from which an explant has been taken.

❐ explantation

The removal of cells, tissues or organs of animals and plants for observation of their growth and development in appropriate culture media.

❐ explosion method

A technique for the genetic transformation of cells, in which the transgene is driven into the target (plant) cells by the sudden vaporization (effected by the application of a pulse of high voltage) of a water droplet containing the DNA and gold particles.

❐ export

The removal of a compound from a cell by active transport.

❐ express

To transcribe and translate a gene.

❐ expressed sequence tag

Partially sequenced cDNA clone. Because the read length of a standard DNA sequencing reaction is shorter than the majority of cDNA clones, full length sequence can only be obtained by further manipulations. For the purposes of (1) assigning putative function to a cDNA and (2) designing PCR primers to extract the genomic DNA equivalent to the cDNA, full length sequence is usually unnecessary. By restricting sequencing to a single run, large numbers of cDNAs can be characterized at the EST level.

❐ expression library

A cDNA library that has been inserted into a bacterial host cell engineered to express transgenes.

❐ expression system

Combination of host and vector which provides a genetic context for making a cloned gene functional, i.e. produce peptide, in the host cell.

❐ expression vector

A cloning vector that has been constructed in such a way that, after insertion of a DNA molecule, its coding sequence is properly transcribed and the mRNA is translated. The cloned gene is put under the control of a promoter sequence for the initiation of transcription, and often also has a transcription termination sequence at its end.

❑ expressivity

Degree of expression of a trait controlled by a particular gene. The gene may show different degrees of expression in different individuals.

❑ *ex-situ* conservation

The conservation of components of biological diversity outside their natural habitats.

❑ extension

The short single-stranded stretch of nucleotides remaining on a double-stranded DNA molecule, following treatment with a restriction endonuclease which makes a staggered cut. The presence of these unpaired regions make the molecule more easily ligatable, and are thus important in gene cloning. *Synonyms*: protruding end; sticky end; overhang; cohesive end.

❑ extracellular signals

An animal cell depends on multiple extracellular signals. Each cell type displays a set of receptor proteins that enables it to respond to a corresponding set of signal molecules produced by other cells. These signal molecules work in combinations to regulate the behavior of the cell. As shown here, many cells require multiple signals to survive, additional signals to divide, and still other signals to differentiate. If deprived of appropriate signals, most cells undergo a form of cell suicide known as programmed cell death, or apoptosis.

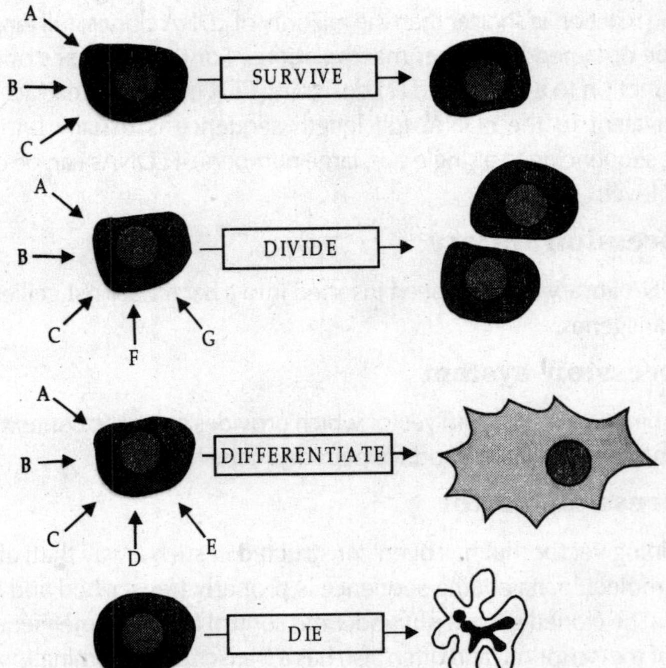

▫ extrachromosomal

In eukaryotes, non-nuclear DNA, present in cytoplasm organelles such as mitochondria and chloroplasts. In prokaryotes, non-chromosomal DNA, i.e. plasmids.

▫ extranuclear genes

Genes residing elsewhere than in the nucleus (e.g. in mitochondria, chloroplasts or plastids).

▫ exude

Slowly discharge liquid material (such as tannins or oxidized polyphenols from plant material) through pores or cuts, or by diffusion into the medium.

F

f factor

Abbreviation for fertility factor. A bacterial plasmid that confers the ability to function as a genetic donor in conjugation.

f_1

Abbreviation for filial generation 1. The initial hybrid generation resulting from a cross between two parents.

f_1, f_2, f_n

Subsequent hybrid generations, counting from the F_1. Thus, for example, F_4 describes the progeny of the F_3, which is the progeny of the F_2 generation, where all progeny are derived from intercrossing or self-fertilization.

f_2

The second hybrid generation, produced either by intercrossing two F_1 individuals, or by self-fertilizing an F_1 individual.

f_{ab}

A product of hydrolysis of an IgG antibody, consisting of the variable region with some of the constant region of a heavy chain, and an entire light chain. Contains a single antigen-binding site.

factorial mating

A mating scheme in which each male parent is mated with each female parent. Made possible in animals by means of *in vitro* embryo production. Such a mating scheme substantially reduces the rate of inbreeding in a selection programme.

facultative anaerobe

An organism that will grow under either aerobic or anaerobic conditions.

false negative

A negative assay result that should have been positive.

false positive

A positive assay result that should have been negative.

❐ families of small organic molecules

The four main families of small organic molecules in cells. They form the monomeric building blocks, or subunits, for most of the macromolecules and other assemblies of the cell. Some, like the sugars and the fatty acids, are also energy sources.

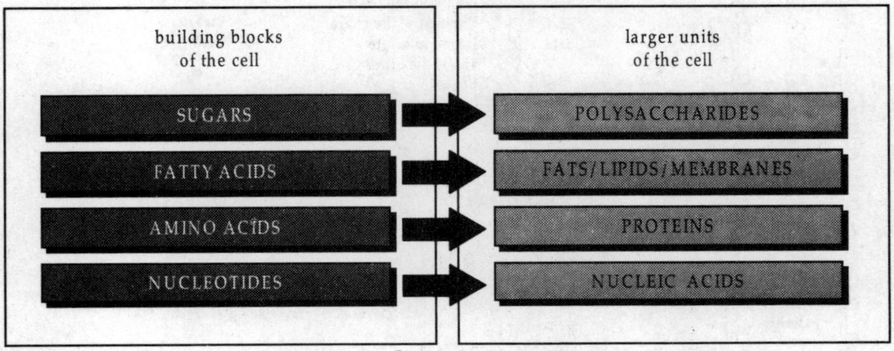

❐ farm animal genetic resources

Those animal species that are used, or may be used, for the production of food and agriculture, and the populations within each of them. Within each species, these populations can be classified as wild and feral populations, landraces and primary populations, standardized breeds, selected lines, and any conserved genetic material.

❐ farmers' privilege

Rights to hold germplasm, covered by plant variety protection, as a seed source for subsequent seasons. Considered as optional for governments to include in their legislation.

❐ farmers' rights

Rights first recognized by Resolution 5 of the 1989 FAO Conference as "rights arising from the past, present and future contributions of farmers in the conservation, improvement and the making available of plant genetic resources"; this item became an attachment to the 'International Undertaking on Plant Genetic Resources'. The binding 'International Treaty of Plant Genetic Resources for Food and Agriculture' that resulted from the renegotiations of the Undertaking makes provision for the Farmers' Rights in Article 9.

❐ fatty acids and other lipids part 1

A chemical unit that occurs naturally, either singly or combined, and consists of strongly linked carbon and hydrogen atoms in a chain-like structure. The end of the chain contains a reactive acid group made up of carbon, hydrogen, and oxygen. This acid group permits reaction with glycerol to make the fatty acid a fundamental unit of the triglyceride fat molecule. A natural fat is a mixture of triglyceride fat molecules. Fatty acids found in soybeans include palmitic, stearic, oleic, linoleic, and linolenic.

COMMON FATTY ACIDS

These are carboxylic acids with long hydrocarbon tails.

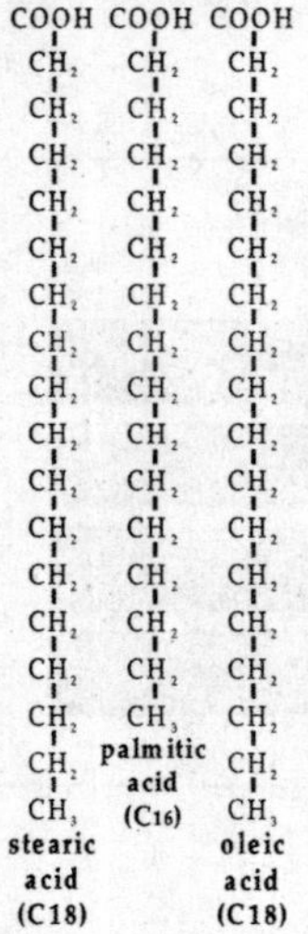

Hundreds of different kinds of fatty acids exist. Some have one or more double bonds in their hydrocarbon tail and are said to be **unsaturated**. Fatty acids with no double bonds are **saturated**.

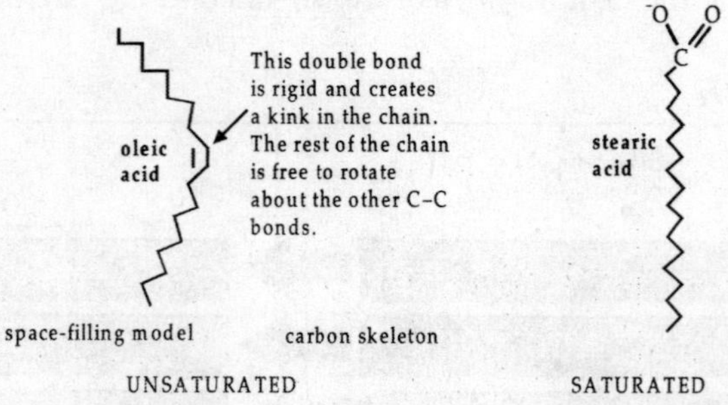

TRIACYLGLYCEROLS

Fatty acids are stored as an energy reserve (fats and oils) through an ester linkage to **glycerol** to form **triacylglycerols**.

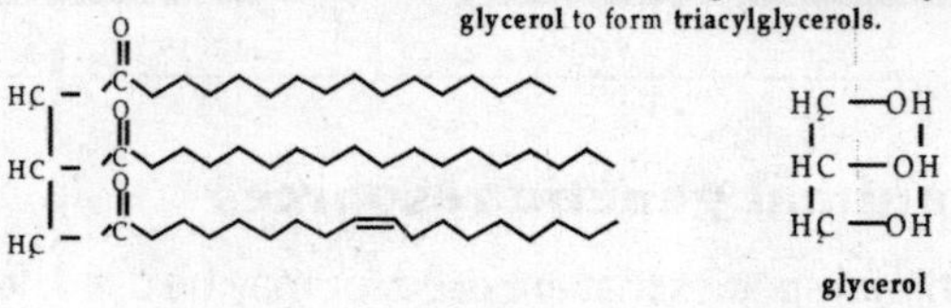

PHOSPHOLIPIDS

Phospholipids are the major constituents of cell membranes.

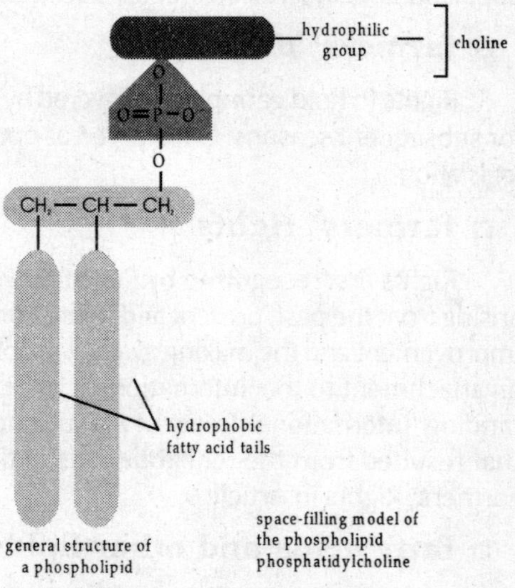

In phospholipids two of the –OH groups in glycerol are linked to fatty acids, while the third –OH group is linked to phosphoric acid. The phosphate is further linked to one of a variety of small polar groups (alcohols).

CARBOXYL GROUP

If free, the carboxyl group of a fatty acid will be ionized.

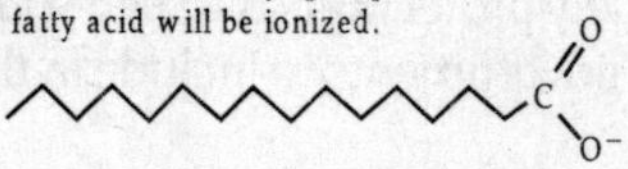

But more usually it is linked to other groups to form either **esters**

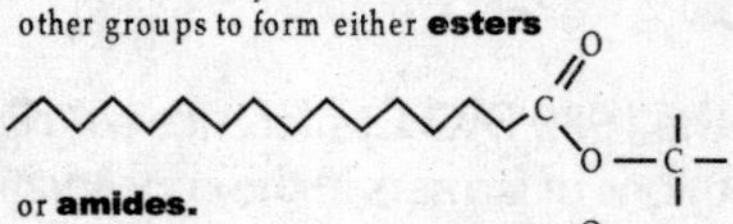

or **amides.**

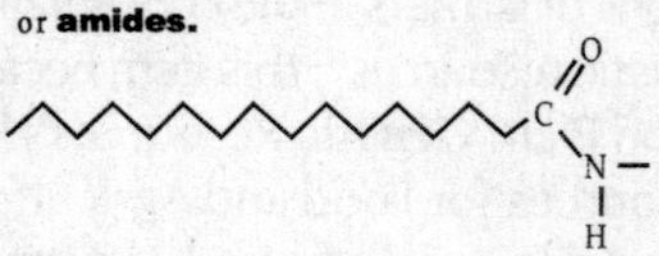

▫ fatty acids and other lipids part 2

An outline of some of the types of sugars commonly found in cells.

LIPID AGGREGATES

Fatty acids have a hydrophilic head and a hydrophobic tail.

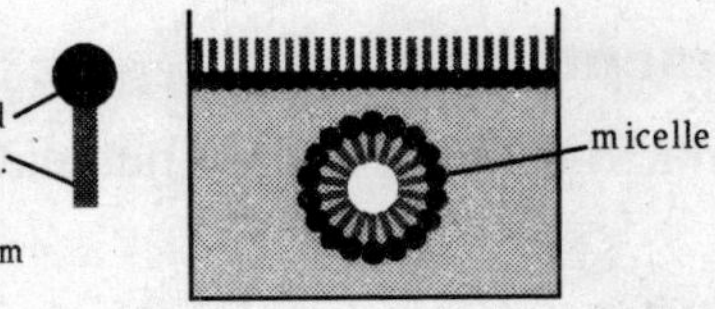

In water they can form a surface film or form small micelles.

Their derivatives can form larger aggregates held together by hydrophobic forces: **Triglycerides** form large spherical fat droplets in the cell cytoplasm.

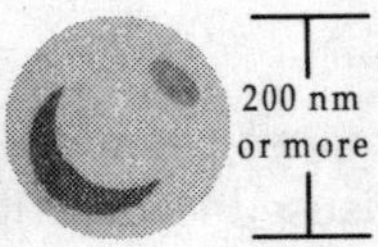

Phospholipids and **glycolipids** form self-sealing lipid bilayers that are the basis for all cellular membranes.

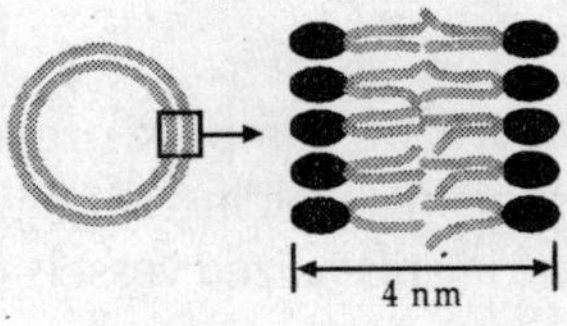

POLYISOPRENOIDS

long chain polymers of isoprene

dolichol phosphate—used to carry activated sugars in the membrane-associated synthesis of glycoproteins and some polysaccharides

OTHER LIPIDS

Lipids are defined as the water-insoluble molecules in cells that are soluble in organic solvents. Two other common types of lipids are steroids and polyisoprenoids. Both are made from isoprene units.

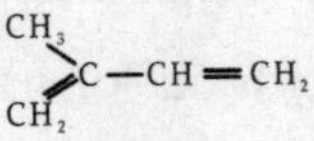

STEROIDS

Steroids have a common multiple-ring structure.

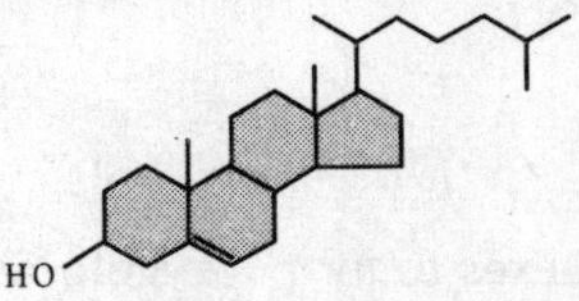

cholesterol—found in many membranes

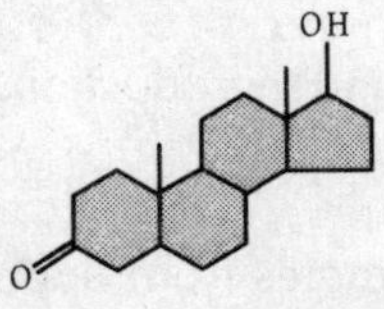

testosterone—male steroid hormone

GLYCOLIPIDS

Like phospholipids, these compounds are composed of a hydrophobic region, containing two long hydrocarbon tails, and a polar region, which, however, contains one or more sugar residues and no phosphate.

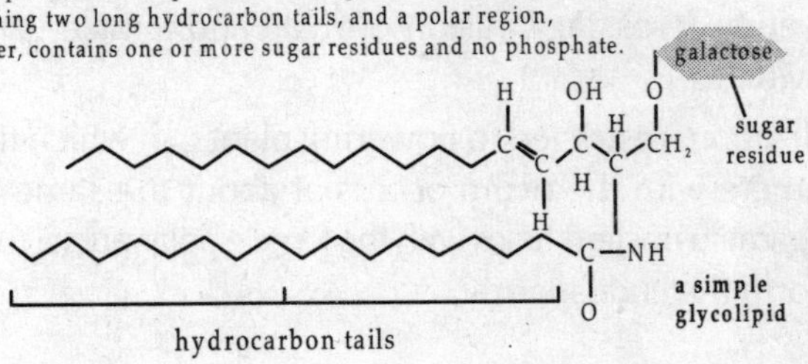

a simple glycolipid

❐ f_c

A product of hydrolysis of an IgG antibody, consisting of parts of the constant regions of two heavy chains held together by a disulphide bridge, but excluding the antigen-binding regions, and also excluding the light chains.

❐ fed-batch fermentation

Culture of cells or micro-organisms where nutrients are added periodically to the bioreactor.

❐ feedback inhibition

The process by which the accumulated end product of a biochemical pathway stops synthesis of that product. The effect is that a late metabolite of a synthetic pathway regulates the synthesis of an earlier step of the pathway.

❐ fermentation

The anaerobic breakdown of complex organic substances, especially carbohydrates, by micro-organisms, yielding energy. Often misused to describe large-scale aerobic cell culture in specialized vessels (fermenters, bioreactors) for secondary product synthesis.

❐ fermentation substrates

Materials used as food for growing micro-organisms. The fermentation substrates and the trace materials needed, together with chemicals added to make the fermentation easier, form the culture medium.

❐ fertile

Capable of breeding and reproduction.

❐ fertilization

The union of two gametes from opposite sexes to form a zygote. Typically, each gamete contains a haploid set of chromosomes. Hence the zygotic nucleus contains a diploid set of chromosomes. Several categories can be distinguished:

1. Self-fertilization (selfing): fusion of male and female gametes from the same individual.
2. Cross-fertilization (crossing): fusion of male and female gametes from different individuals.
3. Double fertilization; restricted to flowering plants, in which the fusion of one male gamete with the ovum occurs at about the same time as the second male gamete nucleus fuses with the female polar nuclei (or secondary nucleus) to form the endosperm.

❐ fertilizer

Any substance that is added to soil in order to increase its productivity. Fertilizers can be of biological origin (e.g. composts), or they can be synthetic (artificial fertilizer).

❐ feulgen staining

A histochemical stain by which the distribution of DNA in the chromosomes of dividing cell nuclei can be observed.

❐ fia

Abbreviation for fluorescence immunoassay.

❐ fibril

A microscopic to sub-microscopic cellulose thread that is part of the cellulose matrix of plant cell walls.

❐ fibroblasts

Irregularly shaped, branching cells distributed throughout vertebrate connective tissue. A cell type which is readily cultured *in vitro*.

❐ fibrous root

Root system in which both primary and lateral roots have approximately equal diameters.

❐ filter bioreactor

A cell culture system, in which cells are grown on a fine mesh of an inert material, which allows the culture medium to flow past it but retains the cells. This is similar in idea to membrane and hollow fibre reactors, but can be much easier to set up, being similar to conventional tower bioreactors, but with the mesh replacing the central reactor space.

❐ filter sterilization

Process of removing microbial contaminants from a liquid by passing through a filter with pores too small to allow the passage of micro-organisms and spores.

❐ filtration

1. Separation of solids from liquids by using a porous material that only allows passage of the liquid or of solids smaller than the pore size of the filter. The material passing the filter forms the filtrate.
2. Removal of cell aggregates to obtain a filtrate of single cells that can be utilized as plating inocula.

❐ fish - fluorescence in situ hybridization

A physical mapping approach that uses fluorescein tags to detect hybridization of probes with metaphase chromosomes and with the less-condensed somatic interphase chromatin. FISH involves the preparation of short sequences of single-stranded DNA, called probes, which are complementary to the DNA sequences the researchers wish to paint and examine. These probes hybridize, or bind, to the complementary DNA and, because they are labeled with fluorescent tags, allow researchers to see the location of those sequences of DNA. Unlike most other techniques used to study chromosomes, which require that the cells be actively dividing, FISH can also be performed on nondividing cells, making it a highly versatile procedure.

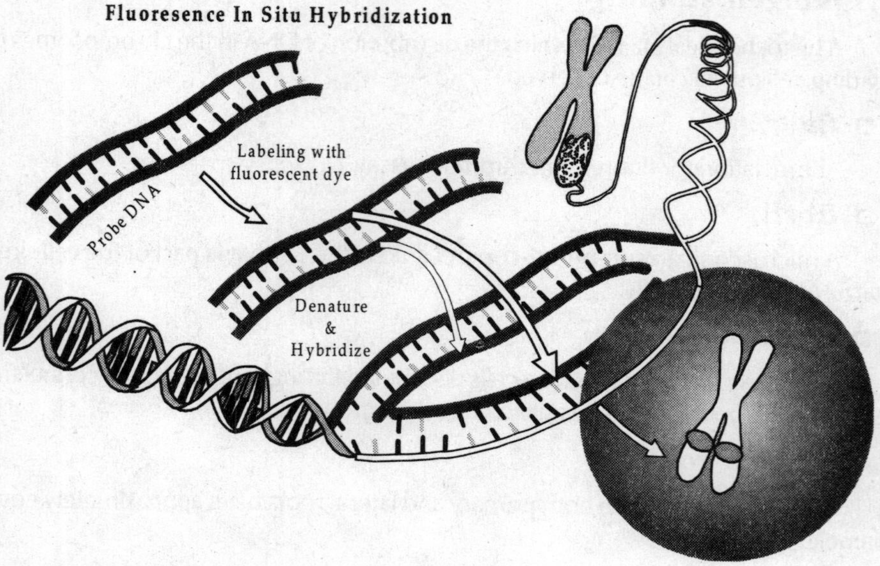

❒ fission

Asexual reproduction involving the division of a single-celled individual into two daughter single-celled individuals of approximately equal size.

❒ fitness

The survival value and the reproductive capability of an individual, compared to that of competitor individuals of the same or other species within a population or an environment.

❒ fixation

The situation in which only one allele for a given gene/locus is present in a population. This can occur as a result of direct selection where the allele delivers a greater level of fitness; because of indirect selection, where the locus is linked to a gene that is subject to direct selection; or because of genetic drift.

❒ flaming

A technique for sterilizing instruments, to remove live micro-organism contaminants. The instrument is dipped in alcohol, and the alcohol remaining on the instrument is ignited, thereby heat-sterilizing the surface.

❒ flanking region

The DNA sequences extending either side of a specific sequence.

❒ flavin adenine dinucleotide

A co-enzyme important in various biochemical reactions. It comprises a phosphorylated vitamin B_2 (riboflavin) molecule linked to AMP, and functions as a hydrogen acceptor in dehydrogenation reactions. The reduced form is oxidized back to FAD by the electron transport chain, generating two molecules of ATP per molecule of reduced FADH.

❑ floating ball anologies for enzyme catalysis

(A) A barrier dam is lowered to represent enzyme catalysis. The ball represents a potential enzyme substrate (compound X) that is bouncing up and down in energy level due to constant encounters with waves (an analogy for the thermal bombardment of the substrate with the surrounding water molecules). When the barrier (activation energy) is lowered significantly, it allows the energetically favorable movement of the ball (the substrate) downhill. (B) The four walls of the box represent the activation energy barriers for four different chemical reactions that are all energetically favorable, in the sense that the products are at lower energy levels than the substrates. In the left-hand box, none of these reactions occurs because even the largest waves are not large enough to surmount any of the energy barriers. In the right-hand box, enzyme catalysis lowers the activation energy for reaction number 1 only; now the jostling of the waves allows passage of the molecule over this energy barrier only, inducing reaction 1. (C) A branching river with a set of barrier dams serves to illustrate how a series of enzyme-catalyzed reactions determines the exact reaction pathway followed by each molecule inside the cell.

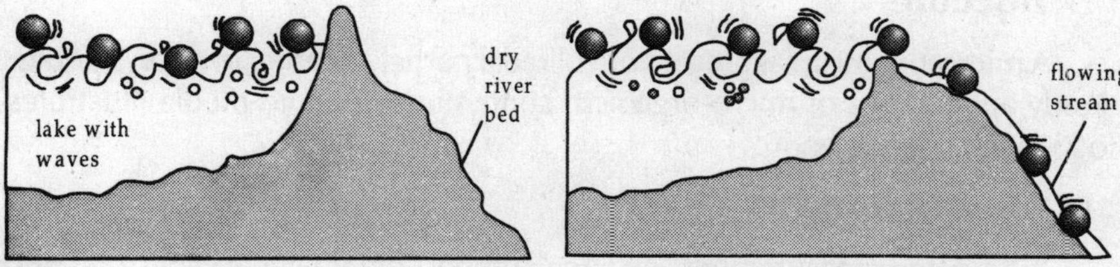

(A)

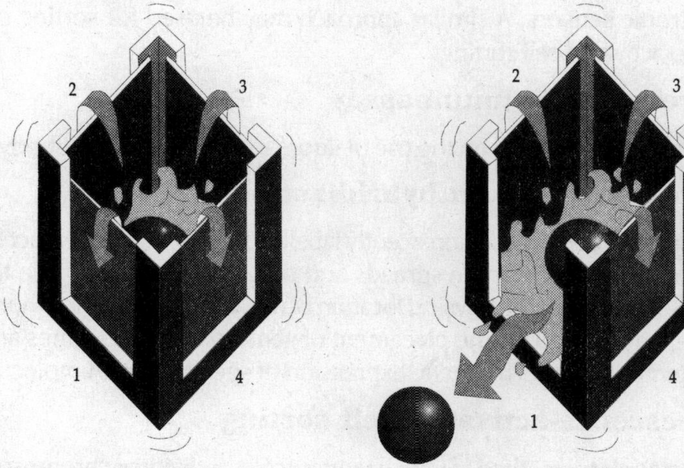

(B) uncatalyzed enzyme catalysis of reaction

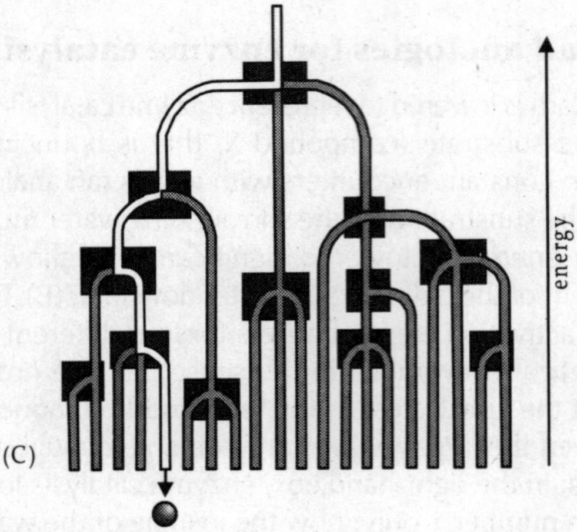

▫ flocculant

A chemical agent that causes small particles to aggregate (flocculate).

▫ floccule

A micro-organism aggregate or colloidal particle floating in or on a liquid. The cloudy appearance of micro-organism contaminated liquid media illustrates the flocculation phenomenon.

▫ flow cytometry

Automated measurements on large numbers of individual cells or other small biological materials, made as the cells flow one by one in a fluid stream past optical and/or electronic sensors. A similar approach may be used for sorting cells - see fluorescence-activated cell sorting.

▫ fluorescence immunoassay

An immunoassay based on the use of fluorescence-labelled antibody.

▫ fluorescence *in situ* hybridization

Hybridization of cloned, fluorescently labelled DNA or RNA, to intact biological materials, notably chromosome spreads and thin tissue sections. The technique allows the visualization of the physical location of nucleic acid sequences homologous to the probe, and is used for the placement of genes on chromosomes and for the spatial and temporal pattern of gene expression of specific mRNA molecules.

▫ fluorescence-activated cell sorting

A flow cytometry method in which targets (cells, individual chromosomes etc.) are labelled with a fluorescent dye, which is excited by a laser beam. Differences in the fluorescence signal emitted are used as a criterion for sorting the material. A specific application is in sperm sexing.

▢ fluorescent probe

A probe which is labelled with a fluorescent dye, so that the signal emitted can be captured by photometric methods.

▢ foetus

Pre-natal stage of a viviparous animal, between the embryonic stage and birth. *AlteRNAtive spelling*: fetus. fog Fine particles of liquid suspended in the air, such as of water in a fog chamber used for acclimatizing recent *ex vitro* transplants. *See*: mist propagation.

▢ fold-back

The structure of a double-stranded DNA molecule formed when a molecule containing an inverted repeat sequence is denatured and then allowed to re-anneal at low DNA concentrations. Under these conditions, the repeated sequence self-anneals to form a double-stranded region within each of the separated strands of the original molecule.

▢ folded genome

The condensed state of the chromosomal DNA of a bacterium. The DNA is segregated into domains, and each domain is independently negatively super oiled.

▢ follicle

An enclosing cluster of cells that protects and nourishes a cell or structure within. Thus a follicle in the ovary contains a developing egg cell, while a hair follicle envelops the root of hair.

▢ follicle stimulating hormone

A hormone, secreted by the anterior pituitary gland in mammals, that stimulates the ripening of the specialized structures in the ovary (Graafian follicles) that produce ova in female mammals; and in males, the formation of sperm in the testis. FSH is a major constituent of fertility drugs.

▢ food processing enzyme

Enzyme used to control food texture, flavour, appearance, or nutritional value. Amylases break down complex polysaccharides to simpler sugars; proteases tenderize meat proteins. A prominent target of food biotechnology is to develop novel food enzymes which can improve the quality of processed foods.

▢ forced cloning

The insertion of foreign DNA into a cloning vector in a predetermined orientation.

▢ foreign dna exogenous

DNA that is incorporated into a host genome.

□ forms of cell signaling

Forms of cell signaling. (A) Hormones produced in endocrine glands are secreted into the bloodstream and can be distributed widely throughout the body. (B) Paracrine signals are released by cells into the extracellular medium in their neighborhood and act locally. (C) Neuronal signals are transmitted along axons to remote target cells. (D) Contact-dependent signaling requires cells to be in direct membrane-to-membrane contact with each other. Many of the same types of signal molecules are used in endocrine, paracrine, and neuronal signaling. The crucial differences lie in the speed and selectivity with which the signals are delivered to their targets.

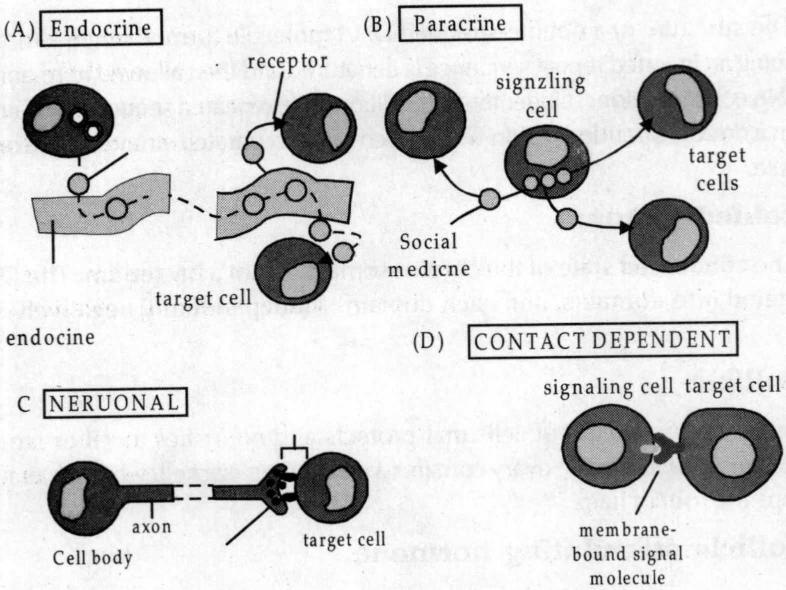

□ forskolin

A medicinal, diterpenoid, compound exclusive to plant roots and used in the preparation of drugs for the treatment of cardiomyopathy, glaucoma and certain cancers.

□ fortify

To add strengthening components or beneficial ingredients to a nutrient medium.

□ forward mutation

A mutation from the wild type to the mutant type.

□ fouling

The coating or plugging (by materials or micro-organisms) of equipment, thus preventing it from functioning properly.

❐ founder animal

An organism that carries a transgene in its germ line and can be used in matings to establish a pure-breeding transgenic line, or one that acts as a breeding stock for transgenic animals.

❐ founder principle

The possibility that a new, isolated population, initiated by a small number of individuals taken from a parent population, may be genetically different from the parent population, because the founding individuals might not be typical of the parent population.

❐ four-base cutter

A type II restriction endonuclease with a four-nucleotide recognition site. Because any particular sequence of four bases occurs more frequently by chance than one of six bases, four-base cutters cleave more frequently than six-base cutters, and therefore generate, on average, smaller restriction fragments. *Synonyms*: four-base-pair-cutter, four-cutter.

❐ fractionation

The separation in components of a complex mixture of molecules.

❐ frameshift mutation

A mutation that changes the reading frame of a DNA, either by the insertion or the deletion of nucleotides. Because of the triplet nature of codons, this occurs if the number of nucleotides involved is not a multiple of three.

❐ free-living conditions

Natural or greenhouse conditions experienced by plantlets upon transfer from *in vitro* conditions to soil. Prior to transfer, nutrients were supplied by the culture medium, but following transfer, plantlets must take up nutrients from soil and synthesize their own food supply.

❐ free water

The cellular water released into the intercellular spaces when tissue is frozen and thawed.

❐ freeze-dry

The removal of water as vapour from frozen material under vacuum. Used to measure water content and to preserve samples, particularly spores. Unlike oven-drying, bound water remains associated with the specimen. fresh weight The weight, including the water content, of a specimen. *Synonym*: wet weight.

❐ friable

A term commonly used to describe a crumb-like callus. In this state, the callus is easily dissected and readily dispersed into single cells or clumps of cells in solution.

❑ functional food

A foodstuff that provides a health benefit beyond basic nutrition, demonstrating specific health or medical benefits, including the prevention and treatment of disease.

❑ functional genomics

The field of research, that aims to determine patterns of gene expression and interaction in the genome, based on the knowledge of extensive or complete genomic sequence of an organism.

❑ functions of plasma membrane proteins

Membranes are vital because they separate the cell from the outside world. They also separate compartments inside the cell to protect important processes and events. Cellular membranes have diverse functions in the different regions and organelles of a cell. However, at the electron microscopic level, they share a common structure following routine preparative steps. The figure shows the typical **"Unit" membrane** which resembles a railroad track with two dense lines separated by a clear space. This figure actually shows two adjacent plasma membranes, both of which have the **"unit membrane"** structure.

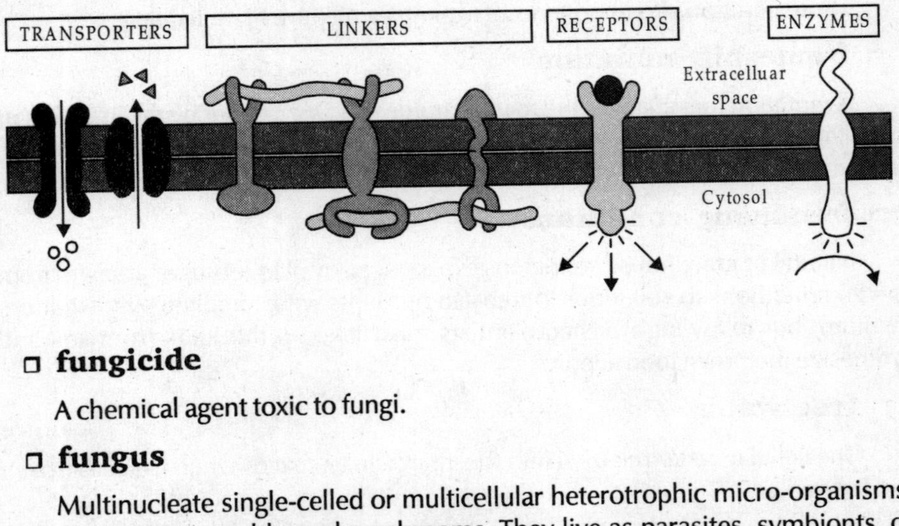

❑ fungicide

A chemical agent toxic to fungi.

❑ fungus

Multinucleate single-celled or multicellular heterotrophic micro-organisms, including yeasts, moulds, and mushrooms. They live as parasites, symbionts, or saprophytes. Lacking any vascular tissues (unlike plants), their cell walls are made of chitin or other non-cellulose compounds.

❑ *Fusarium* spp.

A group of fungal pathogens of many economic crop species, particularly cereals, where severe infestation leads to losses in both grain yield and quality. The latter can be a particularly serious problem as many of these fungi produce mycotoxins, some of which are dangerous to both livestock and human health. Specific strains are also employed on an industrial scale to produce protein for human consumption.

▫ fusion biopharmaceuticals

Fusion proteins with pharmaceutical properties. Their advantages are:

1. Synergistic activities in one molecule, i.e. when the molecule binds to its target, it can perform more than one function simultaneously;
2. An adverse effect or poor stability of one part of the molecule may be offset by the properties of the other; and
3. One part of the molecule can act as a targeting mechanism for the active protein.

▫ fusion protein

A polypeptide translated from a chimeric gene. The different genes are joined so that their coding sequences are in the same reading frame, and the resulting construct is transcribed and translated as a single gene, producing a single protein. These are used for a number of purposes, including:

1. To add an affinity tag to a protein;
2. To produce a protein with the combined characteristics of two natural proteins;
3. To produce a protein where two different activities are physically linked.

▫ fusion toxin

A fusion protein that consists of a toxic protein domain plus a cell receptor binding domain. The latter delivers the toxin directly to the target cell, thus sparing other healthy tissues from the effect of the toxin.

▫ fusogenic agent

Any chemical or virus, etc., that causes cells to fuse together.

G

❐ g

Abbreviation for guanine.

❐ g cap

The 5'-terminal methylated guanine nucleoside that is present on many eukaryotic mRNAs. It is joined to the mRNA, via a 5'?5' phosphodiester bond, after transcription.

❐ g protein

Proteins found on the inner surface of the plasma membrane, which bind to the guanine nucleotides, GTP and GDP. They transmit signals from outside the membrane, via trans-membrane (G-protein-coupled) receptors, to adenylate cyclase, which catalyses the formation of the second messenger, cyclic AMP, inside the cell.

❐ galactomannan

A gum in which the structural chain is made up of D-mannose units with 1?4 linkages. The ratio of galactose to mannose is 1:2.

❐ gall

A tumorous growth in plants.

❐ gamete

A mature reproductive cell which is capable of fusing with a cell of similar origin but of opposite sex to form a zygote from which a new organism can develop. Gametes normally have a haploid chromosome content. In animals, a gamete is a sperm or egg; in plants, it is pollen, spermatic nucleus, or ovum.

❐ gamete and embryo storage

Storage of ova, sperm or fertilized embryos outside their original source. Almost invariably this means cryopreservation.

❐ gametic (phase) disequilibrium

In relation to any two loci, the occurrence of haplotypes (gametes) at a frequency other than that predicted from the product of the respective allele frequencies.

❑ gametic (phase) equilibrium

In relation to any two loci, the occurrence of haplotypes (gametes) at a frequency equal to the product of the frequency of the two relevant alleles. For example, *A* and *B* are in gametic equilibrium if the frequency of A_iB_i gametes equals the product of the frequencies of alleles A_i and B_i.

❑ gametoclone

A plant regenerated from a tissue culture originating from gametic tissue.

❑ gametogenesis

The process of the formation of gametes.

❑ gametophyte

The phase of the plant life cycle that carries the gamete producing organs. In flowering plants, the pollen grain is the male gametophyte and the embryo sac is the female gametophyte.

❑ gametophytic incompatibility

A phenomenon in plants, in which a pollen grain is genetically incapable of fertilizing a particular egg, because both gametes carry the identical allele at an incompatibility locus (usually denoted *S*). This is a mechanism for forcing crossfertilization.

❑ gap

A missing section on one of the strands of double-stranded DNA. The DNA will therefore have a single-stranded region.

❑ gapped dna

A double-stranded DNA molecule with one or more internal single-stranded regions.

❑ gas transfer

The rate at which gases are transferred from gas into solution, an important parameter in fermentation systems because it controls the rate at which the organism can metabolize. Efficient gas transfer can be achieved in several ways, including the use of small bubbles, from which gas dissolves faster than from larger ones, due to their larger surface area per unit of volume; or spreading the liquid out, for example in a thin sheet, or in a thin permeable tube, as in hollow fibre bioreactor.

❑ gastrula

An early animal embryo consisting of two layers of cells; an embryological stage following the blastula.

❑ gc island

A segment of double-stranded DNA that is rich in GC base pairs. This type of sequence is characteristic of eukaryotic genomic regions with a high gene content.

❐ gdp

Abbreviation for guanosine 5'-diphosphate.

❐ gel

A jelly-like solid, used widely as a matrix for the electrophoresis of macromolecules, for encapsulation, and to solidify media for cell cultures.

❐ gelatin

A glutinous, proteinaceous gelling and solidifying agent. Gelatin is produced by the partial hydrolysis (via boiling) of collagen, found in the connective tissues of many farm animals. Used to gel or solidify nutrient solutions for tissue culture, and as a food additive.

❐ gelatinization

The swelling of starch when added to hot water. Hydrolysis causes the molecule to lose structure, and technically gelatinization is not complete until there is no structure left at all.

❐ gel filtration

A method of protein or DNA purification, where differences in size are used to separate the components of a complex mixture.

❐ gelrite™

The brand name of a *Pseudomonas*-derived refined polysaccharide used as a gelling agent and agar substitute.

❐ gem

Abbreviation for genetically engineered micro-organism. *See*: genetically modified organism.

❐ gene

The unit of heredity transmitted from generation to generation during sexual or asexual reproduction. More generally, the term is used in relation to the transmission and inheritance of particular identifiable traits. The simplest gene consists a segment of nucleic acid that encodes an individual protein or RNA.

❐ gene addition

The addition of a functional copy of a gene to the genome of an organism.

❐ gene amplification

The selective production of multiple copies of one gene without a proportional increase in others.

❐ gene bank

1. The physical location where collections of genetic material in the form of seeds, tissues or reproductive cells of plants or animals are stored.
2. Field gene bank: A facility established for the *ex situ* storage and maintenance, using horticultural techniques, of individual plants. Used for species whose

seeds are recalcitrant, or for clonally propagated species of agricultural importance, e.g. apple varieties.

3. A collection of cloned DNA fragments from a single genome. Ideally the bank should contain cloned representatives of all the DNA sequences in the genome.

❑ gene cloning

The synthesis of multiple copies of a chosen DNA sequence using a bacterial cell or another organism as a host. The gene of interest is inserted into a vector, and the resulting recombinant DNA molecule is amplified in an appropriate host cell.

❑ gene conversion

A process, often associated with recombination, during which one allele is replicated at the expense of another, leading to non-Mendelian segregation ratios.

❑ gene expression

The process by which a gene produces mRNA and protein, and hence exerts its effect on the phenotype of an organism.

❑ gene flow

The spread of genes from one breeding population to another (usually) related population by migration, thereby generating changes in allele frequency.

❑ gene imprinting

The differential expression of a single gene according to its parental origin.

❑ gene insertion

The incorporation of one or more copies of a gene into a chromosome.

❑ gene interaction

The modification of the action of one gene by another, non-allelic gene.

❑ gene modification

Chemical change to a gene's DNA sequence.

❑ gene pool

1. The sum of all genetic information in a breeding population at a given time.
2. In plant genetic resources, use is made of the terms 'primary', 'secondary' and 'tertiary' gene pools. In general, members of the primary gene pool are inter-fertile; those of the secondary can be crossed with those in the primary gene pool under special circumstances; but to introgress variation from the tertiary gene pool, special techniques are required to achieve crossing.

❑ gene regulation

The process of controlling the synthesis or suppression of gene products in specific cells or tissues.

❐ gene replacement

The incorporation of a transgene into a chromosome at its normal location by homologous recombination, thus replacing the copy of the gene originally present at the locus.

❐ gene resources conservation

The conservation of species, populations, individuals or parts of individuals, by *in situ* or *ex situ* methods, to provide a diversity of genetic materials for present and future generations.

❐ gene therapy

The proposed treatment of an inherited disease by the transformation of an affected individual with a wild-type copy of the defective gene causing the disorder. In germ-line (or heritable) gene therapy, reproductive cells are transformed; in somatic-cell (or non-inheritable) gene therapy, cells other than reproductive ones are modified.

❐ gene tracking

Following the inheritance of a particular gene from generation to generation.

❐ gene translocation

The movement of a gene from one chromosomal location to another.

❐ genera

Plural form of genus.

❐ generally regarded as safe

Designation given to foods, drugs, and other materials with a long-term history of not causing illness to humans, even though formal toxicity testing may not been conducted. Certain host organisms for recombinant DNA experimentation have recently been given this status.

❐ generative nucleus

In many flowering plants, shed pollen is two-celled (in others it is three-celled or has a variable number). Before pollen is shed, the male gametophyte divides mitotically to give a generative and a vegetative nucleus. The former is the progenitor of the sperm cells.

❐ genes

A gene can be defined as a region of DNA that controls a hereditary characteristic. It usually corresponds to a sequence used in the production of a specific protein or RNA.

A gene carries biological information in a form that must be copied and transmitted from each cell to all its progeny. This includes the entire functional unit: coding DNA sequences, non-coding regulatory DNA sequences, and introns.

Genes can be as short as 1000 base pairs or as long as several hundred thousand base pairs. It can even be carried by more than one chromosome.

The estimate for the number of genes in humans has decreased as our knowledge has increased. As of 2001, humans are thought to have between 30,000 and 40,000 genes.

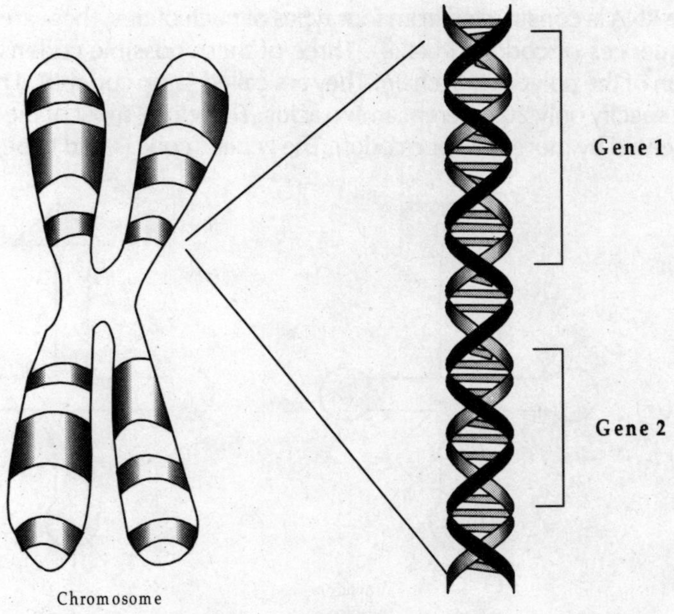

Genes

❐ genet

The individual(s) descended vegetatively from a single sexually produced zygote, including all entities derived from it. All these individuals are genetically identical to one another (barring mutation).

❐ genetic assimilation

Eventual extinction of a natural species as massive gene flow occurs from a related species.

❐ genetic code

Amino acids specified by each codon sequence on mRNA. Key for the below table:

Ala: Alanine	Cys: Cysteine	Asp: Aspartic acid	Glu: Glutamic acid
Phe: Phenylalanine	Gly: Glycine	His: Histidine	Ile: Isoleucine
Lys: Lysine	Leu: Leucine	Met: Methionine	Asn: Asparagine
Pro: Proline	Gln: Glutamine	Arg: Arginine	Ser: Serine
Thr: Threonine	Val: Valine	Trp: Tryptophane	Tyr: Tyrosisne

A = adenine, G = guanine, C = cytosine, T = thymine, U = uracil

DNA transfers information to mRNA in the form of a code defined by a sequence of nucleotides bases. During protein synthesis, ribosomes move along the mRNA molecule and "read" its sequence three nucleotides at a time (codon) from the 5' end to the 3' end. Each amino acid is specified by the mRNA's codon, and then pairs with a sequence of three complementary nucleotides carried by a particular tRNA (anticodon).

Since RNA is constructed from four types of nucleotides, there are 64 possible triplet sequences or codons (4x4x4). Three of these possible codons specify the termination of the polypeptide chain. They are called "stop codons". That leaves 61 codons to specify only 20 different amino acids. Therefore, most of the amino acids are represented by more than one codon. The genetic code is said to be degenerate.

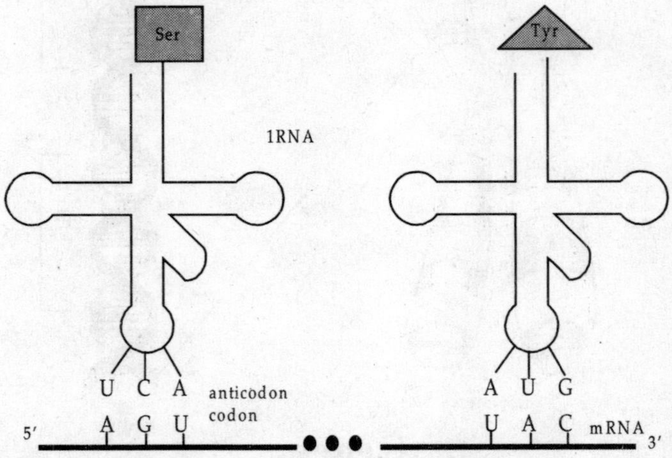

2nd Base in codon

1st Base in codon	U	C	A	G	3rd Base in codon
U	Phe Phe Leu Leu	Ser Ser Ser Ser	Tyr Tyr STOP STOP	Cys Cys STOP Trp	U C A G
C	Leu Leu Leu Leu	Pro Pro Pro Pro	His His Gln Gln	Arg Arg Arg Arg	U C A G
A	Ile Ile Ile Met	Thr Thr Thr Thr	Asn Asn Lys Lys	Ser Ser Arg Arg	U C A G
G	Val Val Val Val	Ala Ala Ala Ala	Asp Asp Glu Glu	Glu Glu Glu Glu	U C A G

The Genetic Code

❐ genetic complementation

When two DNA molecules that are in the same cell together produce a function that neither DNA molecule can supply on its own.

❐ genetic disease

A disease caused by an abnormality in the genetic material, which could be at the level of DNA sequence at a locus, or at the level of karyotype. Usually refers to inherited diseases, although somatic mutations can also cause disease without being inherited.

❐ genetic distance

A measure of the genetic similarity between any pair of populations. This is measured on the basis of variation in a combination of phenotypic traits, allele frequencies or DNA sequences. For example, the genetic distance between two populations having the same allele frequencies at a particular locus, and based solely on that locus, is zero.

❐ genetic distancing

The collection of the data on phenotypic traits, marker allele frequencies or DNA sequences for two or more populations, and estimation of the genetic distances between each pair of populations.

❐ genetic diversity

The heritable variation within and among populations which is created, enhanced or maintained by evolutionary or selective forces.

❐ genetic drift

Change in allele frequency from one generation to another within a population, due to the sampling of finite numbers of genes that is inevitable in all finite-sized populations. The smaller the population, the greater is the genetic drift, with the result that some alleles are lost, and genetic diversity is reduced. Thus minimization of genetic drift is an important consideration for conservation of genetic resources.

❐ genetic engineering

Modifying genotype, and hence phenotype, by transgenesis.

❐ genetic equilibrium

The maintenance of a steady state with respect to allele frequencies in a group of interbreeding organisms.

❐ genetic erosion

The loss over time of allelic diversity, particularly in farmed organisms, caused by either natural or man-made processes.

❐ genetic gain

The increase in productivity achieved following a change in gene frequency effected by selection.

▫ genetic heterogeneity

Occurs where the genetic determination of a given phenotype differs between individuals.

▫ genetic immunization

Delivery to a host organism of a cloned gene that encodes an antigen. After the cloned gene is expressed, it elicits an antibody response that protects the organism from infection by the relevant pathogen.

▫ genetic information

It contained in a nucleotide base sequence in chromosomal DNA or RNA.

▫ genetic map

The linear array of genes on a chromosome, based on recombination frequencies (linkage map) or physical location (physical or chromosomal map).

▫ genetic marker

A DNA sequence used to identify a particular location (locus) on a particular chromosome.

▫ genetic pollution

Uncontrolled spread of genetic information (frequently referring to transgenes) into the genomes of organisms in which such genes are not present in nature.

▫ genetic relatedness

A quantitative estimate of the proportion of genes, ®, shared between the genomes of any two individuals, groups or populations, e.g. $r = 0.5$ for full siblings and parent offspring pairs.

▫ genetic resources

genetic material of actual or potential value.

▫ genetic selection

The process of selecting genes, cells, clones, etc., within populations or between populations or species. Genetic selection usually results in differential survival rates of the various genotypes, reflecting many variables, including the selection pressure and genetic variability present in populations.

▫ genetic use restriction technology

A proposed technology applying transgenesis to genetically compromise the fertility or the performance of saved seed of a cultivar or of second generation animals. The intention is to protect the market for the seed producer or to prevent undesired escape of genes. Two types of GURTs have been patented: variety-level GURT (V-GURT), which produces sterile progeny, and trait-specific GURT (T-GURT), in which only the added value transgenic trait is genetically protected.

❐ genetic variation

Differences between individuals attributable to differences in genotype.

❐ genetically engineered organism

Occasional alteRNAtive term for genetically modified organism.

❐ genetically modified organism

An organism that has been transformed by the insertion of one or more transgenes.

❐ genetics

The science of heredity.

❐ genome

1. The entire complement of genetic material (genes plus non-coding sequences) present in each cell of an organism, virus or organelle.
2. The complete set of chromosomes (hence of genes) inherited as a unit from one parent.

❐ genomic library

A clone library specifically constructed from restriction fragments of the genomic DNA of an organism.

❐ genomics

The research strategy that uses molecular characterization and cloning of whole genomes to understand the structure, function and evolution of genes and to answer fundamental biological questions. *See*: bio-informatics, functional genomics and proteomics.

❐ genotype

1. The genetic constitution of an organism.
2. The allelic constitution at a particular locus, e.g. *Aa* or *aa*.
3. The sum effect of all loci that contribute to the expression of a trait.

❐ genus

A group of closely related species, whose perceived relationship is typically based on physical resemblance, now often supplemented with DNA sequence data.

❐ geo

Abbreviation for genetically engineered organism.

❐ geotropism

A growth curvature induced by gravity.

❐ germ

1. The botanical term for a plant embryo.
2. Colloquial: a disease-causing micro-organism.

❒ germ cell

A member of a cell lineage (the germ line) leading to the production of gametes. In mammals, germ cells are found in the germinal epithelium of the ovaries and testes.

❒ germ cell gene therapy

The repair or replacement of a defective gene within the gamete-forming tissues, resulting in a heritable change in an organism's genetic constitution.

❒ germ layer

The layers of cells in an animal embryo at the gastrula stage, from which the various organs of the animal's body will be derived.

❒ germ line

A lineage of cells which, during the development of an organism, are set aside as potential gamete-forming tissues. The location, nature and time of formation of potential gamete-forming tissues are species specific, and may vary greatly from one species to another.

❒ germ line gene therapy

The delivery of a gene or genes to a fertilized egg or an early embryonic cell. The transferred gene(s) is present in all or some of the nuclei of the cells of the mature individual, including possibly the reproductive cells, and alters the phenotype of the individual that develops.

❒ germicide

Any chemical agent used to control or kill any pathogenic and non-pathogenic micro-organisms.

❒ germinal epithelium

1. A layer of epithelial cells on the surface of the ovary that are continuous with the mesothelium.
2. The layer of epithelial cells lining the seminiferous tubules of the testis, which gives rise to spermatogonia.

❒ germination

1. The initial stages in the growth of a seed to form a seedling.
2. The growth of spores (fungal or algal) and pollen grains.

❒ germplasm

1. An individual, group of individuals or a clone representing a genotype, variety, species or culture, held in an *in situ* or *ex situ* collection.
2. Original meaning, now no longer in use: the genetic material that forms the physical basis of inheritance and which is transmitted from one generation to the next by means of the germ cells.

❐ gestation

The period between conception (fertilization of the egg) to parturition (birth) spent *in utero* by the foetus of viviparous animals.

❐ gfp

Abbreviation for green fluorescent protein.

❐ GH

Abbreviation for growth hormone.

❐ gibberellins

A class of plant growth regulators which are active in the elongation, enhancement of flower, fruit and leaf size, germination, vernalization and other physiological processes.

❐ gland

A specialized group of cells or a single cell in animals or plants that secretes a specific substance. The two types of animal glands are: endocrine, which secrete directly into the blood vessels; and exocrine, which secrete through a duct or network of ducts into a body cavity or onto the body surface.

❐ glaucous

A surface with a waxy, white coating. In most cases, this waxy covering can be rubbed off.

❐ globulins

Common class of proteins in blood, eggs and milk, and seeds. Characterized by their slight solubility in water but are freely soluble in dilute salt solutions. Gamma-globulins are defined further by their electrophoretic behaviour, and include the immunoglobulins.

❐ glp

Abbreviation for good laboratory practice.

❐ glucocorticoid

A steroid hormone that regulates gene expression in higher animals.

❐ glucose invertase

An enzyme that catalyses the hydrolysis of sucrose into its component monosaccharides, glucose and fructose.

❐ glucose isomerase

An enzyme that catalyses the interconversion of the two sugars, glucose and fructose. As fructose is a lower energy compound compared with glucose, a mixture of glucose and fructose with the enzyme will end up almost entirely as fructose.

❐ glucosinolates

A class of molecules produced in the seeds and green tissue of a range of plants, in particular brassicas. Their natural role is thought to be involved in plant-insect interactions. Their importance in plant breeding is largely because of their negative influence on taste and their positive effect on the prevention of cancers of the alimentary tract.

❐ gluten

A mixture of two seed storage protein classes, gliadin and glutenin, found in the endosperm of cereal (particularly wheat) grain. High levels of gluten impart elasticity to dough, and thus the composition of wheat glutens largely determines whether a specific flour is suitable for biscuit or bread making. Sensitivity of the lining of the intestine to gluten in some humans results in coeliac disease, a condition that requires a gluten-free diet.

❐ glycoalkaloids

A group of modified alkaloids, including solanine and tomatine, having a range of toxic effects in humans and other species. They are of particular significance in food plants from the *Solanaceae*.

❐ glycoform

One of several structures possible for a given glycoprotein, determined by the type and position of attachment of the component oligosaccharide(s). Certain glycoforms may exhibit different biological activities from one another because the oligosaccharide units mediate interactions with other cell components.

❐ glycolysis

Each of the 10 steps shown is catalyzed by a different enzyme. Note that step 4 cleaves a six-carbon sugar into two three-carbon sugars, so that the number of molecules at every stage after this doubles. As indicated, step 6 begins the energy generation phase of glycolysis, which causes the net synthesis of ATP and NADH molecules.

STEP 1

Glucose is phosphorylated by ATP to form a sugar phosphate. The negative charge of the phosphate prevents passage of the sugar phosphate through the plasma membrane, trapping glucose inside the cell.

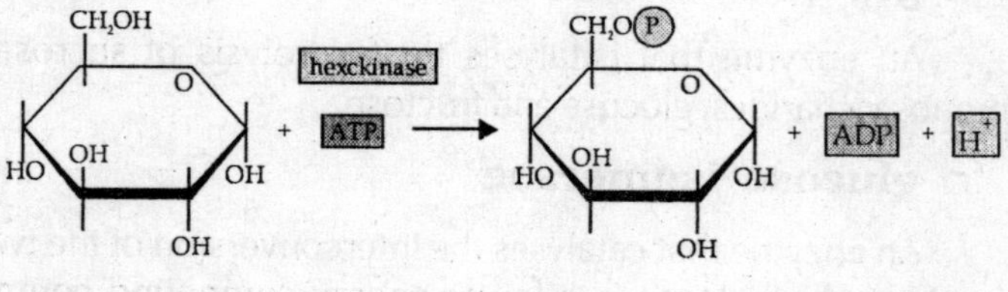

STEP 2

A readily reversible rearrangement of the chemical structure (isomerization) moves the carbonyl oxygen from carbon 1 to carbon 2, forming a ketose from an aldose sugar.

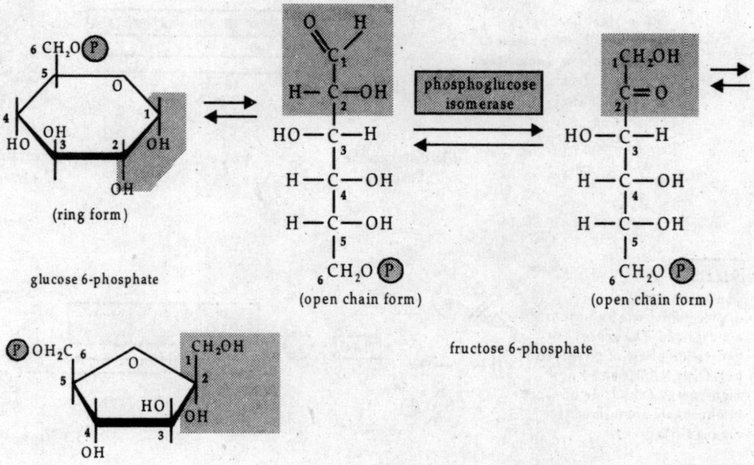

STEP 3

The new hydroxyl group on carbon 1 is phosphorylated by ATP, in preparation for the formation of two three-carbon sugar phosphates. The entry of sugars into glycolysis is controlled at this step, through regulation of the enzyme *phosphofructokinase*.

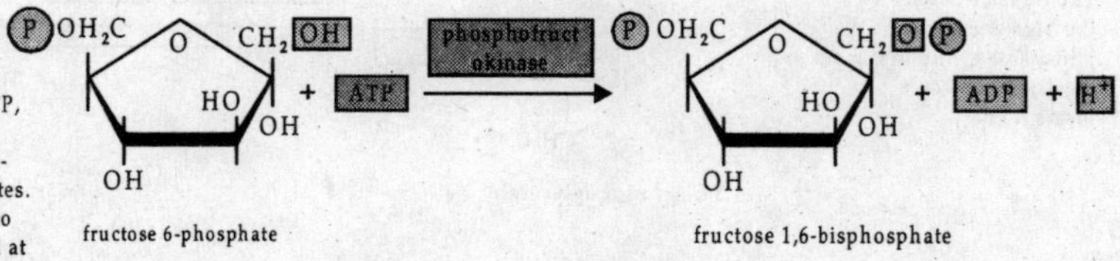

STEP 4

The sixcarbon sugar is cleaved to produce two three-carbon molecules. Only the glyceraldehyde 3-phosphate can proceed immediately through glycolysis.

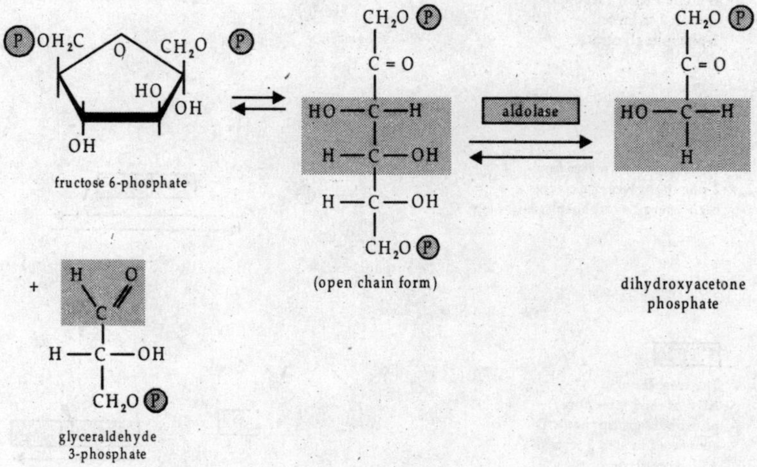

STEP 5

The other product of step 4, dihydroxyacetone phosphate, is isomerized to form glyceraldehyde 3-phosphate.

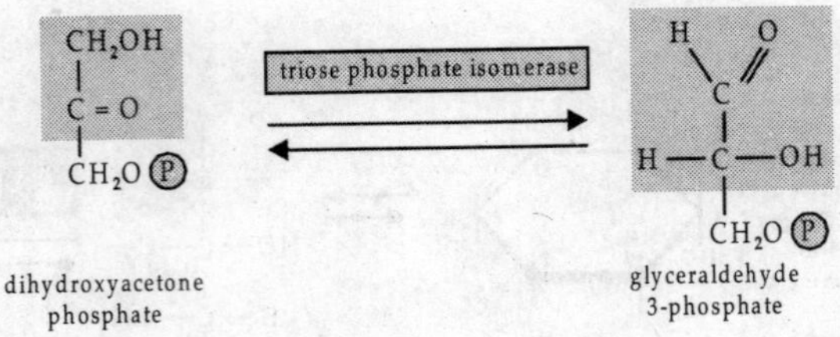

STEP 6

The two molecules of glyceraldehyde 3-phosphate are oxidized. The energy generation phase of glycolysis begins, as NADH and a new high-energy anhydride linkage to phosphate are formed (see Figure 4–5).

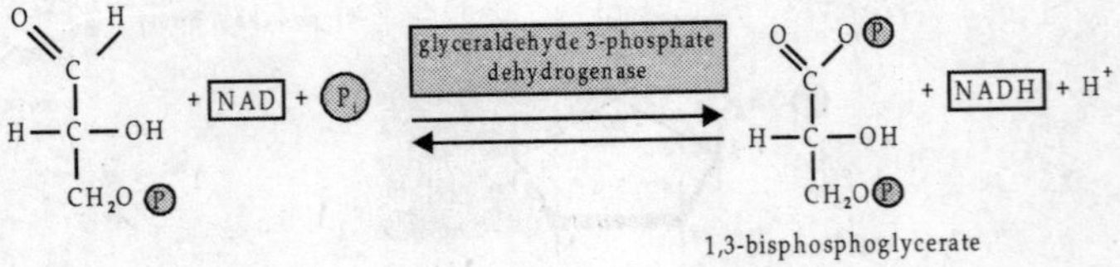

STEP 7

The transfer to ADP of the high energy phosphate group that was generated in step 6 forms ATP.

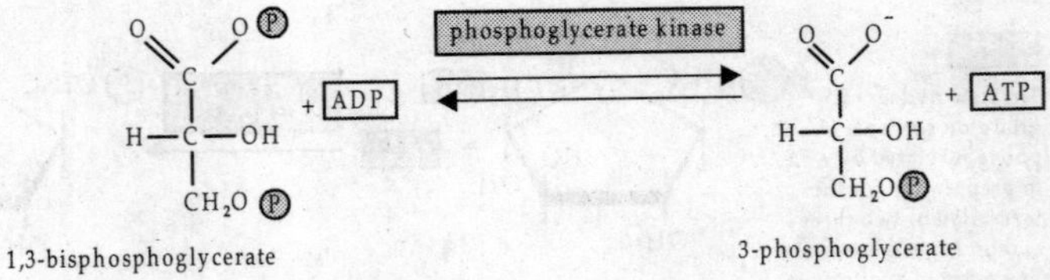

STEP 8

The remaining phosphate ester linkage in 3-phosphoglycerate, which has a relatively low free energy of hydrolysis, is moved from carbon 3 to carbon 2 to form 2-phosphoglycerate.

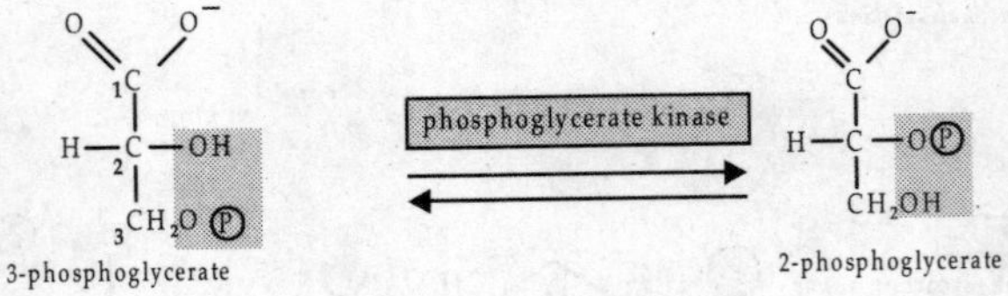

STEP 9

The removal of water from 2-phosphoglycerate creates a high-energy enolphosphate linkage.

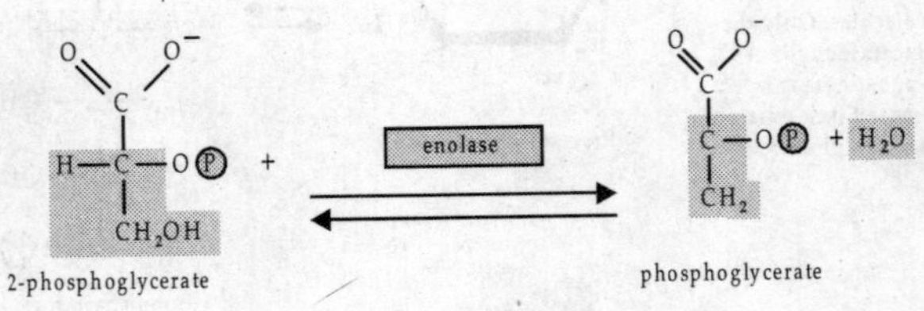

STEP 10

The transfer to ADP of the high-energy phosphate group that was generated in step 9 forms ATP, completing glycolysis.

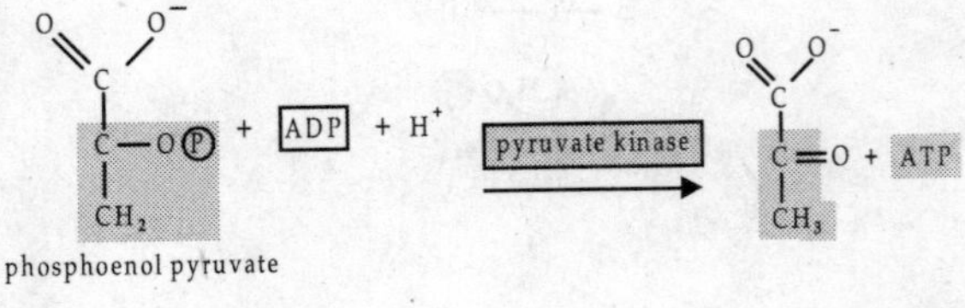

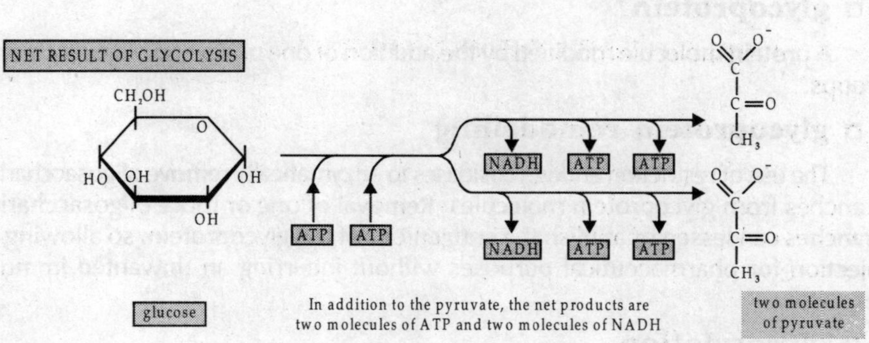
NET RESULT OF GLYCOLYSIS
CH2OH
O
HO
OH
OH
OH
ATP
ATP
NADH
ATP
ATP
NADH
ATP
ATP
C=O
CH3
glucose
In addition to the pyruvate, the net products are
two molecules of ATP and two molecules of NADH
two molecules
of pyruvate

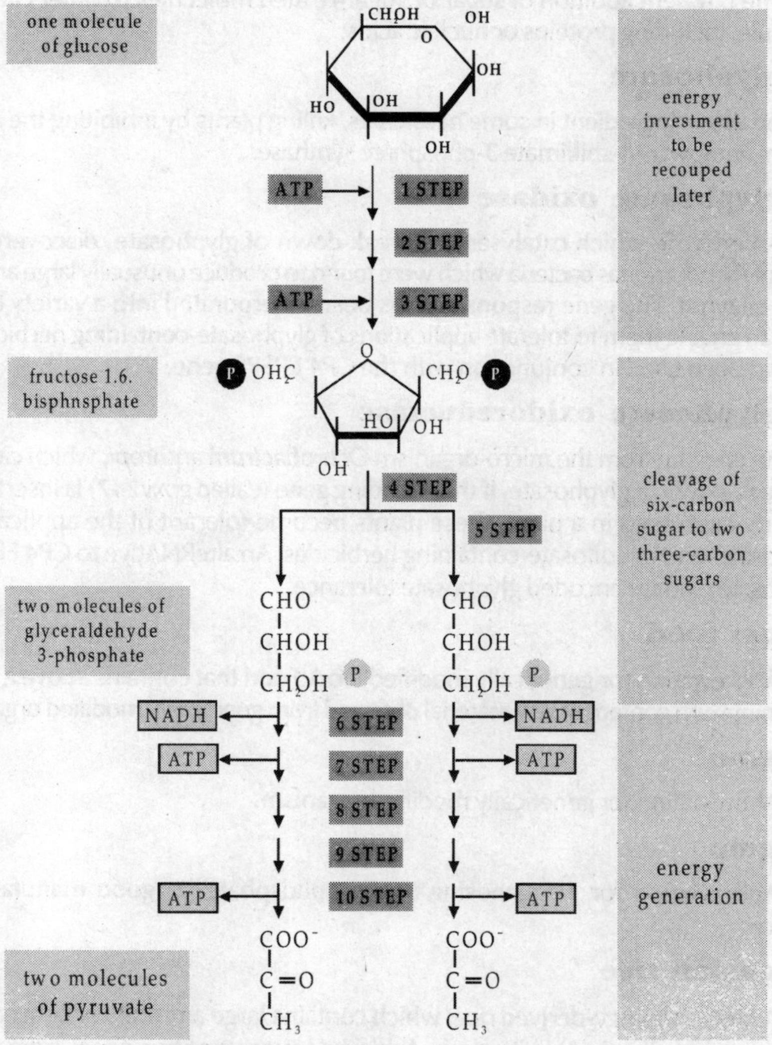
one molecule
of glucose
energy
investment
to be
recouped
later
ATP
1 STEP
2 STEP
ATP
3 STEP
fructose 1.6.
bisphnsphate
cleavage of
six-carbon
sugar to two
three-carbon
sugars
4 STEP
5 STEP
two molecules of
glyceraldehyde
3-phosphate
CHO
CHOH
NADH
6 STEP
NADH
ATP
7 STEP
ATP
8 STEP
9 STEP
energy
generation
ATP
10 STEP
ATP
COO⁻
C=O
CH3
two molecules
of pyruvate

❑ glycoprotein

A protein molecule modified by the addition of one or several oligosaccharide groups.

❑ glycoprotein remodelling

The use of restriction endoglycosidases to enzymatically remove oligosaccharide branches from glycoprotein molecules. Removal of one or more oligosaccharide branches can lessen or abolish the antigenicity of the glycoprotein, so allowing its injection for pharmaceutical purposes without incurring an unwanted immune response.

❑ glycosylation

The covalent addition of sugar or sugar-related molecules to other classes of molecule, including proteins or nucleic acids.

❑ glyphosate

An active ingredient in some herbicides, killing plants by inhibiting the activity of plant enolpyruvyl-shikimate 3-phosphate synthase.

❑ glyphosate oxidase

An enzyme which catalyses the break-down of glyphosate, discovered in a strain of *Pseudomonas* bacteria which were found to produce unusually large amounts of the enzyme. The gene responsible has been incorporated into a variety of crop plants to enable them to tolerate applications of glyphosate-containing herbicides. It has also been used in conjunction with the CP4 EPSPS gene.

❑ glyphosate oxidoreductase

An enzyme from the micro-organism *Ochrobactrum anthropi*, which catalyses the break-down of glyphosate. If the encoding gene (called *goxv247*) is inserted and properly expressed in a plant, these plants become tolerant of the application of glyphosate- and/or sulfosate-containing herbicides. An alteRNAtive to CP4 EPSPS or glyphosate oxidase encoded glyphosate tolerance.

❑ gm food

Abbreviation for genetically modified food. Food that contains above a certain legal minimum content of raw material obtained from genetically modified organisms.

❑ gmo

Abbreviation for genetically modified organism.

❑ gmp

Abbreviation for 1. guanosine 5'-monophosphate. 2. good manufacturing practice.

❑ golden rice

A biotechnology-derived rice, which contains large amounts of beta carotene (a precursor of vitamin A) in its seeds. Achieved by inserting two genes from daffodil and one from the bacterium *Erwinia uredovora*.

❑ golgi apparatus

An assembly of vesicles and folded membranes within the cytoplasm of plant and animal cells that stores and transports secretory products (such as enzymes and hormones) and plays a role in formation of a cell wall (when this is present).

❑ gonad

One of the (usually paired) animal organs that produce reproductive cells (gametes). The most important gonads are the male testis, which produces spermatozoa, and the female ovary, which produces ova (egg cells). The gonads also produce hormones that control secondary sexual characteristics.

❑ good laboratory practice

Written codes of practice designed to reduce to a minimum the chance of procedural or instrument problems which could adversely affect a research project or other laboratory work.

❑ good manufacturing practice

Codes of practice designed to reduce to a minimum the chance of procedural or instrument/manufacturing plant problems which could adversely affect a manufactured product.

❑ graft

1. Verb. To place a detached branch or bud (scion) in close cambial contact with a rooted stem (rootstock) in such a manner that scion and rootstock unite to form a single plant.
2. Noun. Colloquial synonym for scion.

❑ graft chimera

A plant which is a mosaic of two sorts of tissue differing in genetic constitution and assumed to have arisen as the result of a nuclear fusion following grafting.

❑ graft hybrid

An individual formed from graft (2) and stock showing the characteristics of both progenitors.

❑ graft inoculation test

A test based on the use of a suspected viral carrier which is grafted to an indicator plant. If symptoms appear in the indicator plant, the viral assay is positive.

❑ graft union

The point at which a scion from one plant is joined to a rootstock from another plant.

❑ grafting

The process of making a graft (1).

▫ graft-versus-host disease

The rejection of transplanted organs by the recipient's immune system, due to attack of the recipient's T lymphocytes on the transplanted organ caused by differences in major histocompatibility complex proteins.

▫ gram staining

A technique to distinguish between two major bacterial groups, based on whether or not their cell wall retains the Gram stain. Gram-positive bacteria are stained dark purple, while Gram-negative bacteria are only faintly coloured. Stain retention is determined by the structure of the cell wall.

▫ granum

Structure within the chloroplasts, appear as green granules with the light microscope and as a series of parallel lamellae with the electron microscope. They contain the chlorophyll and carotenoid pigments directly involved in photosynthesis.

▫ gras

Abbreviation for generally regarded as safe.

▫ gratuitous inducer

A substance that can induce transcription of a gene or genes, but is not a substrate for the induced enzyme(s).

▫ green fluorescent protein

A protein derived from a species of jelly fish, that fluoresces when exposed to ultra violet light. Its encoding gene has been isolated and is replacing GUS as a reporter gene in plant transgenesis, since it can be assayed non-destructively in real time.

▫ green revolution

Name given to the dramatic increase in crop productivity during the third quarter of the 20th century, as a result of integrated advances in genetics and plant breeding, agronomy, and pest and disease control.

▫ Gro-luxa

A wide-spectrum fluorescent lamp suitable for artificial light for plant growth.

▫ growth cabinet

An enclosed space in which environmental conditions can be controlled. The degree of control over temperature, light and humidity is a function of the quality of the cabinet.

▫ growth factor

Any of various chemicals, particularly polypeptides, that have a variety of important roles in the stimulation of new cell growth and cell maintenance. They bind to the cell surface on receptors. Specific growth factors can cause new cell proliferation.

▫ growth hormone

A group of hormones, secreted by the mammalian pituitary gland, that stimulates protein synthesis and growth of the long bones in the legs and arms. They also promote the breakdown and use of fats as an energy source, rather than glucose.

▫ growth inhibitor

Any substance inhibiting the growth of an organism. The inhibitory effect can range from mild inhibition (growth retardation) to severe inhibition or death (toxic reaction). The concentration of the inhibitor, the length of exposure to it, and the relative susceptibility of the organisms exposed to the inhibitor, are all important factors which determine the extent of the inhibitory effect.

▫ growth phase

Each of the characteristic periods in the growth curve of a bacterial culture, as indicated by the shape of a graph of viable cell number versus time, namely: lag phase; logarithmic phase; stationary phase; death phase.

▫ growth rate

Change in an organism's mass per unit of time.

▫ growth regulator

A synthetic or natural compound that at low concentrations elicits and controls growth responses in a manner similar to hormones.

▫ growth retardant

A chemical that selectively interferes with normal hormonal promotion of growth and other physiological processes, but without appreciable toxic effects.

▫ growth ring

Rings visible in a cross-section of a woody stem, such as a tree trunk. Each ring represents the xylem formed in one year as a result of fluctuating activity of the vascular cambium.

▫ growth substance

Any organic substance, other than a nutrient, that is synthesized by plants and regulates growth and development. They are usually made in a particular region, such as the shoot tip, and transported to other regions, where they take effect.

▫ gtp

Abbreviation for guanosine 5'-triphosphate, a nucleotide which is important as a ligand for G proteins and as a direct precursor molecule for RNA synthesis.

▫ guanine

One of the bases found in DNA and RNA.

❐ guanosine

The (ribo)nucleoside resulting from the combination of the base guanine (G) and a D-ribose sugar. The corresponding deoxyribonucleoside is called deoxyguanosine.

❐ guanosine triphosphate (guanosine 5-triphosphate)

Abbreviation: GTP.

❐ guanylic acid

Synonym for guanosine monophosphate (abbreviation: GMP), a (ribo)nucleotide containing the nucleoside guanosine. The corresponding deoxyribonucleotide is called deoxyguanylic acid.

❐ guard cell

Specialized epidermal cells found in pairs around a stoma. Their function is to control the opening and closing of the stoma through changes in turgor.

❐ guide RNA

An RNA molecule that contain sequences that function as a template during RNA editing.

❐ guide sequence

An RNA molecule (or a part of it) which hybridizes with eukaryotic mRNA and aids in the splicing of intron sequences. Guide sequences may be either external (EGS) or internal (IGS) to the RNA being processed and may hybridize with either intron or exon sequences close to the splice junction.

❐ gurt

Abbreviation for genetic use restriction technology.

❐ gus

Abbreviation for beta-glucuronidase.

❐ gus gene

An *E. coli* gene that encodes for production of beta-glucuronidase (GUS). Because this activity is absent in plants, the gene is commonly utilized as a reporter gene to detect the occurrence of transformation events.

❐ gymnosperm

A class of plant (e.g. conifers) whose ovules and the seeds into which they develop are borne unprotected, rather than enclosed in ovaries, as are those of the flowering plants, the (angiosperms).

❐ gynandromorph

An individual in which one part of the body is female and another part is male; a sex mosaic.

❐ gynogenesis

Female parthenogenesis: after fertilization of the ovum, the male nucleus is eliminated and the haploid (gynogenetic) individual possesses the maternal genome only.

H

❐ h

Prefix used to designate the human form of an enzyme. For example, hGH is human growth hormone.

❐ habituation

The phenomenon whereby, after a number of sub-cultures, cells can grow without the addition to the tissue culture medium of previously obligatory factors. Such cells are then autonomous.

❐ hac

Abbreviation for human artificial chromosome.

❐ haemoglobin

Protein containing iron, located in erythrocytes of vertebrates; important for the transportation of oxygen to the cells of the body.

❐ haemolymph

The mixture of blood and other fluids in the body cavity of an invertebrate.

❐ haemophilia

A sex-linked hereditary bleeding disorder in which it takes a long time for the blood to clot and abnormal bleeding occurs. This disease affects mostly males.

❐ hairpin loop

A region in one strand of a polynucleotide which, due to an inverted repeat in the sequence, may under appropriate conditions fold back on itself and form a limited segment of double-stranded DNA with a loop at one end.

❐ hairy root culture

A culture consisting of highly branched roots of a plant. A plant tissue is treated with the bacterium *Agrobacterium rhizogenes* containing the Ri plasmid, which causes the explant to grow highly branched roots from the sites of infection. Transgenes engineered into the plasmid can be expressed in these cultures.

▢ hairy root disease

A disease of broad-leaved plants, where a proliferation of root-like tissue is formed from the stem. Hairy root disease is a tumorous state similar to crown gall, and is induced by the bacterium *Agrobacterium rhizogenes,* when containing an Ri plasmid.

▢ halophyte

A plant species adapted to soils containing a concentration of salt that is toxic to most plant species.

▢ hammerling's acetabularia

Hammerling's experiment with the single celled green algae, *Acetabularia*, showed that the nucleus of a cell contains the genetic information that directs cellular development.

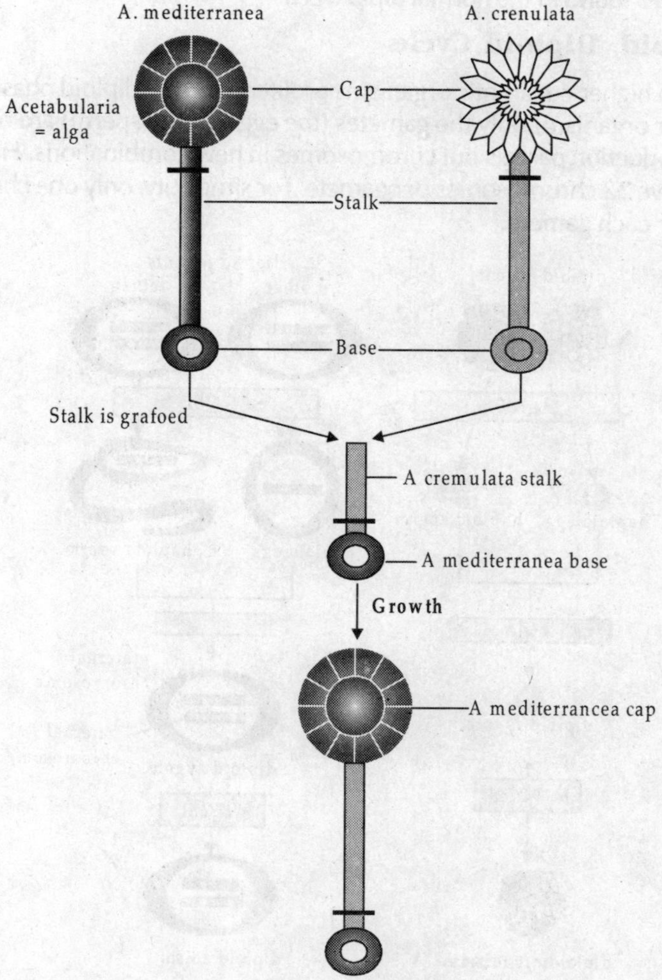

HAMMERLING'S ACETABLARIA

A. mediterranea has a smooth, disc shaped cap, while *A. crenulata* has a branched, flower-like cap. Each *Acetabularia* cell is composed of three segments: the "foot" or base which contains the nucleus, the "stalk," and the "cap."

In his experiments, Hammerling grafted the stalk of one species of *Acetabularia* onto the foot of another species. In all cases, the cap that eventually developed on the grafted cell matched the species of the foot rather than that of the stalk. In this example, the cap that is allowed to grow on the grafted stalk looks like the base species one... *A. mediterranea*.

This experiment shows that the base is responsible for the type of cap that grows. The nucleus that contains genetic information is in the base, so the nucleus directs cellular development.

❐ haploid

A cell or organism containing one of each of the pairs of homologous chromosomes found in the normal diploid cell.

❐ haploid, Diploid Cycle

Cells in higher eucaryotic organisms proliferate in the diploid phase to form a multicellular organism; only the gametes (the egg and the sperm) are haploid. (B) Sexual reproduction parcels out chromosomes in new combinations. Humans, for example, have 23 chromosomes per gamete. For simplicity, only one chromosome is shown for each gamete.

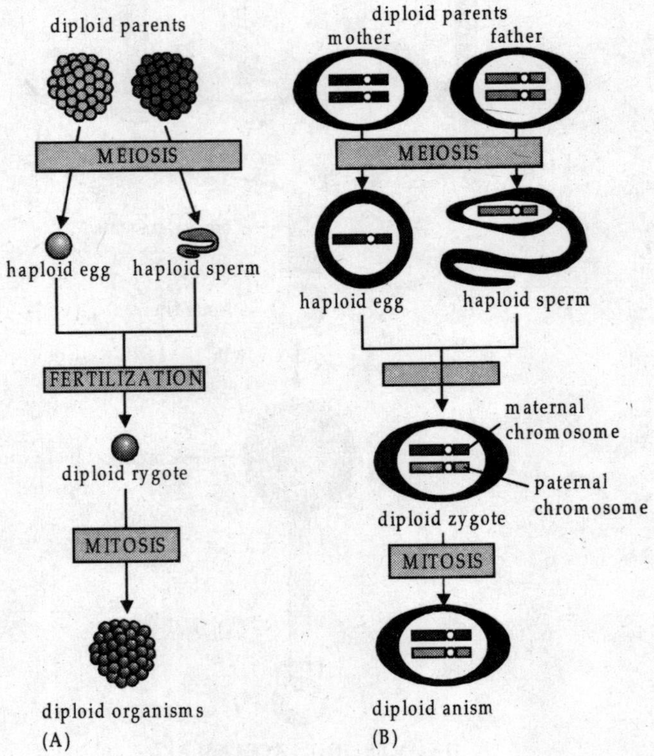

❐ haplotype

A specific allelic constitution at a number of loci within a defined linkage block.

❐ hapten

A small molecule, which by itself is not an antigen, but which as a part of a larger structure when linked to a carrier protein, can serve as an antigenic determinant.

❐ haptoglobin

A serum protein that interacts with haemoglobin during recycling of the iron molecule of haemoglobin.

❐ hardening off

Adapting glasshouse or controlled environment grown plants to outdoor conditions by reducing availability of water, lowering the temperature, increasing light intensity, or reducing the nutrient supply. The hardening-off process conditions plants for survival when transplanted outdoors.

❐ Hardy-Weinberg equilibrium

The frequencies of genotypes at a locus resulting from random mating at that locus; for two alleles, A_1 and A_2, with respective frequencies in a population of p and q, the Hardy-Weinberg equilibrium frequencies are p^2 A_1A_1; $2pq$ A_1A_2; q^2 A_2A_2. Departure from these frequencies is an indication that random mating is not occurring.

❐ harvesting

1. The process involved in gathering ripened crops.
2. The collection of cells from cell cultures or of organs from donors for the purpose of transplantation.

❐ heat shock protein

A class of protein chaperones which are typically over-expressed as a response to heat stress. Two such proteins - HSP 90 and HSP 70 - have a role in ensuring that crucial proteins are folded into the correct conformation.

❐ helix

A structure with a spiral shape. The normal state of double-stranded DNA is in the form of a double helix.

❐ helminth

A class of parasitic worms, especially those which are internal parasites of man and animals.

❐ helper cell

T cells that assist in stimulating B and T lymphocytes to develop into antibody-producing plasma cells and killer T cells, respectively.

❐ helper plasmid

A plasmid that provides a function or functions to another plasmid in the same cell.

❐ helper virus

A virus that provides a function or functions to another virus in the same cell.

❐ hemicellulase

An enzyme that catalyses the degradation of hemicellulose.

❐ hemicellulose

Any cellulose-like carbohydrate, but excluding cellulose itself. Together with pectin and lignin, hemicelluloses form the cell wall matrix.

❐ hemizygous

The condition in which genes are present only once in the genotype and not in pairs. Occurs for all genes in haploids, for all genes located in the differential segments of the sex chromosomes in diploids, and in various aneuploids and deletion mutant heterozygotes.

❐ hepa filter

Abbreviation for high efficiency particulate air filter. A filter capable of excluding particles larger than 0.3ìm. HEPA filters are used in laminar air flow cabinets to ensure that the air is pathogen-free.

❐ herbicide

A substance that is toxic to plants; the active ingredient in agrochemicals intended to kill specific unwanted plants, especially weeds.

❐ herbicide resistance

The ability of a plant to remain unaffected by the application of a herbicide.

❐ heredity

Resemblance among individuals related by descent; transmission of traits from parents to offspring.

❐ heritability

The degree to which a given trait is controlled by inheritance, as opposed to being controlled by non-genetic factors. *See*: broad-sense heritability; narrow-sense heritability.

❐ hermaphrodite

1. An animal that has both male and female reproductive organs, or a mixture of male and female attributes.
2. A plant whose flowers contain both stamen and carpels.

❐ hershey-chase blender experiment

Side by side experiments are performed with separate bacteriophage (virus) cultures in which either the protein capsule is labeled with radioactive sulfur or the DNA core is labeled with radioactive phosphorus.

1. The radioactively labeled phages are allowed to infect bacteria.
2. Agitation in a blender dislodges phage particles from bacterial cells.
3. Centrifugation concentrates cells, separating them from the phage particles left in the supernatant.

Results:

1. Radioactive sulfur is found predominantly in the supernatant.
2. Radioactive phosphorus is found predominantly in the cell fraction, from which a new generation of infective phage can be isolated.

Conclusion: The active component of the bacteriophage that transmits the infective characteristic is the DNA. There is a clear correlation between DNA and genetic information.

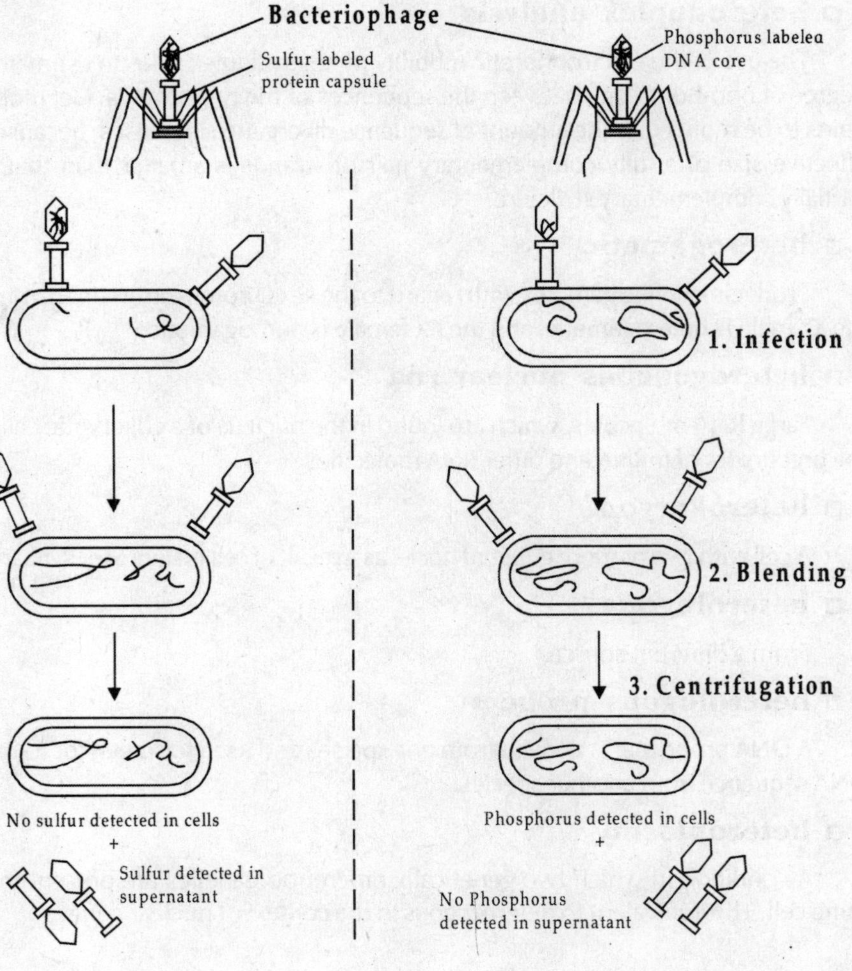

▫ heteroallele

A gene having mutations at two or more different sites.

▫ heterochromatin

Regions of chromosomes that remain contracted during interphase and therefore stain more intensely in cytological preparations. These regions have a high content of repetitive DNA, and a low content of genes; thus they are for the most part genetically inactive.

▫ heteroduplex

A double-stranded DNA molecule or DNA-RNA hybrid, where each strand is of a different origin. Where the two DNAs differ significantly in sequence, single-stranded regions will be revealed when the heteroduplex is observed under the electron microscope. A map of homologous and non-homologous regions of the two molecules may thereby be constructed (heteroduplex mapping). *Synonym*: hybrid DNA (DNA/RNA).

▫ heteroduplex analysis

The use of the electrophoretic mobility of heteroduplex DNA to estimate the degree of non-homology between the sequences of the two strands. Gel mobility tends to be reduced as the amount of sequence divergence increases, because the effective size of a fully complementary pair of strands is smaller than that of a partially complementary structure.

▫ heterogametic

Producing unlike gametes with regard to the sex chromosomes. In mammals, the XY male is heterogametic, and the XX female is homogametic.

▫ heterogeneous nuclear rna

Large RNA molecules, which are found in the nucleus of a eukaryotic cell and the precursors of mRNA and other RNA molecules.

▫ heterokaryon

A cell with two or more different nuclei as a result of cell fusion. *See*: synkaryon.

▫ heterologous

From a different source.

▫ heterologous probe

A DNA probe that is derived from one species and used to screen for a similar DNA sequence from another species.

▫ heteroplasmy

A condition in which two genetically different organelles are present in the same cell. The equivalent to heterozygous in the context of nuclear genes.

❒ heteroploid

Cells with nuclei containing chromosome numbers other than diploid.

❒ heteropyknosis

The property of certain chromosomes, or of their parts, to remain more dense during the cell cycle and to stain more intensely than other chromosomes or parts.

❒ heterotroph

Organism non capable of self-nourishment utilizing carbon dioxide or carbonates as the sole source of carbon and obtaining energy from radiant energy or from the oxidation of inorganic elements, or compounds such as iron, sulphur, hydrogen, ammonium and nitrites.

❒ heterozygote

An individual with non-identical alleles for a particular gene or genes. The condition is termed "heterozygous".

❒ hfr

High-frequency recombination strain of *Escherichia coli*; in these strains, the F factor (plasmid) is integrated into the bacterial chromosome.

❒ hgh

Abbreviation for human growth hormone.

❒ high throughput screening

Automated systems designed to process large numbers of assays, especially in the context of genotyping.

❒ histocompatibility

The degree to which tissue from one organism is tolerated by the immune system of another organism.

❒ histoglobulin

The peptides present on the surface of nucleate cells, responsible for the differences between genetically non-identical individuals that cause rejection of tissue grafts between such individuals. Products of the major histocompatibility complex genes.

❒ histology

Science that deals with the microscopic structure of animal and plant tissues.

❒ histone

Group of water-soluble proteins rich in basic amino acids, closely associated with DNA in plant and animal chromatin. Histones are involved in the coiling of DNA in chromosomes and in the regulation of gene activity.

❐ hiv (human immunodeficiency virus)

AIDS was first reported in 1981 in the United States and has since become a major epidemic, killing nearly 12 million people and infecting more than 30 million others worldwide. The disease is caused by HIV, a virus that destroys the body's ability to fight infections and certain cancers.

The agent that leads to the development of the acquired immunodeficiency syndrome (AIDS). Individuals infected with the virus have an increased risk of developing lymphoma. The lymphomas are of the B cell type and may involve the brain or be widespread at the time of occurrence. Patients carrying the virus become immunologically impaired and are susceptible to unusual infections that are caused by microbes that usually do not infect healthy persons.

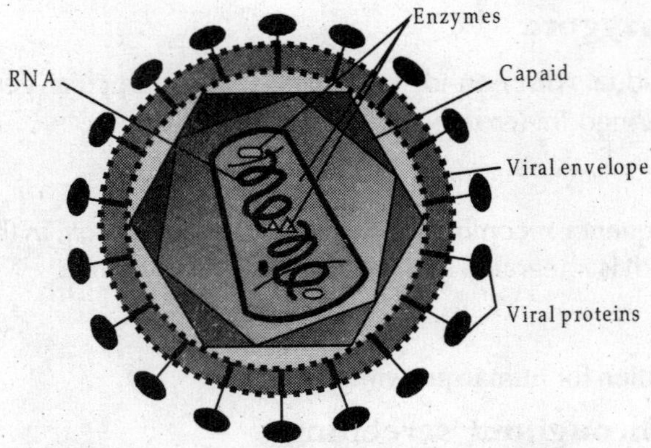

Structure of Human Immunodeficiency Virus (HIV)

❐ hla

Abbreviation for human-leukocyte-antigen system.

❐ hnrna

Abbreviation for heterogeneous nuclear RNA.

❐ hogness box

Synonym for TATA box.

❐ hollow fibre

A tube of porous material, with an internal diameter of a fraction of a millimetre, making its ratio of surface area to internal volume very large. Employed as filters or in bioreactors as a method of retaining cells while allowing the easy removal of spent medium and/or products.

❐ holometabolous

An insect that undergoes complete metamorphosis to the adult from a morphologically distinct larval stage.

❐ homeobox

A highly conserved 180 bp DNA sequence that controls body part-, organ- or tissue-specific gene expression, most particularly involved in segmentation in animals (e.g. development of antennae or legs of *Drosophila melanogaster*), but also in a variety of other eukaryotes. It encodes a DNA-binding region, the homeodomain, which acts as a transcription factor.

❐ homeotic genes

Genes that act in concert to determine fundamental patterns of development.

❐ homeotic mutation

A mutation that causes a body part to develop in an inappropriate position in an organism, such as the mutation in *Drosophila melanogaster* that causes legs to develop on the head in place of antennae.

❐ homoallele

One of a number of otherwise identical alleles which differ at the same site in their sequence. Homoalleles are inherited as strict alteRNAtives; but heteroalleles, in principle, could through recombination create a genotype which contains a 'double' variant.

❐ homodimer

A protein comprising two identical polypeptide chains, or a dimer of identical residues.

❐ homoduplex DNA

A double-stranded fully complementary DNA molecule.

❐ homoeologous

Referring to chromosomes which are descended from a common progenitor, but which have evolved to be no longer fully homologous. Homoeologous chromosomes have similar gene content to one another, but are structurally altered in subtle ways to inhibit, and sometimes completely prevent their pairing with one another at meiosis.

❐ homogametic

Producing similar gametes with regard to the sex chromosomes. In mammals, the female is homogametic (XX), and the male is heterogametic (XY).

❐ homogenotization

An allele replacement technique, in which a bacterial cell is transformed with a plasmid containing an altered copy of the target sequence, and a double recombination event effects the substitution of the wild type allele by the altered one. An antibiotic resistance gene is usually fused to the altered copy in the plasmid, so that double recombinants can be selected.

homokaryon

A cell with two or more identical nuclei as a result of fusion.

homologous

1. General definition: from the same source, or having the same evolutionary function or structure.
2. Of chromosomes: identical with respect to gene content and linear ordering. Homologous chromosomes pair and recombine with one another at meiosis.
3. Of DNA/proteins: identical, or nearly identical nucleotide/amino acid sequence.

homologous recombination

The exchange of DNA fragments between the two non-sister chromatids of the same chromosome in the course of meiosis.

homology

1. The degree of identity between individuals, or characters.
2. The degree of identity of sequence (nucleotide or amino acid) between a number of DNA or polypeptide molecules.

homomultimer

A protein consisting of a number of identical subunits.

homoplasmy

The condition in which all copies of an organelle in a cell are genetically identical.

homopolymer

A polymer (nucleic acid, polypeptide, etc.) which contains only one kind of residue (e.g. the polynucleotide GGGGGGGG...).

homozygote

An individual that has two copies of the same allele for a given gene on its two homologous chromosomes. The condition is termed "homozygous".

hormone

A specific chemical, produced in one part of a plant or animal body, and transported to another part where, at low concentrations, it promotes, inhibits or quantitatively modifies a biological process.

host

An organism that contains another organism or a cloning vector.

host-specific toxin

A metabolite, produced by a pathogen, and which is responsible for the adverse effects of the pathogen. The toxin has a host specificity equivalent to that of the pathogen. Utilized for *in vitro* selection experiments to screen for tolerance or resistance to the pathogen.

❑ human artificial chromosome

Analogous to yeast artificial chromosome, a construct comprising a human centromere and telomeres, which would allow for the cloning of very large fragments of DNA, and their transfer into human cells for the purpose of gene therapy. Not yet established as a working technology, although some partial success has been reported.

❑ human chromosomes

Chromosome: a very long DNA molecule and associated proteins, that carry portions of the hereditary information of an organism.

a. Structure of a chromosome (Typical metaphase chromosome): A chromosome is formed from a single DNA molecule that contains many genes. A chromosomal DNA molecule contains three specific nucleotide sequences which are required for replication: a DNA *replication origin*; a *centromere* to attach the DNA to the mitotic spindle.; a *telomere* located at each end of the linear chromosome.

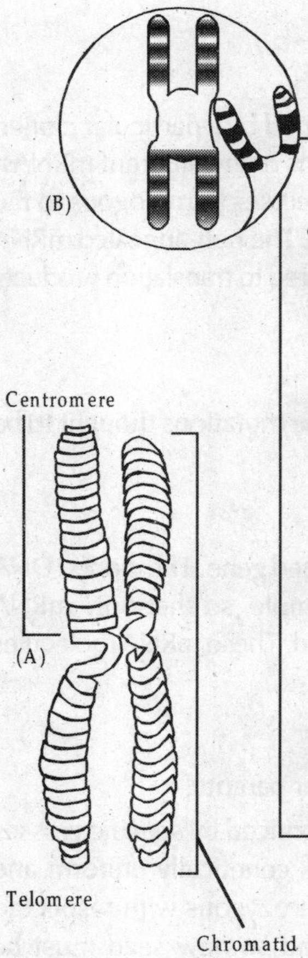

HUMAN CHROMOSOMES

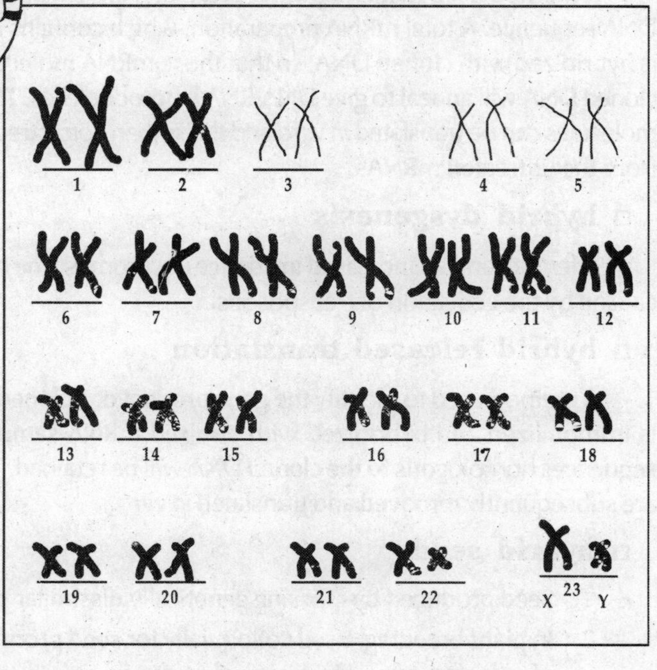

The DNA molecule is highly condensed. The human DNA helix occupy too much space in the cell. Small proteins are responsible for packing the DNA into units called nucleosomes.

b. Stained chromosomes: Chromosomes are stained with A-T (G bands) and G-C (R bands) base pair specific dyes. When they are stained, the mitotic chromosomes have a banded structure that unambiguously identifies each chromosome of a karyotype. Each band contains millions of DNA nucleotide pairs which do not correspond to any functional structure.

c. Karyotype of a male: The human haploid genome contains 3,000,000,000 DNA nucleotide pairs, divided among twenty two (22) pairs of autosomes and one pair of sex chromosomes.

▫ hup$^+$

Abbreviation for hydrogen-uptake positive.

▫ hybrid

1. The offspring of two genetically unlike parents.
2. Of DNA molecules, *see*: heteroduplex.

▫ hybrid arrested translation

A method used to identify what protein(s) are encoded by a particular cloned DNA sequence. A total mRNA preparation, which contains many different mRNAs, is hybridized with cloned DNA, so that those mRNA molecules homologous to the cloned DNA will anneal to give DNA/RNA heteroduplexes. The non-annealed mRNA molecules can be translated *in vitro* and this is then compared to translation products from the untreated mRNAs.

▫ hybrid dysgenesis

Infertility and an increased incidence of chromosome mutations thought to be caused by the activation of transposons.

▫ hybrid released translation

A method used to identify the gene product of a cloned gene. The cloned DNA is immobilized and hybridized with a mixed mRNA sample, so that only mRNA sequences homologous to the cloned DNA will be retained. These mRNA molecules are subsequently removed and translated *in vitro*.

▫ hybrid seed

1. Seed produced by crossing genetically dissimilar parents.
2. In plant breeding, used colloquially for seed produced by specific crosses of selected pure lines, such that the F_1 crop is genetically uniform and displays hybrid vigour. As the F_1 plants are heterozygous with respect to many genes, the crop does not breed true and so new seed must be purchased each season.

❐ hybrid selection

The process of choosing individuals possessing desired characteristics from among a hybrid population.

❐ hybrid vigour

The extent to which a hybrid individual outperforms both its parents with respect to one or many traits. The genetic basis of hybrid vigour is not well understood, but the phenomenon is widespread, particularly in inbreeding plant species.

❐ hybridization

1. The process of forming a hybrid by cross pollination of plants or by mating animals of different types.
2. The production of offspring of genetically different parents, normally from sexual reproduction, but also asexually by the fusion of protoplasts or by transformation.
3. The pairing of two DNA strands, often from different sources, by hydrogen bonding between complementary nucleotides.

❐ hybridoma

A synthetic hybrid cell, derived by fusing a B lymphocyte with a tumour cell. The former secretes a single antibody, while the latter confers the property of growing indefinitely in tissue culture. The underlying technology behind the monoclonal antibody.

❐ hydrogen-uptake positive

A term describing a micro-organism that is capable of assimilating (or taking up) hydrogen gas.

❐ hydrolysis

A chemical reaction in which water is added across a covalent bond, often cleaving the molecule into two. Occurs for example when polynucleotides, polypeptides, and polysaccharides are broken into their component monomers. Thus sucrose can be hydrolysed to glucose and fructose; and proteins to individual amino acids.

❐ hydrophobic interaction

An interaction between a hydrophobic ('water-hating') part of a molecule and an aqueous environment. Particularly significant in establishing the conformation of molecules in solution, and thus their biological activity. Many enzymes have a structure where the polypeptide chain is folded to form a hydrophobic core and a hydrophilic ('water-loving') surface.

❐ hydroponics

The growing of plants without soil. Plants are fed with an aerated solution of nutrients, and the roots are either supported within an inert matrix, or are freely floating in the nutrient solution.

❐ hygromycin

An antibiotic used as selective agent in bacterial and transgenic plant cell cultures.

❐ hyperploid

The situation in which a particular chromosome or chromosome segment is present in more than the normal number.

❐ hypersensitive response

1. A specific reaction of a plant to attack by a pathogen. The plant cells surrounding the point of infection rapidly die and dry out, so that pathogen spread within the plant is prevented. Often associated with the interaction of race-specific *R* genes with a matching pathogen avirulence.
2. The abnormal response of an animal to the presence of a particular antigen.

❐ hypersensitive site

Regions in the DNA that are highly susceptible to digestion with restriction endonucleases.

❐ hypertonic

A solution with an osmotic potential greater than that of living cells. Treatment with such solutions leads to the plasmolysis of cells.

❐ hypervariable region

The parts of both the heavy and light chains of an antibody molecule that enable it to bind to a specific site on an antigen.

❐ hypervariable segment

A region of a protein that varies considerably between strains or individuals.

❐ hypocotyl

The portion of an embryo or seedling below the cotyledons. A transitional area between stem and root.

❐ hypomorph

A mutation that reduces, but does not completely abolish gene expression.

❐ hypoplastic

Defective and reduced growth or development (e.g. dwarfing and stunting in plants) resulting from an abnormal condition, for example disease or nutritional stress.

❐ hypoploid

The situation in which a particular chromosome or chromosome segment is present in less than the normal number.

▫ hypothalamic peptides

Peptides generated in the vertebrate forebrain and concerned with regulating the body's physiological state.

▫ hypotonic

Osmotic potential less than that of living cells. Cells placed in a hypotonic solution will absorb water and display swelling and turgidity.

I

❐ idiotype

An identifying property or characteristic of an item or system.

1. A plant form expected on physiological grounds to represent an optimal type for the environment in which the plant is to be grown.
2. A classification of antibody molecules according to the antigenicity of the variable regions. Each idiotype is unique to a particular immunoglobulin raised to a particular antigen.

❐ igs

Abbreviation for:

1. internal guide sequence.
2. intergenic spacer.

❐ imaginal disc

A mass of cells in the larvae of *Drosophila melanogaster* and other holometabolous insects that gives rise to particular adult organs, such as antennae, eyes or wings.

❐ imbibition

1. The absorption of liquids or vapours into the ultramicroscopic spaces or pores found in materials.
2. The initial water uptake by seeds prior to germination.

❐ immediate early gene

A viral gene that is expressed promptly after infection.

❐ immobilized cells

Cells entrapped in matrices such as alginate, polyacrylamide and agarose, for use in membrane and filter bioreactors.

❐ immortalization

The genetic transformation of a cell type into a cell line which can proliferate indefinitely.

▫ immortalizing oncogene

A gene that, upon transfection, enables a primary cell to grow indefinitely in culture.

▫ immune response

The processes, including the synthesis of antibodies, that are used by vertebrates to respond to the presence of a foreign antigen. *See*: primary immune response; secondary immune response.

▫ immunity

The lack of susceptibility of an animal or plant to infection by a particular pathogen, or to the harmful effects of their toxins.

▫ immunization

The production of immunity in an individual by artificial means. Active immunization involves the introduction, either orally or by infection, of specially treated bacteria, viruses or their toxins so as to stimulate the production of antibodies.

▫ immunoaffinity chromatography

A purification technique in which an antibody is bound to a matrix and is used to isolate a protein from a complex mixture.

▫ immunoassay

An detection system for a particular molecule, which exploits the specific binding of an antibody raised against it. For measurement, the antibody can incorporate a radioactive or fluorescent label, or be linked to an enzyme which catalyses an easily monitored reaction such as a change in colour.

▫ immunogenicity

The ability to elicit an immune response.

▫ *immunoprophylaxis*

The process of active or passive immunization. Active immunization with vaccines leads to long-term protection through mobilizing the organism's own immune system. Passive immunization is achieved by injection of antibody against a specific pathogen obtained either by fractionation of blood from an individual previously infected with the pathogen, or by monoclonal antibody technology.

▫ immunosensor

A biosensor having an antibody as the biological element.

▫ immunosuppression

The suppression of the immune response. Necessary following organ transplants from a genetically different donor in order to prevent the host rejecting the grafted organ.

❒ immunosuppressor

A substance, an agent or a condition that prevents or diminishes the immune response.

❒ immunotherapy

The use of an antibody or a fusion protein containing the antigen binding site of an antibody to cure a disease or enhance the well-being of a patient. *Synonym*: immunochemical control.

❒ immunotoxin

Protein drugs consisting of an antibody joined to a toxin molecule. Made by either chemically linking the molecules, or by fusing the genes for the toxin and the antibody, and expressing a fusion protein. The antibody portion of the molecule targets the toxin.

❒ impeller

An agitator that is used for mixing the contents of a bioreactor.

❒ import and export

Cells can import materials from the external medium by capturing them in vesicles that pinch off from the plasma membrane. The vesicles ultimately fuse with lysosomes, where intracellular digestion occurs. By a converse process, cells export materials that they have synthesized in intracellular compartments: the materials are stored in the intracellular vesicles and released to the exterior when these vesicles fuse with the plasma membrane.

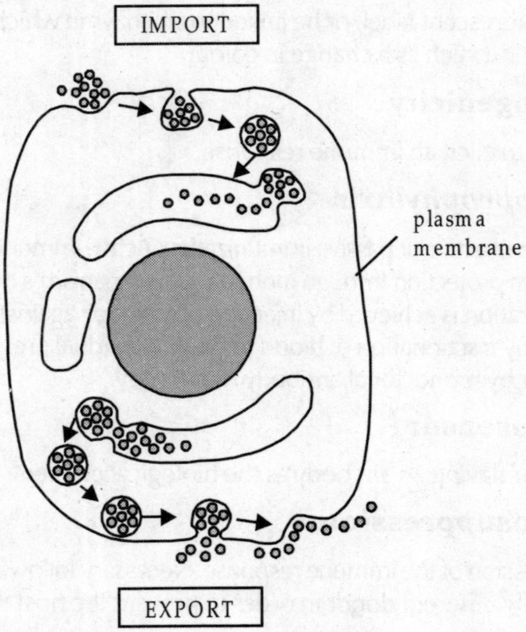

□ importance of combinatorial gene control

A hypothetical scheme illustrating how combinations of a few gene regulatory proteins can generate many different cell types during development. In this simple scheme a ³decision² to make a new gene regulatory protein (shown as a numbered circle) is made after each cell division. Repetition of this simple rule enables eight cell types (A through H) to be created using only three different regulatory proteins. Each of these hypothetical cell types would then express different genes, as dictated by the combination of gene regulatory proteins that are present within it.

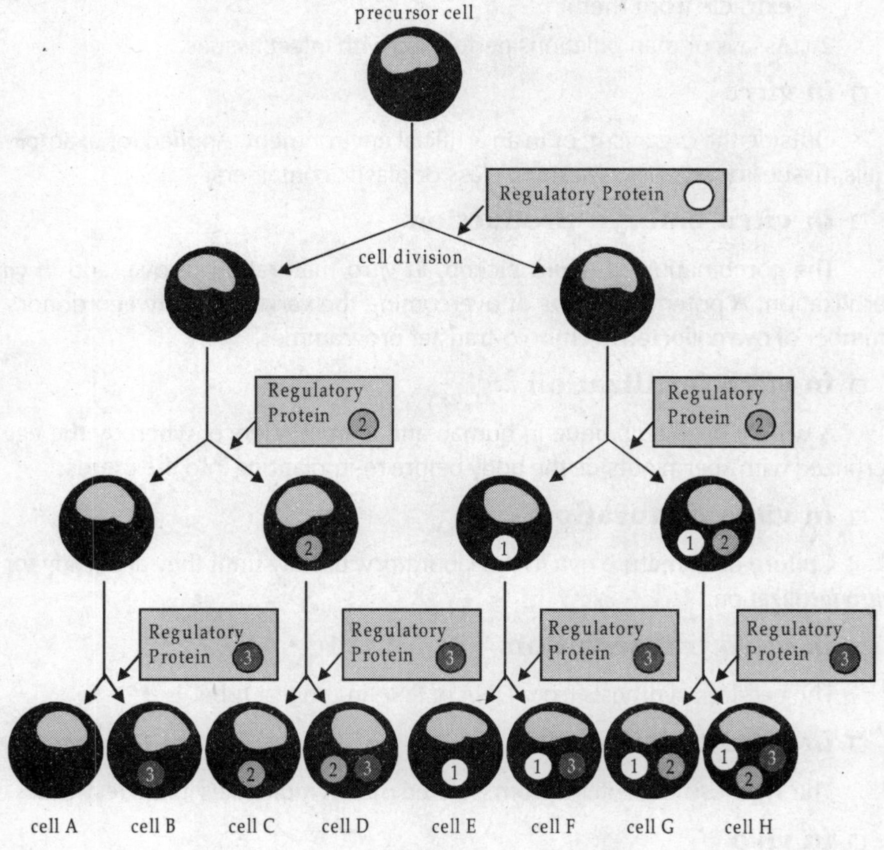

□ *in silico*

In a computer file. In the present context, the use of data bases of DNA and protein sequence to help answer biological questions. This is growing area of biology as the amount of genomics and proteomics data continues to grow.

□ *in situ* colony hybridization

A procedure for screening bacterial colonies or plaques growing on plates or membranes for the presence of specific DNA sequences by the hybridization of nucleic acid probes to the DNA molecules present in these colonies or plaques.

□ *in situ* hybridization

The visualization of *in vivo* location of macromolecules (particularly polynucleotides and polypeptides) by the histological staining of tissue sections or cytological preparations via labelled probes/antibodies.

□ *in situ*

In the natural place or in the original place.

1. **Experimental treatments performed on cells or** tissue **rather than on extracts from them.**
2. Assays or manipulations performed with intact tissues.

□ *in vitro*

Outside the organism, or in an artificial environment. Applied for example to cells, tissues or organs cultured in glass or plastic containers.

□ *in vitro* embryo production

The combination of ovum pickup, *in vitro* maturation of ova, and *in vitro* fertilization. A potential means of overcoming the variability between donors in number of ova collected in embryo-transfer programmes.

□ *in vitro* fertilization

A widely used technique in human and animal science, whereby the egg is fertilized with sperm outside the body before re-implanting into the uterus.

□ *in vitro* maturation

Culture of immature ova in the laboratory, usually until they are ready for *in vitro* fertilization.

□ *in vitro* transcription

The cell-free synthesis from DNA of RNA in the test tube.

□ *in vitro* translation

The synthesis of proteins from isolated mRNA molecules in the test-tube.

□ *in vivo*

The natural conditions in which organisms reside. Refers to biological processes that take place within a living organism or cell under normal conditions.

□ *in vivo* gene therapy

The delivery of a gene or genes to a tissue or organ of a complete living individual to alleviate a genetic disorder.

□ inactivated agent

A virus, bacterium or other organism that has been treated to prevent it from causing a disease.

□ inbred line

The product of inbreeding, i.e. the intercrossing of individuals that have ancestors in common. In plants and laboratory animals, it refers to populations resulting from at least 6 generations of selfing or 20 generations of brother-sister mating, so that they have become, for all practical purposes, completely homozygous. In farm animals, the term is sometimes used to describe populations that have resulted from several generations of the mating of close relatives, without having reached complete homozygosity.

□ inbreeding

Matings between individuals that have one or more ancestors in common, the extreme condition being self-fertilization, which occurs naturally in many plants and some primitive animals.

□ inbreeding depression

The reduction in vigour over generations of inbreeding. This affects species which are normally outbreeding and highly heterozygous.

□ inclusion body

A protein that is overproduced in a recombinant bacterium and forms a crystalline structure inside the bacterial cell.

□ incompatibility

1. Genetically or physiologically determined prevention of intermating.
2. A physiological interaction resulting in graft rejection or failure.
3. A function of a related group of plasmids. Incompatible plasmids share similar replication functions, and this leads to the exclusion of one or the other plasmid if they are present in the same cell. Plasmids belonging to one incompatibility group are very closely related.

□ incompatibility group

Plasmids must belong to different incompatibility groups to co-exist within the same cell. A plasmid cloning vector should always belong to an incompatibility group different from that of the host bacterium's endogenous plasmids.

□ incomplete dominance

A gene action in which heterozygotes have a phenotype that is different from either homozygote, and is usually intermediate between them.

□ incomplete penetrance

Where the phenotype does not allow perfect prediction of the genotype as a result of interference in gene expression by the environment.

□ incubation

1. The hatching of eggs by means of heat, either natural or artificial.

2. Period between infection and appearance of symptoms induced by a pathogen.
3. The culture of cells and organisms.

▫ incubator

An apparatus in which environmental conditions (light, photoperiod, temperature, humidity, etc.) are fully controlled; used for hatching eggs, multiplying micro-organisms, culturing plants, etc.

▫ indehiscent

Describing a fruit or fruiting body that does not open to release its seeds or spores when ripe.

▫ independent assortment

The random distribution during meiosis of alleles (at different genes) to the gametes that is the case when the genes in question are located on different chromosomes or are unlinked on the same chromosome.

▫ indeterminate growth

The condition in which the terminal bud persists and produces successive lateral branches over an indefinite period.

▫ indirect embryogenesis

Plant embryo formation on callus tissues derived from explants, including zygotic or somatic embryos and seedlings.

▫ indirect organogenesis

Plant organ formation on callus tissues derived from explants. *Opposite*: direct organogenesis.

▫ inducer

A low-molecular-weight compound or a physical agent that associates with a repressor protein to produce a complex that can no longer bind to the operator. Thus, the presence of the inducer turns on the expression of the gene(s) controlled by the operator.

▫ inducible

A gene or gene product whose transcription or synthesis is increased by exposure of the cells to an inducer or to a condition, e.g. heat.

▫ inducible enzyme

An enzyme that is synthesized only in the presence of the substrate that acts as an inducer.

▫ inducible gene

A gene that is expressed only in the presence of a specific metabolite, the inducer.

❑ inducible promoter

The activation of a promoter in response to either the presence of a particular compound, i.e. the inducer, or to a defined external condition, e.g. elevated temperature.

❑ induction media

1. Media used to induce the formation of organs or other structures.
2. **Media causing variation or mutation in the tissues exposed to it.**

❑ induction of the lac operon

Gene transcription can be switched on and off by gene regulation proteins. The lac operon in E.Coli is an example of that dual control. Glucose and lactose levels control the initiation of transcription of the lac operon, i.e. whether the lac operon is switched "ON" or "OFF".

1. In an E. Coli cell growing in the absence of lactose, a repressor protein binds to the operator, preventing RNA polymerase from transcribing the lac operon's genes. The operon is OFF.

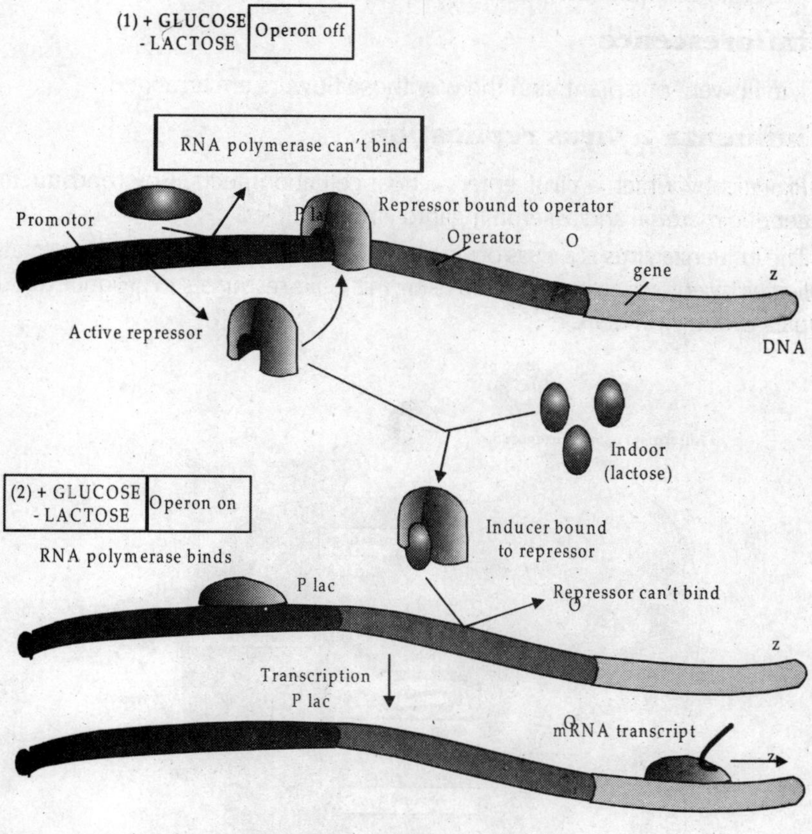

Induction of the Lac Operon

2. When the inducer, lactose, is added, it binds to the repressor and changes the repressor's shape so as to eliminate binding to the operator. As long as the operator remains free of the repressor, RNA polymerase that recognizes the promoter can transcribe the operon's structural genes into mRNA. The operon is ON.

❐ induction

The act or process of causing some specific effect to occur; for example the transcription of a specific gene or operon, or the production of a protein by an organism after it is exposed to a specific stimulus.

❐ infection

The successful colonization of any living organism by a pathogen.

❐ infectious agent

Synonym of pathogen.

❐ infiltrate

The entry of liquid into pores or other spaces.

❐ inflorescence

The flowers of a plant, and the way those flowers are arranged.

❐ influenza a virus replication

Process by which a virus enters a host cell and infects it by reproducing its own genetic material and assembling into virus particles.

The influenza virus is a class of viruses containing RNA as its hereditary material. It replicates by entering a host cell and using this cell's resources to produce hundreds of copies of the viral RNA.

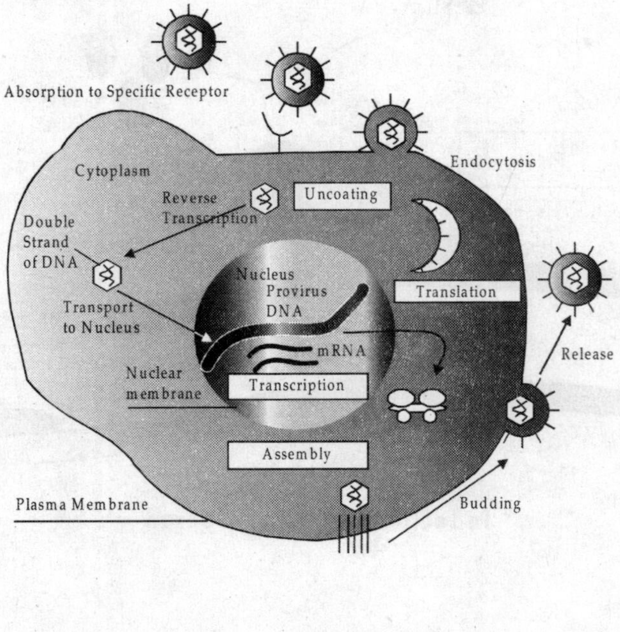

The virus attaches to the outside of the host cell and its RNA enters into the cell. The viral genes are transcribed and translated by the cell's enzymes and ribosomes. In this way, the virus takes over the cell's productivity. Now, instead of producing only new cellular material, the cell produces hundreds of new virus particles. The new virus particles are eventually released from the cell and drift off, and some may land on a host cell of their own to pirate.

❐ inheritance

The transmission of genes and phenotypes from generation to generation.

❐ inhibitor

1. Any substance or object that retards a chemical reaction.
2. A metabolite or modifier gene that interferes with a reaction or with the expression of another gene.

❐ initial

Cells in a meristem that remain permanently capable of differentiation, and which develop into tissues of particular structure and function.

❐ initiation

Causing something to start.

1. Early steps or stages of a tissue culture process.
2. Early stages of biosynthesis.

❐ initiation factor

Soluble protein required for the initiation of translation.

❐ initiation of transcription by rna polymerase

In order to begin transcription, RNA polymerase requires a number of general transcription factors (called TFIIA, TFIIB, and so on). (A) The promoter contains a DNA sequence called the TATA box, which is located 25 nucleotides away from the site where transcription is initiated. (B) The TATA box is recognized and bound by transcription factor TFIID, which then enables the adjacent binding of TFIIB (C). For simplicity the DNA distortion produced by the binding of TFIID (D) The rest of the general transcription factors as well as the RNA polymerase itself assemble at the promoter. (E) TFIIH then uses ATP to phosphorylate RNA polymerase II, changing its conformation so that the polymerase is released from the complex and is able to start transcribing. As shown, the site of phosphorylation is a long polypeptide tail that extends from the polymerase molecule.

The exact order in which the general transcription factors assemble on the promoter is not known with certainty. One view holds that the general factors assemble off the DNA with the polymerase and that this whole assembly then binds to the DNA in a single step. The general transcription factors have been highly conserved in evolution; some of those from human cells can be replaced in biochemical experiments by the corresponding factors from simple yeasts.

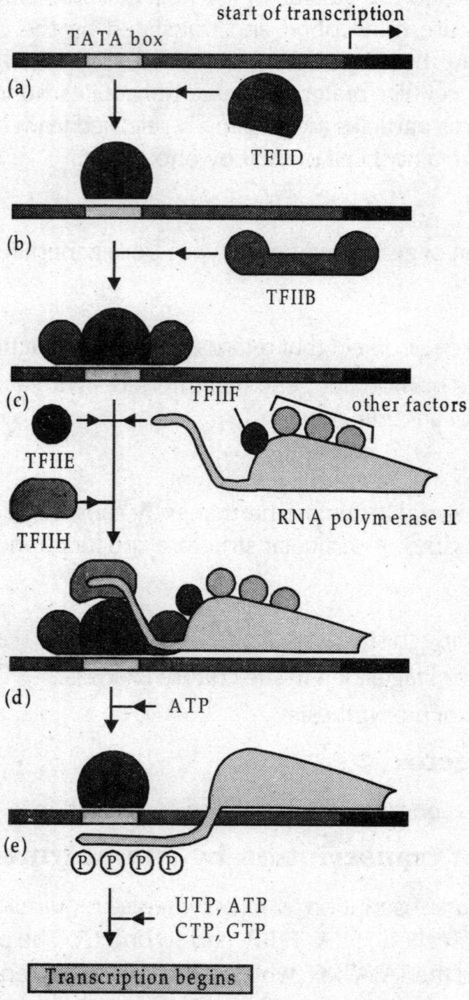

◻ inoculate

Deliberately introduce, in contrast to contamination.

1. In bacteriology, tissue culture, etc., placing an inoculum into (or onto) medium to initiate a culture.
2. In immunology, to carry out immunization.
3. In plant pathology, application of pathogen spores etc. on to plants under conditions where infection should result in the absence of resistance.

◻ inoculation cabinet

Small room or cabinet for inoculation (of tissue or micro-organism cultures) operations, often with a current of sterile air to carry contaminants away from the work area.

inoculum

1. A small piece of tissue cut from callus, or an explant from a tissue or organ, or a small amount of cell material from a suspension culture, transferred into fresh medium for continued growth of the culture.
2. Microbial spores or parts (such as mycelium).
3. Vaccine. inorganic compound Historically, chemicals that could not be derived from living processes. In modern usage, chemicals that do not contain carbon, although carbonates and a few other simple carbon compounds are generally regarded as inorganic.

inositol

A cyclic acid (hexahydroxycyclohexane) that is a constituent of certain cell phosphoglycerides. It is a nutrient frequently referred to as a "vitamin" in plant tissue culture. Also acts as a growth factor in some animals and micro-organisms.

inositol lipid

A membrane-anchored phospholipid that transduces hormonal signals by stimulating the release of any of several chemical messengers.

insecticide

A substance that kills insects.

insert

1. To incorporate a DNA molecule into a cloning vector; also used as a noun to describe such a DNA molecule.
2. To introduce a gene or gene construct into a new genomic site or into a new genome.

insertion

A type of mutation in which one or more nucleotides is inserted into a DNA sequence. Small insertions within a gene can alter the reading frame, and thus the amino acid sequence of the encoded protein.

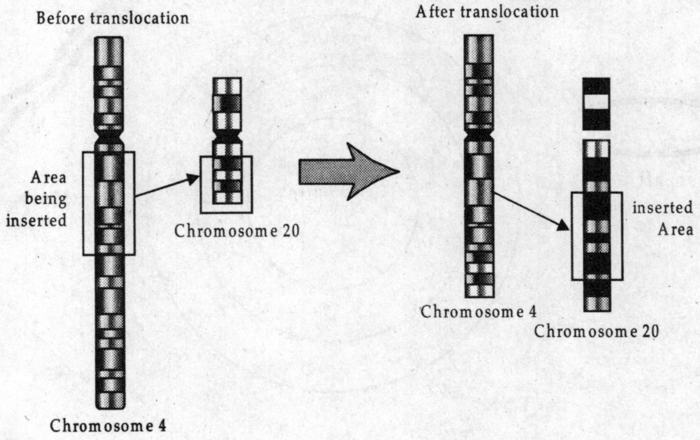

□ inserting dna into plasmid

Plasmids are similar to viruses, but lack a protein coat and cannot move from cell to cell in the same fashion as a virus.

Plasmid vectors are small circular molecules of double stranded DNA derived from natural plasmids that occur in bacterial cells. A piece of DNA can be inserted into a plasmid if both the circular plasmid and the source of DNA have recognition sites for the same restriction endonuclease.

The plasmid and the foreign DNA are cut by this restriction endonuclease (EcoRI in this example) producing intermediates with sticky and complementary ends. Those two intermediates recombine by base-pairing and are linked by the action of DNA ligase. A new plasmid containing the foreign DNA as an insert is obtained. A few mismatches occur, producing an undesirable recombinant.

The new plasmid can be introduced into bacterial cells that can produce many copies of the inserted DNA. This technique is called DNA cloning.

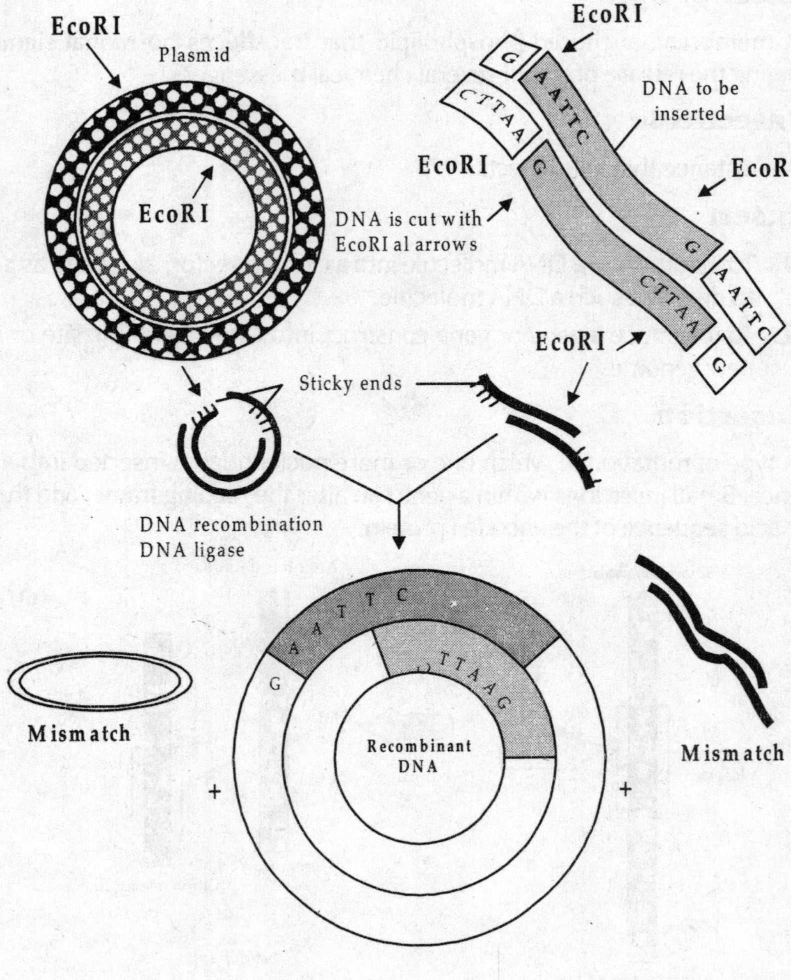

▫ insertion element

Generic term for DNA sequences found in bacteria capable of genome insertion. Postulated to be responsible for site-specific phage and plasmid integration.

▫ insertion mutation

Changes in the base sequence of a DNA molecule resulting from the random integration of DNA from another source.

▫ insertion site

1. A unique restriction site in a vector DNA molecule into which foreign DNA can readily be inserted. This is achieved by treating both the vector and the insert with the relevant restriction endonuclease and then ligating the two different molecules, both having the same sticky ends.
2. The position of integration of a transposon.

▫ *in-situ* conservation

The conservation of ecosystems and natural habitats and the maintenance and recovery of viable populations of species in their natural surroundings and, in the case of domesticated or cultivated species, in the surroundings where they have developed their distinctive properties.

▫ instability

A lack of consistent phenotype, usually as a result of uncontrolled genetic changes. These may be due to transposon activity, or in cell lines, to changes in karyotype.

▫ insulin

A peptide hormone secreted by the Langerhans islets of the pancreas, and that regulates the level of sugar in the blood.

▫ integrating vector

A vector that is designed to integrate cloned DNA into the host's chromosomal DNA.

▫ integration

The recombination process which inserts a small DNA molecule (usually by homologous recombination) into a larger one. If the molecules are circular, integration involves only a single crossing-over; if linear, then two crossings-over are required.

▫ integration-excision region

The portion of bacteriophage lambda (ë) DNA that enables ë-DNA to be inserted into a specific site in the *E. coli* bacteriophage lambda chromosome or excised from this site.

▫ integument

One of the layers that enclosed the ovule, and is the precursor of the seed coat.

❑ intellectual property rights

The legal framework, which includes patenting and plant variety protection, by which inventors control the commercial application of their work.

❑ intensifying screen

A plastic sheet impregnated with a rare-earth compound, such as calcium tungstate, which reacts to radiation by emitting light. When placed on one side of a piece of X-ray film with a radioactive sample on the other side, the intensifying screen will capture some of the radioactive energy which has passed through the film, exposing the X-ray film and so enhancing the sensitivity of the detection. Often used in Southern and northern blotting procedures.

❑ intercalary

1. Meristematic tissue or growth not restricted to the apex of an organ, i.e. growth at nodes.
2. Referring to internal segments of a chromosomes (i.e. not at the ends).

❑ intercalary growth

A pattern of stem elongation typical of grasses. Elongation proceeds from the lower internodes to the upper internodes through the differentiation of meristematic tissue at the base of each internode.

❑ intercalating agent

A chemical capable of inserting between adjacent base pairs in a double-stranded nucleic acid. A prominent example is ethidium bromide.

❑ intercellular space

The pore space between cells, especially typical of leaf tissues.

❑ interconversion of atp and adp

An adenosine-derived nucleotide, $C_{10}H_{16}N_5O_{13}P_3$, that contains high-energy phosphate bonds and is used to transport energy to cells for biochemical processes, including muscle contraction and enzymatic metabolism, through its hydrolysis to ADP. ATP is hydrolyzed to AMP when it is incorporated into DNA or RNA.

Adenosine Diphosphate consists of one adenosine molecule and 2 inorganic phosphate molecules. The bonds that connect the phosphate molecules have a high potential energy state, and therefore is ideal to be used in biological processes.

The interconversion of ATP and ADP the two outermost phosphates of ATP are held to the rest of the molecule by high-energy phosphoanhydride bonds and are readily transferred. Water can be added to ATP to form ADP and inorganic phosphate (Pi). This hydrolysis of the terminal phosphate of ATP yields between 11 and 13 kcal/mole of usable energy, depending on the intracellular conditions. The large negative delta G of this reaction arises from a number of factors. Release of the terminal phosphate group removes an unfavorable repulsion between adjacent negative charges; in addition, the inorganic phosphate ion (Pi) released is stabilized by resonance and by favorable hydrogen-bond formation with water.

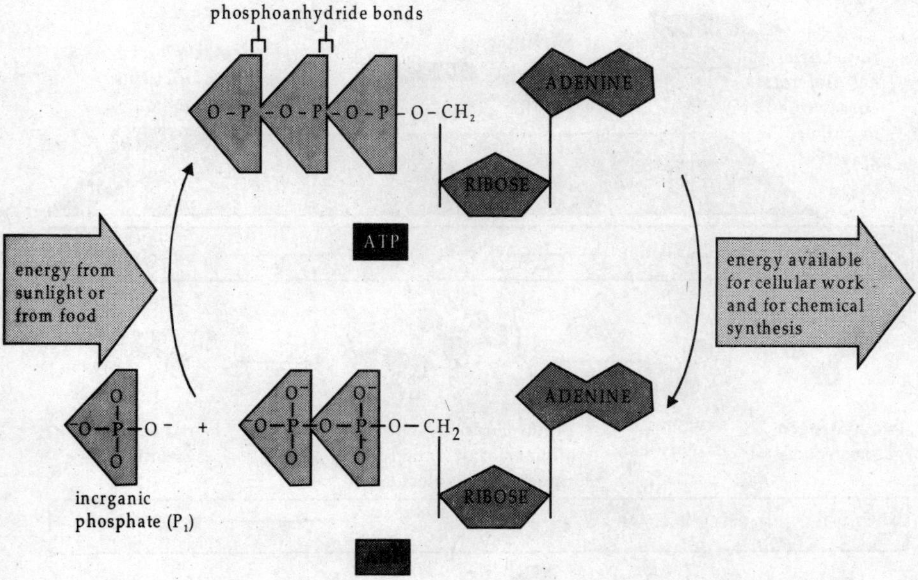

❒ interconversions of energy

All energy forms are, in principle, interconvertible. In all these processes the total amount of energy is conserved; thus, for example, from the height and weight of the brick in the first example, we can predict exactly how much heat will be released when it hits the floor. In the second example, note that the large amount of chemical bond energy released when water is formed is initially converted to very rapid thermal motions in the two new water molecules; but collisions with other molecules almost instantaneously spread this kinetic energy evenly throughout the surroundings (heat transfer), making the new molecules indistinguishable from all the rest.

Transformation of Energy: Carbohydrate Oxidation

$$C_6H_{12}O_6 + 6O_2 \underset{2}{\overset{1}{\rightleftharpoons}} 6CO_2 + 6H_2O + \text{energy} \rightarrow \text{Heat}$$

VO_2 | VGO_2 | ATP synthesis → Splitting of ATP for muscular controction

1. Cellular respiration
2. Photosynthesis

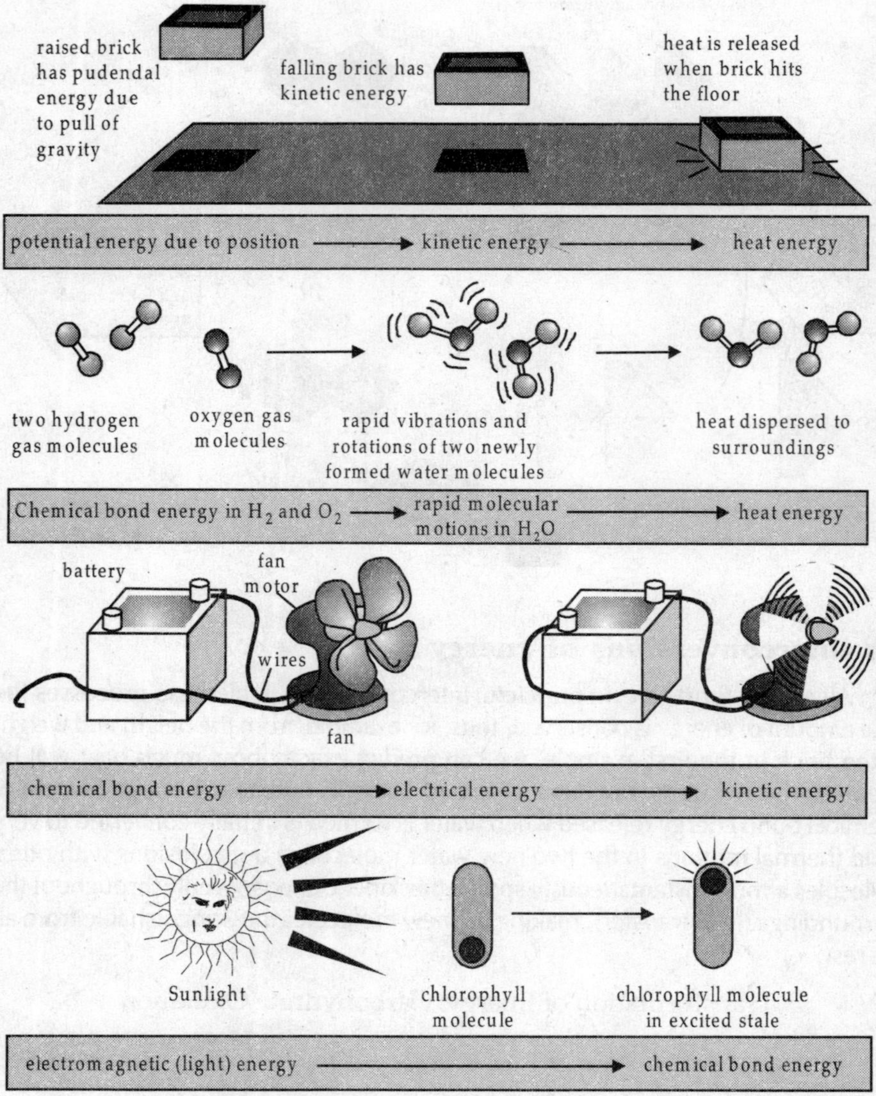

❐ interfascicular cambium

Cambium that arises between vascular bundles.

❐ interference

The effect of one crossing over event in altering the probability of another crossing over event occurring at a nearby location. This probability can be either increased (positive interference) or decreased (negative interference), but the latter is the more usual.

❑ interferon

One of a group of small proteins synthesized by certain T cells of vertebrates, which inhibit virus replication. There are three types of interferon in humans.

❑ intergeneric cross

A hybrid made between parents belonging to two different genera.

❑ intergenic regions

Non-coding DNA located between genes; this comprises a variable but considerable proportion of all eukaryotic genomic DNA, and its function is largely unknown.

❑ intergenic spacer

Non-coding DNA separating tandemly arranged copies of a repeated gene sequence (typically ribosomal DNA). Of particular interest because, unlike the coding sequence itself, the spacers show high levels of interspecific sequence polymorphism, and are thus useful as assays for species identification.

❑ interleukin

A group of proteins that transmit signals between immune cells and are necessary for mounting normal immune responses.

❑ internal membranes and cytosol

A variety of membrane-bounded compartments exist within eukaryotic cells, each specialized to perform a different function. (B) The rest of the cell, excluding all these organelles, is called the cytosol, and is the site of many vital cellular activities.

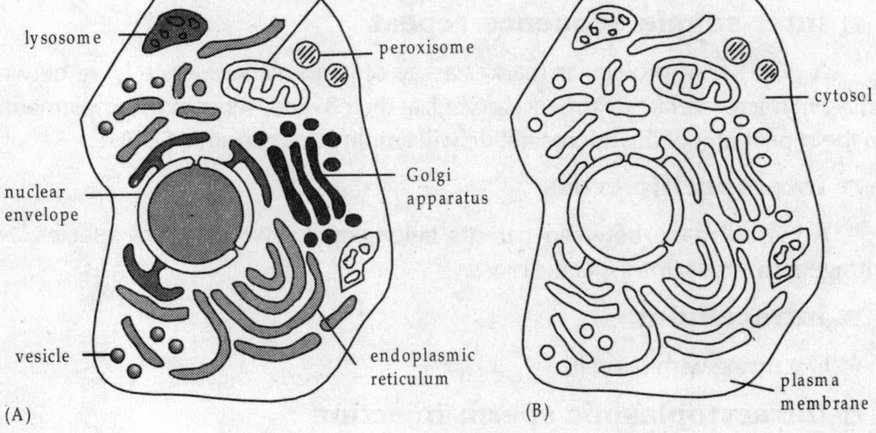

❑ internal transcribed spacer

Non-coding regions separating the individual components of the ribosomal DNA units. These regions show much more sequence polymorphism than the genic regions themselves, and therefore, like the intergenic spacers, are useful source of genetic markers for the ribosomal DNA locus.

❐ international treaty on plant genetic resources for food and agriculture

The international treaty resulting from the revision of the International Undertaking on Plant Genetic Resources was adopted by the 2001 FAO Conference as a binding international instrument to enter into force after ratification by 40 states. Its objectives are the conservation and sustainable use of plant genetic resources for food and agriculture and equitable sharing of the benefits of this use.

❐ international indertaking on plant genetic resources

The first comprehensive voluntary, international agreement (adopted in 1983) dealing with plant genetic resources for food and agriculture. Designed as an instrument to promote international harmony in matters regarding access to plant genetic resources for food and agriculture.

❐ internode

The region of a stem between two successive nodes.

❐ interphase

The stage in the cell cycle when the cell is not dividing and during part of which DNA replication occurs; it follows telophase of one mitotic division and extends to the beginning of prophase in the next division.

❐ intersex

Synonym of hermaphrodite.

❐ inter-simple sequence repeat

A PCR-based molecular marker assay of genomic sequence lying between adjacent microsatellites. Primers carrying, at their 3'-end, sequence complementary to the repeat unit of the microsatellite will amplify this genomic DNA.

❐ interspecific cross

A hybrid made between parents belonging to two different species. *See*: intrageneric cross, intraspecific cross.

❐ intracellular

Occurring within a cell.

❐ intracytoplasmic sperm injection

The micro-injection of a single sperm into the cytoplasm of a mature oocyte.

❐ intrageneric cross

A hybrid made between parents belonging to two species in the same genus.

❐ intrageneric

Within a genus, such as an intrageneric cross, or intrageneric variation.

❐ intrag complementation

Occurs when wild type phenotype is restored in an F_1 individual made by crossing two independent mutants, carrying different heteroalleles.

❐ intraspecific cross

A hybrid made between parents belonging to the same species. *See*: intrageneric cross, interspecific cross.

❐ intraspecific

Within a species, such as an intraspecific cross, or intraspecific variation.

❐ introgression

The introduction of new alleles or gene(s) into a population from an exotic source, usually another species. This is achieved by repeated backcrossing of the initial hybrid in order to eliminate all genetic changes except for the desired new gene(s).

❐ intron

A segment of the primary transcript of a eukaryotic gene, removed (before the mature mRNA is translated) in a process known as intron splicing. Some eukaryotic genes contain a large number of introns, which make up the bulk of the DNA sequence of the gene. Introns are also found in genes whose RNA transcripts are not translated, namely eukaryotic rRNA and tRNA genes. In these cases the intron sequence does not appear in the functional RNA molecule.

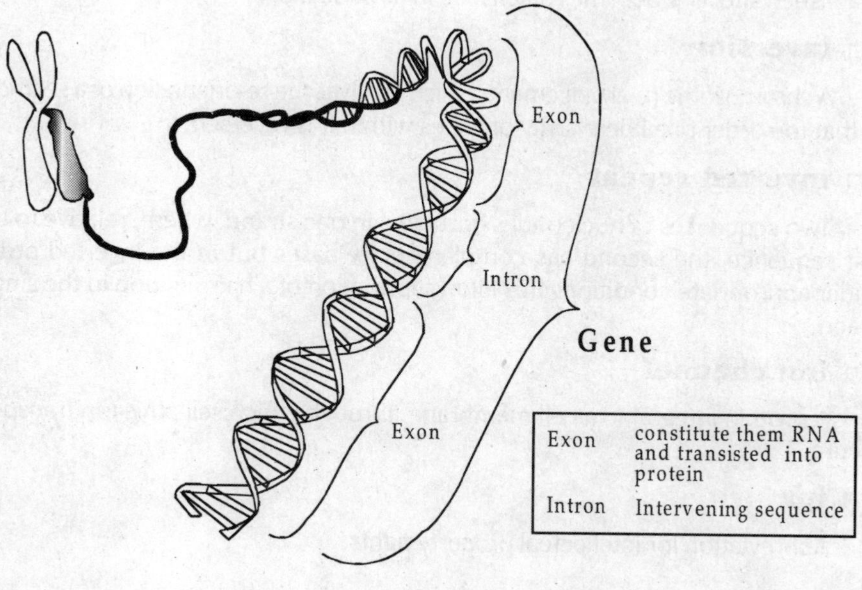

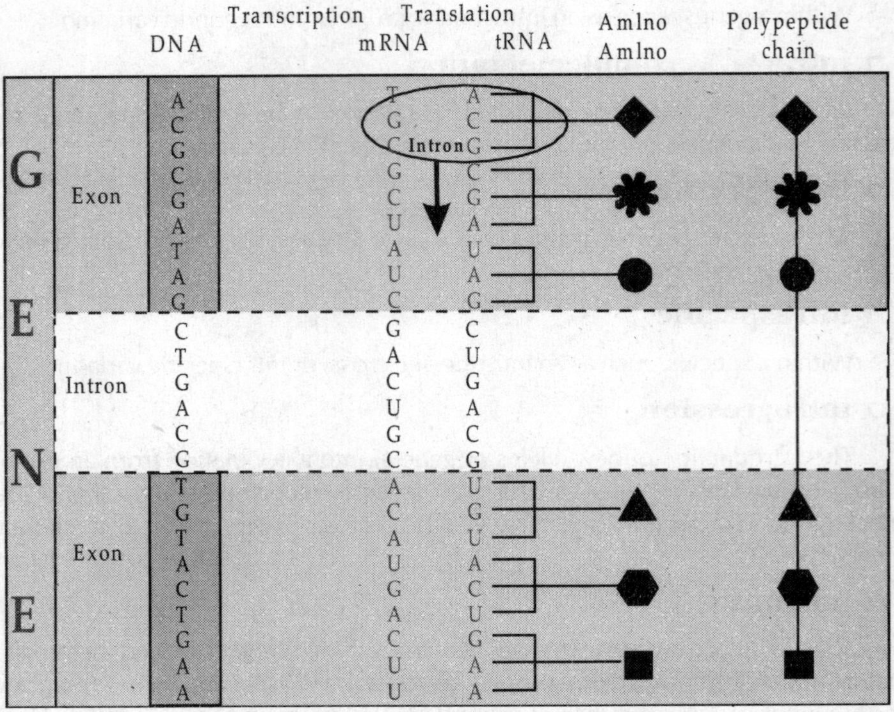

❐ invasiveness

The ability of a plant, particularly a weed, to spread beyond its presently established site, and become established in new locations.

❐ inversion

A chromosome re-arrangement, which involves the re-orientation of a segment so that the order of a linear array of genes within it is reversed.

❐ inverted repeat

Two sequences of nucleotides occurring in one strand, where, relative to the first sequence, the second has complementary bases but in the inverted order. Under appropriate conditions this allows formation of a hairpin loop in the single strand.

❐ ion channel

A protein integral to a cell membrane, through which selective ion transport occurs.

❐ ipr

Abbreviation for intellectual property rights.

❐ iptg

Abbreviation for isopropyl-3-D-thiogalactopyranoside. A synthetic inducer of beta-galactosidase activity in many bacteria. Used in combination with the synthetic chromogenic substrate Xgal to differentiate recombinant from non-recombinant bacterial colonies in cloning strategies using plasmid vectors containing the *lacZ* gene: blue coloured colonies are produced when β-galactosidase activity is not disrupted by an insert; but when it is disrupted, the colonies are white. Hence white colonies are indicative of recombinant plasmids, and blue colonies of non-recombinant ones.

❐ irradiation

Illumination with electromagnetic radiation, typically of sufficiently high energy (low-wavelength UV or gamma, *etc.*) to disrupt biological macromolecules and hence induce mutations.

❐ IS element

Abbreviation for insertion sequence element. A short (800-1400 nucleotide pairs) DNA sequence found in bacteria that is capable of transposing to a new genomic location; DNA sequences contained within an IS element can be transposed along with the IS itself.

❐ isoallele

Multiple similar copies of a gene, usually located at independent positions in the genome, which encode similar gene products and produce the same, or a very similar phenotype.

❐ isochromosome

A chromosome produced following an error in meiosis, in which the two arms are mirror images of each other. The presence of an isochromosome results in the duplication of all genes present on the originating chromosome arm.

❐ isodiametric

Commonly used to describe cells with equal diameters.

❐ iso-electric focusing gel

A variant of gel electrophoresis, in which macromolecules (usually proteins) are separated on the basis of differing iso-electric point, rather than on the basis of size.

❐ isoform

1. A tissue-specific form of a protein.
2. Synonym of isoenzyme.

❐ isogamy

Fusion of gametes of similar size and structure.

❒ isogenic

A group of individuals that possesses the same genotype, irrespective of their being homozygous or heterozygous.

❒ isogenic stock

Strains of organisms that are genetically nearly identical, except with respect to identified genes. Generally produced by repeated backcrossing, or by transformation.

❒ isolating mechanism

The properties of an organism that prevent interbreeding (and therefore exchange of genetic material) between members of different species that inhabit the same geographical area.

❒ isolation medium

An optimum plant tissue culture medium suitable for explant survival, growth and development.

❒ isomer

1. Structural isomers have the same chemical formula but different structures; e.g. leucine and isoleucine.
2. Stereoisomers are different topological forms of an otherwise single chemical structure, due to changes in bond configurations about some axis or plane of symmetry; e.g., D- and L-glucose or cis- and trans-cinnamic acid.

❒ isomerase

A class of enzymes that catalyse the re-arrangement of the atoms within a molecule, thereby converting one isomer into another.

❒ isotonic

Solutions with the same osmotic potential, as a result of same molar concentration. For protoplasts to avoid losing or gaining water, the medium they are suspended in must be isotonic with them.

❒ isotope

One of two or more forms of an element that differ in the number of neutrons carried by the nucleus. Radioactive isotopes (radio-isotopes) are used as probes in many biochemical analyses.

❒ isozyme

A genetic variant of an enzyme. Isozymes for a given enzyme share the same function, but may differ in level of activity, as a result of minor differences in their amino acid sequence. Electrophoretic separation of isozymes has been used to distinguish between individuals and varieties.

❒ issr

Abbreviation for inter-simple sequence repeat.

□ **its**

Abbreviation for internal transcribed spacer.

□ **ivep**

Abbreviation for *in vitro* embryo production.

□ **ivf**

Abbreviation for *in vitro* fertilization.

□ **ivm**

Abbreviation for *in vitro* maturation.

J

❑ jiffy pota

Pots made from wood pulp and peat, commonly used for transplanting tissue culture-derived plants into soil medium.

❑ jivet

Abbreviation for juvenile *in vitro* embryo technology.

❑ jivt

Abbreviation for juvenile *in vitro* embryo technology.

❑ joining segment

A small DNA segment that links genes in order to yield a functional gene encoding an immunoglobulin.

❑ juvenile hormone

A hormone secreted by insects from a pair of endocrine glands close to the brain. Its function is to inhibit metamorphosis in order to maintain the larval features.

❑ juvenile *in vitro* embryo technology

A technology involving collection of immature eggs from young animals, their *in vitro* maturation and fertilization, and the transfer of the resultant embryos into recipient females. The method is designed to achieve rapid generation turnover.

❑ juvenility

Early phase of development in which an organism is incapable of sexual reproduction.

K

kanamycin

An antibiotic of the aminoglycoside family that inhibits translation by binding to the ribosomes. Important as a substrate for selection of plant transformants.

kan^r

Kanamycin-resistance gene.

kappa chain

One of two classes of antibody light chains. The other is a lambda chain.

karyogamy

The fusion of nuclei or nuclear material that occurs at fertilization during sexual reproduction.

karyogram

A diagrammatic representation of the full chromosome set of a species, highlighting characteristic physical features of individual chromosomes.

karyokinesis

The division of a cell nucleus.

karyotype

The chromosome constitution of a cell, an individual, or of a related group of individuals, as defined both by the number and the morphology of the chromosomes, usually in mitotic metaphase; chromosomes arranged in order of length and according to position of centromere; also, the abbreviated formula for the chromosome constitution, such as 47, + 21 for human trisomy-21 (Down's syndrome).

kb

Abbreviation for kilobase (of single-stranded nucleic acid).

kbp

Abbreviation for kilobase pairs (of double-stranded DNA).

❐ **k_{cat}/k_m**

The catalytic efficiency of an enzyme-catalysed reaction. The greater the value of k_{cat}/K_m, the more rapidly and efficiently the substrate is converted into product.

❐ **k_{cat}**

The catalytic rate constant that characterizes an enzyme-catalysed reaction. The larger the k_{cat} value, the faster the conversion of substrate into product.

❐ **k_d**

Abbreviation for dissociation constant. Describes the strength of binding (or affinity) between molecules and their ligands.

❐ **kda**

Abbreviation for kiloDalton. A unit of molecular mass equal to 1000 Dalton.

❐ **killer t cell**

T cells that kill cells displaying recognized antigens.

❐ **kilobase**

A length of single-stranded nucleic acid composed of 1000 bases. One kilobase of single-stranded DNA has a mass of about 330 kiloDalton (exact mass depends on base composition).

❐ **kilobase pairs**

A length of double-stranded DNA composed of 1000 base pairs.

❐ **kinase**

An enzyme that catalyses the transfer of a phosphate group from a high energy state (as in ATP) to another molecule.

❐ **kinetics**

Dynamic processes involving motion. Often used as a suffix to indicate studies involving movement or rates of reactions.

❐ **kinetin**

A cytokinin.

❐ **kinetochore**

Structure at the centromere of eukaryotic chromosomes. The kinetochore consists of inner and outer electron dense plates and a central zone containing repetitive DNA elements. Kinetochores are involved in the control of chromosome movement in cell division.

❐ **kinetosome**

Granular cytoplasmic structure which forms the base of a cilium or flagellum.

☐ kinin

A substance promoting cell division. In plant systems, the prefix cyto- has been added (cytokinin) to distinguish it from kinin in animal systems.

☐ Klenow fragment

A truncated form of DNA polymerase I from *E. coli*, used extensively for the production of synthetic DNA molecules as it retains polymerase and 3'-exonuclease activities, but not 5'-exonuclease activity.

☐ k_m

A dissociation constant that characterizes the binding of an enzyme to a substrate. The smaller the value of K_m, the tighter the binding of the enzyme to the substrate. Also called the Michaelis constant.

☐ knockout

A mutant individual, in which a single functional gene has been replaced by a non-functional form of the gene. Used to understand gene function via the comparison of the phenotypes of wild type and knockouts.

L

❑ label

A compound or atom that is attached to, or incorporated into, another molecule in order to allow detection of the latter's presence. Commonly, labels exploit radioactivity, fluorescence or antigenicity.

❑ labelling

The process of attaching or inserting a label into a molecule. Most often in the context of nucleic acids or proteins.

❑ lactose

A disaccharide sugar produced in milk, composed of one unit each of glucose and galactose.

❑ lag phase

1. The state of apparent inactivity preceding a response to a treatment; also called a latent phase.
2. The initial growth phase, during which cell number remains relatively constant, prior to the onset of rapid cell division.

❑ lagging strand

The strand of DNA that is synthesized discontinuously during replication (because DNA synthesis can proceed only in the 5'?3' direction).

❑ lambda chain

One of two classes of antibody light chains. The other is a kappa chain.

❑ lambda phage

A bacteriophage that infects *E. coli*, commonly used as a cloning vector.

❑ lamella

A structure, plate or vesicle that is formed by two membranes lying parallel to each other.

❑ lamina

Blade or expanded part of a leaf.

▫ laminar air-flow cabinet

Cabinet designed for cell or tissue culture manipulations requiring a sterile environment. Achieved by a continuous, non-turbulent flow of filter-sterilized air over the working area.

▫ laminarin

A storage polysaccharide of the brown algae.

▫ lampbrush chromosome

Large diplotene chromosomes present in oocyte nuclei, and particularly conspicuous in amphibians. These chromosomes have extended regions called loops, which are active sites of transcription.

▫ landrace

In plant genetic resources, an early, cultivated form of a crop species, evolved from a wild population, and generally composed of a heterogeneous mixture of genotypes.

▫ latent agent

A pathogen, usually a virus, present in a host organism without producing any symptoms.

▫ latent bud

An inactive bud not held back by rest or dormant period, but which may start growth if stimulated.

▫ lateral meristem

A meristem giving rise to secondary plant tissues, such as the vascular and cork cambia.

▫ lawn

A uniform and uninterrupted layer of bacterial growth, typically on agar medium, in which individual colonies cannot be observed.

▫ layering

A technique for vegetative propagation, in which new plants produce adventitious roots before being severed from the parent plant.

▫ lCR

Abbreviation for ligase chain reaction.

▫ ld_{50}

Abbreviation for lethal $dose_{50\%}$. The amount of a substance required to kill 50% of the test population. The higher the LD_{50}, the lower the toxicity of the chemical in that specific test.

lead compound

A chemical that has demonstrated promising biological activity in preliminary assays.

leader sequence

A variable length sequence of nucleotides at the 5' end of an mRNA molecule that precedes the AUG initiation codon where translation begins and is not itself translated into protein.

leading strand

The strand of DNA that is synthesized continuously during replication.

leaf blade

The usually flattened portion of the leaf.

leaf bud cutting

A cutting that includes a short section of stem with attached leaf.

leaf margin

The edge of a leaf.

leaf primordium

A lateral outgrowth from the apical meristem, which will become a leaf when fully developed and expanded.

leaf roll

A symptom of some virus diseases, characterized by curling of the leaves. Can also occur as a response to water stress.

leaf scar

Mark left on a stem after leaf abscission.

leaflet

Expanded leaf-like part of a compound leaf.

leaky mutant

A mutant in which the gene product still retains some biological activity.

lectin

A group of plant proteins that can bind to specific oligosaccharides on the surface of cells, causing the cells to clump together.

leptonema

Stage in meiosis immediately preceding synapsis and post DNA replication, in which the chromosomes appear as single, fine, threadlike structures.

❑ l hal allele

A utant form of a gene that, in the homozygous state, is fatal.

❑ l ukocyte

V ite blood cell, up to 0.02 mm in diameter, of which leukocyte are normally 4-11 n llion per millilitre of human blood. There are several kinds, all involved in the body' defence mechanisms. Granulocytes have granules in their cytoplasm; mono ytes ingest and feed on bacteria and other micro-organisms that cause infection; lymph cytes include the B cells that are involved with the production of antibodies.

❑ levels of chromatin packing

This schematic drawing shows some of the orders of chromatin packing thought to giv rise to the highly condensed mitotic chromosome. The folding of naked DNA nto nucleosomes is the best understood level of packing. The structures corresponding to the additional layers of chromosome packing are more speculative.

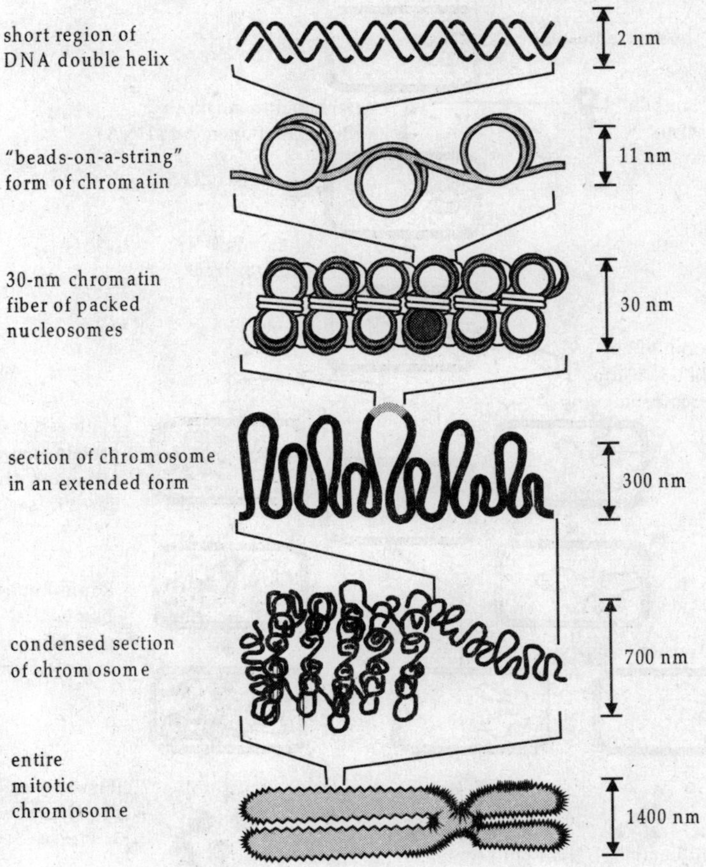

NET RESULT: EACH DNA MOLECULE HAS BEEN PACKAGED INTO A MITOTIC CHROMOSOME THAT IS 50,000x SHORTER THAN ITS EXTENDED LENGTH

▫ library

A collection of cells, usually bacteria or yeast, that have been transformed with recombinant vectors carrying DNA derived from an unrelated organism.

▫ life cycle of a bacteriophage

The linear double-stranded DNA lambda genome contains about 50,000 nucleotide pairs and encodes 50-60 different proteins. When the lambda DNA enters the cell the ends join to form a circular DNA molecule. The bacteriophage can multiply in E. coli by a lytic pathway, which destroys the cell, or it can enter a latent prophage state. Damage to a cell carrying a lambda prophage induces the prophage to exit from the host chromosome and shift to lytic growth. The entrance and exit of the lambda DNA from the bacterial chromosome are site-specific recombination events.

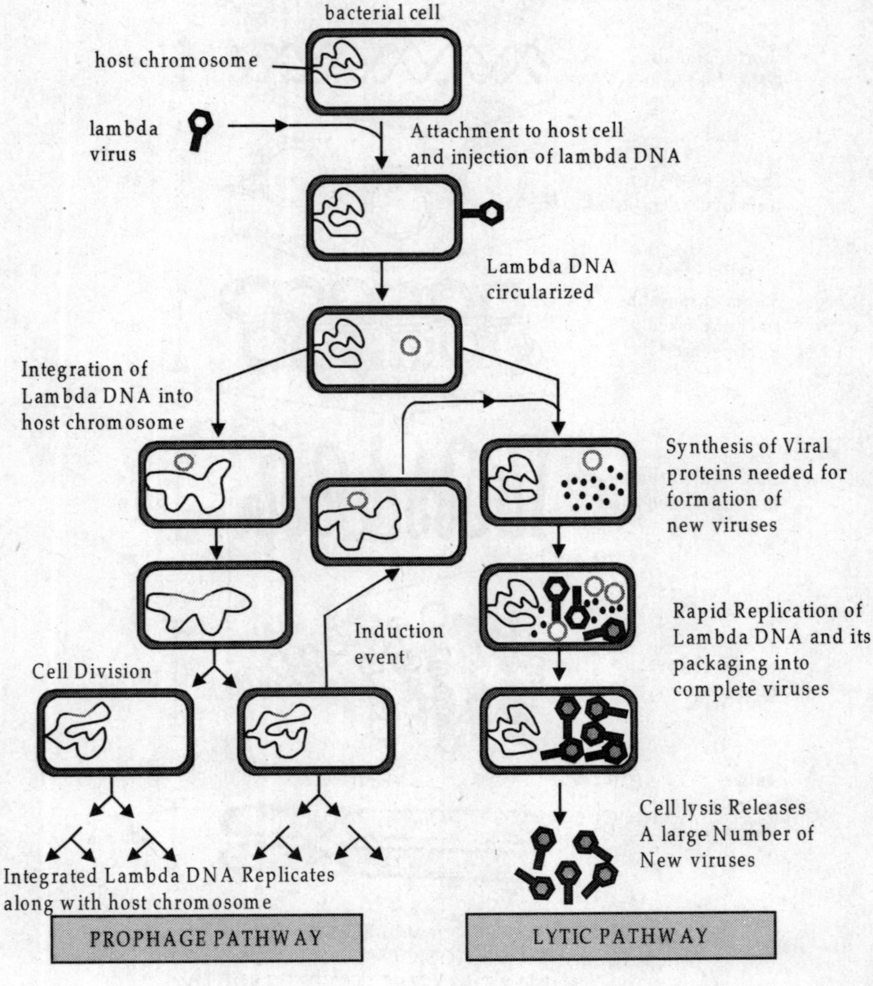

❐ life cycle of a retrovirus

The typical retrovirus genome consists of a single-stranded RNA of about 8500 nucleotides. The enzyme reverse transcriptase is a multifunctional enzyme that first makes a DNA copy of the viral RNA molecule. It then acts as a nuclease to remove the RNA, and then makes a second DNA strand, generating a double-stranded DNA copy of the RNA genome. The integration of this DNA into the host chromosome, catalyzed by a viral protein called integrase, is required for the synthesis of new viral RNA molecules by the host cell RNA polymerase. Retroviruses are examples of enveloped viruses, in which the protein shell is further enclosed by an outer lipid bilayer membrane. The envelope contains proteins that enable the virus to bind to cells, and that aid its entry into a cell. As indicated, the lipid membrane is acquired when the virus is released from the cell by a process of budding from the plasma membrane, taking some of the plasma membrane with it. The budding process is reversed when the virus reinfects a cell.

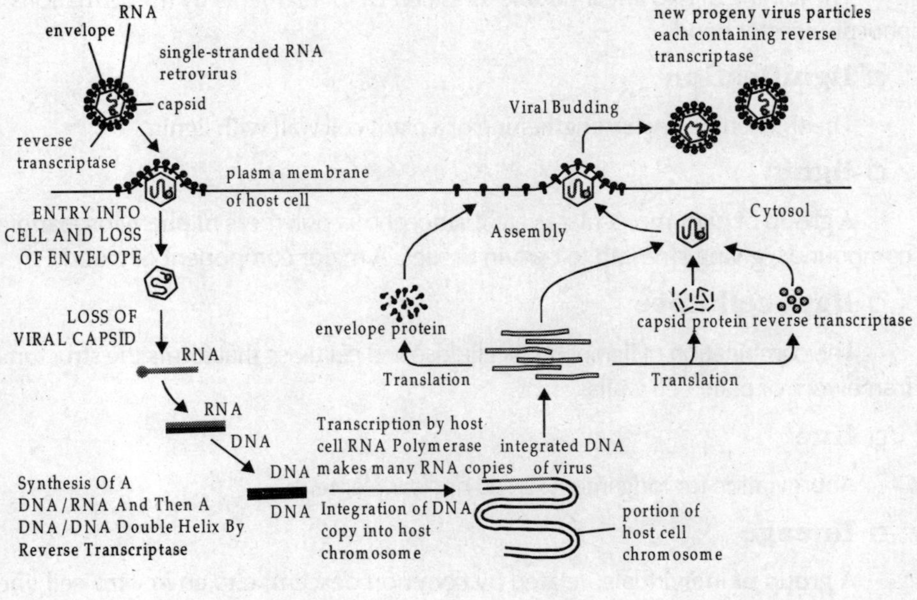

❐ life cycle

The sequence of events from a given developmental stage in one generation to the same stage in the following generation. In sexually reproduced organisms, the starting point is the fusion of gametes to form the zygote.

❐ ligand

A small molecule (e.g. activators, substrates and inhibitors of enzyme activity) bound to a protein by non-covalent forces; an ion or a molecule that binds to another chemical entity to form a larger complex.

❐ ligase chain reaction

A technique for the detection and amplification of target DNA sequences. Two oligonucleotides are synthesized which between them are complementary to the entire target sequence, one to the 5'-side and one to the 3'-side. If the target sequence is present in the DNA sample under examination, the oligonucleotides will bind to it with their ends abutting in the centre, and a heat-stable ligase will join them into a complete polynucleotide. No ligation occurs if the target sequence is absent or if the match between synthetic oligonucleotides and target sequence is imperfect in the region where they abut. At a high temperature, the new polynucleotide dissociates from the original DNA template, and upon cooling, it and the original DNA serve as templates for a second cycle of hybridization, ligation and thermal dissociation. At each cycle there is a doubling of the number of new complete polynucleotides.

❐ ligate, ligation

The joining of two linear double-stranded DNA fragments by the formation of phosphodiester bonds.

❐ lignification

The thickening and strengthening of a plant cell wall with lignin.

❐ lignin

A group of high-molecular-weight amorphous polymers of phenylpropanoid compounds, giving strength to certain tissues. A major component of wood.

❐ lignocellulose

The combination of lignin, hemicellulose and cellulose that forms the structural framework of plant cell walls.

❐ line

Abbreviation for long interspersed nuclear element.

❐ lineage

A group of individuals, related by common descent, e.g. an *in vitro* cell line derived from a single cell.

❐ linear phase

The growth phase during cell culture when cell number increases arithmetically. The linear phase follows a period of exponential growth.

❐ linearized vector

A covalently closed circular DNA vector (typically a plasmid) which has been opened by restriction digestion to convert it to a linear molecule. In molecular cloning, DNA to be cloned is mixed with the linearized vector, and treated with ligase to join and recircularize the resulting hybrid molecule.

❑ linkage

The tendency of a set of genes to be inherited together more often than would be expected if they were assorting independently. exists between two genes when they are located sufficiently close to one another on the same chromosome that a proportion of gametes is produced without crossing-over occurring between them.

❑ linkage map

A linear or circular diagram that shows the relative positions of genes on a chromosome as determined by recombination fraction.

❑ linked gene, linked marker

A gene or marker that is linked to another gene or marker.

❑ linker

A synthetic double-stranded oligonucleotide that carries the recognition sequence for one or more restriction endonucleases. The ligation of a linker to each end of a DNA fragment facilitates the preparation of the fragment for cloning into a vector.

❑ lipase

A class of enzymes which break down lipids into their component fatty acids and glycerol. Lipases used in biotechnology are generally digestive, with a role in the break-down of fats in food into their components, so that these can be used to make other materials.

❑ lipid

Any of a group of fats or fat-like compounds insoluble in water and soluble in fat solvents.

❑ lipofection

Delivery into eukaryotic cells of DNA, RNA or other compounds that have been encapsulated in liposomes.

❑ lipopolysaccharide

A compound containing lipid bound to a polysaccharide; often a component of microbial cells walls.

❑ liposome

A synthetic microscopic spherical structure consisting of a phospholipid bilayer membrane containing a user-defined aqueous solution. Liposomes can be used to transport relatively toxic drugs into diseased cells, where they can exert their maximum effect. DNA molecules may be entrapped in, or bound to the surface of, the vesicles, and subsequent fusion of the liposome with the cell membrane will deliver the DNA into the cell. Liposomes have been used to develop an efficient transfection procedure for *Streptomyces* bacteria.

□ liquefaction

Enzymatic digestion (often by alpha-amylase) of gelatinized starch to form lower molecular weight polysaccharides.

□ liquid medium

Culture solution, without a solidifying agent, for *in vitro* cell growth.

□ liquid membrane

Thin films made up of liquids (as opposed to solids) which are stable in another liquid (usually water). Thus the liquid must not dissolve in the water, but nevertheless must be prevented from collapsing into small droplets.

□ liquid nitrogen

Nitrogen gas condensed to a liquid with a boiling point of about -196 °C. Commonly used as a medium for long-term storage of biological materials.

□ litmus paper

A pH indicator paper. It turns red in acidic and blue in alkaline solutions.

□ live recombinant vaccine

A vaccine created by the expression of a pathogen antigen in a non-pathogenic organism.

□ live vaccine

A living, non-virulent form of a pathogenic micro-organism or virus used to elicit an antibody response for the protection against infection by a virulent form of the same pathogen.

□ living modified organism

"Living organism that possess a novel combination of genetic material obtained through the use of modern biotechnology" (Convention on Biological Diversity). Synonym of GMO, but restricted to organisms that can endanger biological diversity.

□ lmo

Abbreviation for living modified organism

□ locus

A site on a chromosome.

□ lod score

The logarithm of the odds of linkage between two loci. Used to measure the statistical support for linkage.

□ log phase

Abbreviation for logarithmic phase.

❐ logarithmic phase

The growth phase in cell culture, during which cell number doubles every 20-30 minutes.

❐ long interspersed nuclear element

Families of common DNA elements, of average length 6.5 kb, which are dispersed at numerous locations within the genome. The human genome contains over 500,000 LINEs (representing ca. 16% of the genome). They appear to represent degenerate copies of transposable elements.

❐ long template

A DNA strand, synthesized during PCR, which has a primer sequence at one end but is extended beyond the site that is complementary to the second primer at the other end.

❐ long terminal repeat

A characteristic sequence of nucleotides that occurs at each end of a retrovirus element that has become integrated into the host genome.

❐ long-day plant

Plants requiring a period of short nights before the switch from vegetative to reproductive growth can be initiated.

❐ loop bioreactor

Fermenters in which material is cycled between a bulk tank and a smaller tank or loop of pipes. The circulation helps to mix the materials and to ensure that gas injected into the fermenter is well distributed in the liquid. Particularly useful for photosynthetic fermentations, where the photosynthesizing organisms are passed through a system of many small transparent pipes, which allow the access of light.

❐ lps

Abbreviation for lipopolysaccharide.

❐ ltr

Abbreviation for long terminal repeat.

❐ luteinizing hormone

A pituitary hormone which causes growth of the yellow body of the ovary and also stimulates activity of the interstitial cells of the testis.

❐ luxury consumption

Nutrient absorption by an organism in excess of that required for optimum growth and productivity.

❒ lyase

A class of enzymes that catalyse either the cleavage of a double bond and the addition of new groups to a substrate, or the formation of a double bond.

❒ lymphocyte

White blood cells that are important components of the immune system of vertebrates.

❒ lymphokine

Generic name for proteins that are released by lymphocytes to act on other cells involved in the immune response. The term includes interleukins and interferons. A sub-class of cytokines.

❒ lymphoma

Cancer originating in the lymph nodes, spleen and other lympho-reticular sites.

❒ lysis

The destruction or breakage of cells either by viruses or by chemical or physical treatment.

❒ lysogen

A bacterial cell whose chromosome contains integrated bacteriophage DNA.

❒ lysogenic

Bacteria or bacteriophages undergoing lysogeny.

❒ lysogenic bacterium

Bacterium harbouring temperate (non-virulent, lysogenic) bacteriophages.

❒ lysogeny

A condition in which a bacteriophage genome (pro-phage) survives within a host bacterium, either as part of the host chromosome or as part of an extrachromosomal element, and does not initiate lysis.

❒ lysosome

A membrane-bound sac within the cytoplasm of animal cells that contain enzymes responsible for the digestion of material in food vacuoles, the dissolution of foreign particles entering the cell and, on the death of the cell, the breaking down of all cell structures. The digestive system of the cell.

❒ lysozyme

A naturally occurring enzyme extracted from egg white protein and other animal and plant sources, which attacks the cell wall of gram-positive bacteria leading to cell lysis and death.

□ lytic

A phase of the virus life cycle during which the virus replicates within the host cell, releasing a new generation of viruses when the infected cell undergoes lysis.

□ lytic cycle

The steps in viral production that lead to cell lysis.

M

❐ m13

A single-stranded DNA bacteriophage used as a vector for DNA sequencing.

❐ m13 strand

The single-stranded DNA molecule that is present in the infective form of bacteriophage M13.

❐ maap

Abbreviation for multiple arbitrary amplicon profiling.

❐ mab

Abbreviation for monoclonal antibody.

❐ macerate

To disintegrate tissue to disrupt cells. Commonly achieved via mechanical shearing, plasmolysis or enzymatic cell wall degradation.

❐ macromolecule

Any high molecular weight molecule. Often used as a synonym for polymers.

❐ macromolecules in cells

In some dimensions macromolecules can reach a millimetre in length; it is not unusual to come across macromolecules that measure as much as a meter when unrolled (a molecule of water, by contrast, is on the order of one millionth of a millimetre). This is the case for many biological macromolecules, like proteins and DNA. Knowing the three-dimensional structure and sequence of the groups of atoms in macromolecules is crucial in understanding how they react biochemically.

The sequences can be determined with the help of biochips, and by using x-rays, for example, it is possible to uncover a macromolecule's three-dimensional structure.

Macromolecules are abundant in cells. The approximate composition of a bacterial cell is shown on next page. The composition of an animal cell is similar.

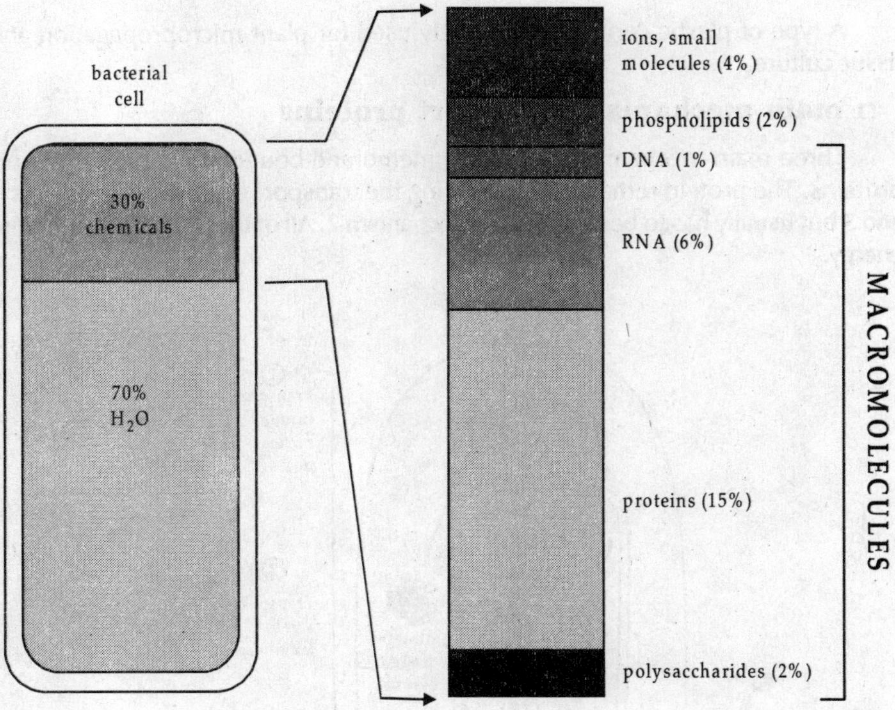

□ macronutrient

A major chemical element essential for normal growth and development. In tissue culture media, macronutrients are those required in concentrations above 0.5 millimole/litre.

□ macrophage

Large white blood cells that ingest foreign substances and display on their surfaces antigens which are recognized by other cells of the immune system.

□ macropropagation

Production of plant clones from growing parts.

□ mad cow disease

Colloquial term for bovine spongiform encephalopathy. *See* proteinaceous infectious particle

□ mads box

A highly conserved DNA sequence motif found in a large family of transcription factors, most of which play important roles in developmental processes. Most prominently, the MADS box genes known in flowering plants are intimately involved in the control of flower morphogenesis.

magenta

A type of plastic container frequently used for plant micropropagation and tissue culture.

main mechanism to import proteins

Three main mechanisms by which membrane-bounded organelles import proteins. The protein remains folded during the transport steps in mechanisms 1 and 3 but usually has to be unfolded in mechanism 2. All of these processes require energy.

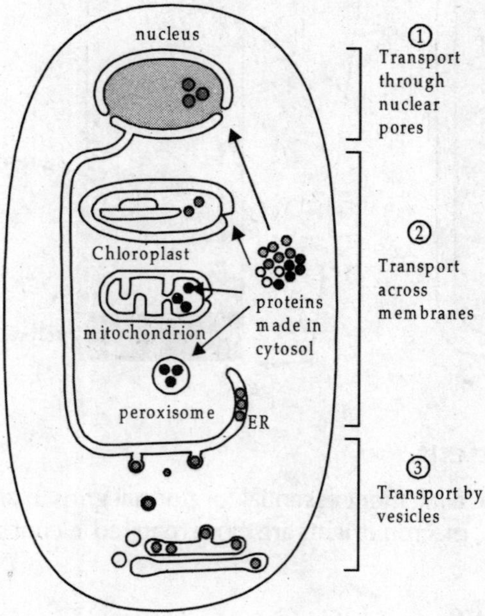

major histocompatibility antigen

A cell-surface protein or glycoprotein that allows the immune system to distinguish foreign or "non-self" from "self". A better term is histoglobulin. These are the antigens that must be matched between donors and recipients during organ and tissue transplants to prevent rejection.

major histocompatibility complex

The large cluster of genes that encode the major histocompatibility antigens in mammals.

malt extract

A mixture of organic compounds prepared from malt, used as a culture medium additive.

❐ malting

Enzymatic reduction of starch to sugars in germinating grain, used in brewing.

❐ mammalian differentiated cell types

Fertilization of mammalian eggs is followed by successive cell divisions and progressive differentiation, first into the early embryo and subsequently into all of the cell types that make up the adult animal. Transfer of a single nucleus at a specific stage of development, to an enucleated unfertilized egg, provided an opportunity to investigate whether cellular differentiation to that stage involved irreversible genetic modification. The first offspring to develop from a differentiated cell were born after nuclear transfer from an embryo-derived cell line that had been induced to become quiescent. Using the same procedure, we now report the birth of live lambs from three new cell populations established from adult mammary gland, fetus and embryo. The fact that a lamb was derived from an adult cell confirms that differentiation of that cell did not involve the irreversible modification of genetic material required for development to term. The birth of lambs from differentiated fetal and adult cells also reinforces previous speculation that by inducing donor cells to become quiescent it will be possible to obtain normal development from a wide variety of differentiated cells.

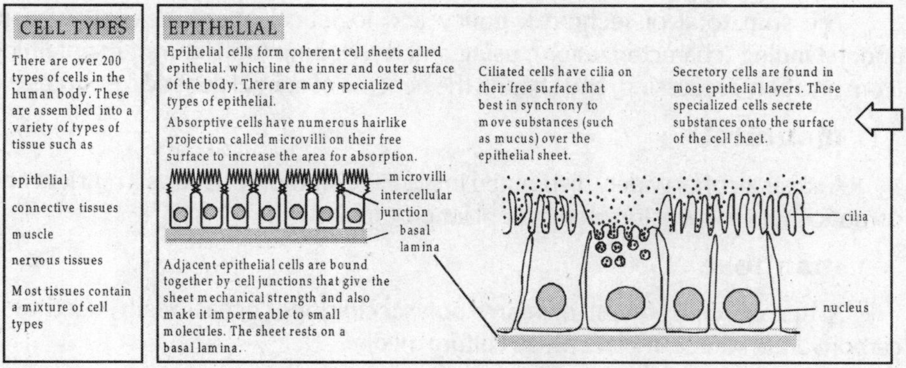

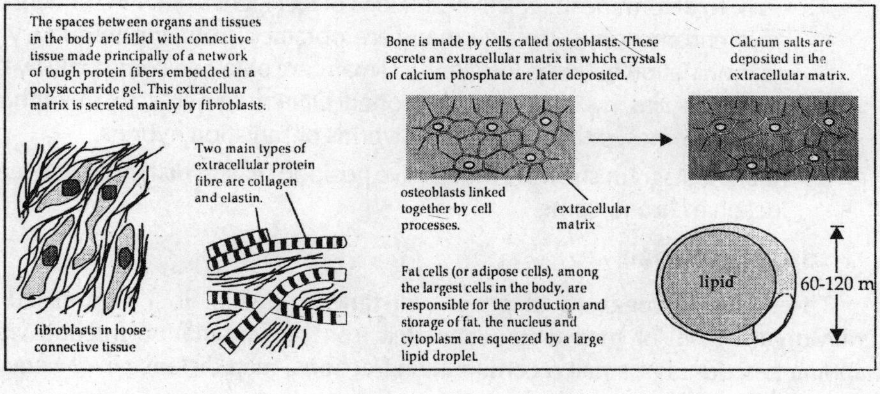

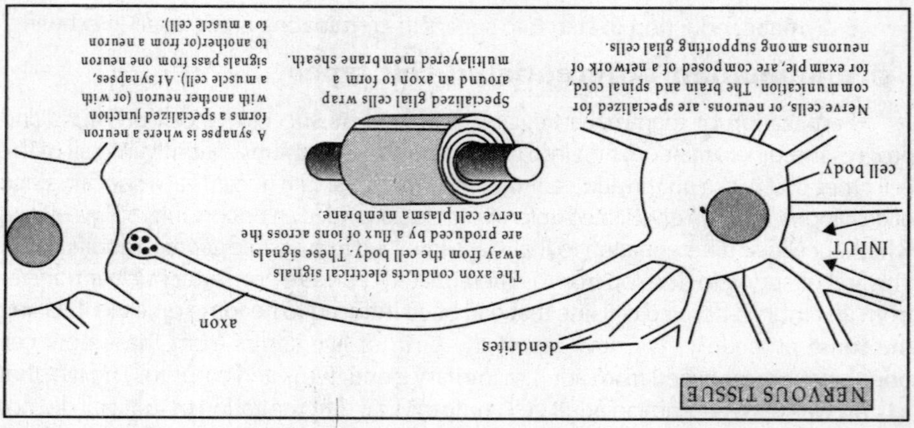

❒ mammary gland

The milk-producing organ of female mammals.

❒ management of farm animal genetic resources

The sum total of technical, policy and logistical operations involved in understanding (characterization), using and developing (utilization), maintaining (conservation), accessing, and sharing the benefits of animal genetic resources.

❒ mannitol

A sugar alcohol widely distributed in plants. Commonly used as a nutrient and osmoticum in suspension media for plant protoplasts.

❒ mannose

A hexose component of many polysaccharides, occasionally used as a carbohydrate source in plant tissue culture media.

❒ map

1. verb: to determine the relative positions of loci (genes or DNA sequences) on a chromosome. Linkage maps are obtained from the frequency of recombination between loci. Physical maps are obtained commonly by the use of *in situ* hybridization of cloned DNA fragments to metaphase chromosomes, or by somatic-cell hybrids or radiation hybrids.
2. noun: a diagram showing the relative position of, and distances between, loci on a chromosome.

❒ map distance

The standard measure of genetic distance between loci, expressed in centiMorgans (cM) or map units. Estimated from recombination fraction via a mapping function. For small recombination fractions, map distance in cM equals the recombination fraction in %.

❐ map unit

One centiMorgan (1 cM) *See*: map distance; crossing-over unit.

❐ mapping function

A mathematical expression relating observed recombination fraction to map distance.

❐ mapping

The construction of a localized (around a gene), or broad-based (whole genome) genetic map. More generally, determining the location of a locus (gene or genetic marker) on a chromosome.

❐ marker

An identifiable DNA sequence that is inherited in Mendelian fashion, and which facilitates the study of inheritance of a trait or a linked gene.

❐ marker gene

A gene of known function or known location, used for marker-assisted selection or genetic studies.

❐ marker peptide

A portion of fusion protein that facilitates its identification or purification.

❐ marker-assisted introgression

The use of DNA markers to increase the speed and efficiency of introgression of a new allele(s) or gene(s) into a breeding population. The markers will be closely linked to the gene(s) in question.

❐ marker-assisted selection

The use of DNA markers to improve response to selection in a population. The markers will be closely linked to one or more target loci, which may often be quantitative trait loci.

❐ mas

Abbreviation for marker-assisted selection.

❐ mass selection

As practised in plant and animal breeding, the selection of a number of individuals, on the basis of their individual phenotypes, to interbreed to form the next generation.

❐ maternal effect

An effect attributable to a genetic contribution of the female parent of the individual being evaluated.

❐ maternal inheritance

Inheritance controlled by non-nuclear genes (e.g. mitochrondria, chloroplast) that are transmitted only through the female line.

❒ matric potential

A water potential component, always of negative value, resulting from the presence of solid (often finely divided) surfaces; primarily responsible for water uptake by a dry seed prior to germination.

❒ maturation

The formation of gametes or spores.

❒ mcs

Abbreviation for multiple cloning site.

❒ mda

Abbreviation for multiple drop array.

❒ mean

In statistics, the arithmetic average; the sum of all measurements or values in a sample divided by the sample size.

❒ means of protein separation

In order to study proteins it is crucial for the biochemists to obtain a sample which contains only the molecule he is interested in. Also in the production of proteins for commercial purposes the demand for purity is high.

A purification scheme might consist of a combination of the following steps:

Crude cell extract

↓

Ammonium sulfate precipitation

↓

Affinity chromatography

↓

Ion exchange chromatography

This particular sequence of steps is of course not applicable in all cases, most often this means that a unique purification protocol has to be developed for each new substance you wish to isolate. Most important is that the steps complement each other and that the degree of purity increases with each step. The number of steps included in the protocol depends on the state of the starting fraction and on how pure you want your substance.

To get a good recovery of the substance, i.e. to minimize losses, it is desirable that each step is as specific as possible. To check purity and yield you may use absorbance measurements, various types of electrophoresis and preferably also some kind of activity measurement. The column chromatography part may be performed in different ways more or less manually. In the course practical we will use manual ways. For more routine purifications this system can be built out to monitor absorbance etc. and quite often one uses a FPLC (Fast Protein Liquid Chromatography) which is a programmable system with more powerful pumps.

PROTEIN SEPARATION

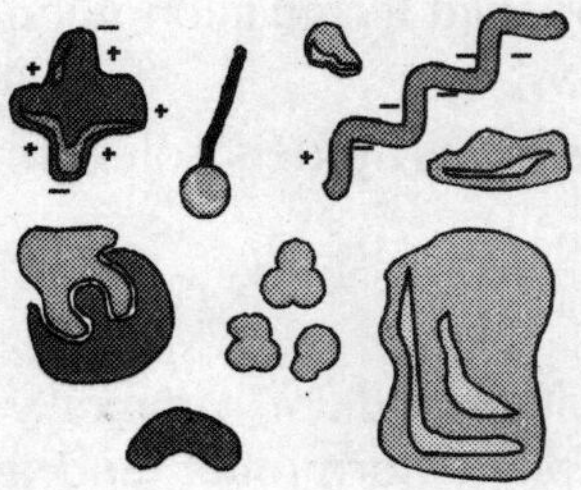

Proteins are very diverse. They differ by size, shape, charge, hydrophobicity, and their affinity for other molecules. All these properties can be exploited to separate them from one another so that they can be studied individually.

THREE KINDS OF CHROMATOGRAPHY

Many types of matrix are available for column chromatography, usually packed in the column in the form of small beads. A typical protein purification strategy might employ in turn each of the three kinds of matrix described below, with a final protein purification of up to 10,000-fold.
Purity can easily be assessed by gel electrophoresis (see *opposite page*).

COLUMN CHROMATOGRAPHY

Proteins are often fractionated by **column chromatography.** A mixture of proteins in solution is applied to the top of a cylindrical column fille D with a permeable solid matrix immersed in solvent. A large amount of solvent is then pumped through the column. Because different proteins are retarded to different extents by their interaction with the matrix, they can be collected separately as they flow out from the bottom. According to the choice of matrix, proteins can be separated according to their charge, their hydrophobicity, their size, or their ability to bind to particular chemical groups (see *below*).

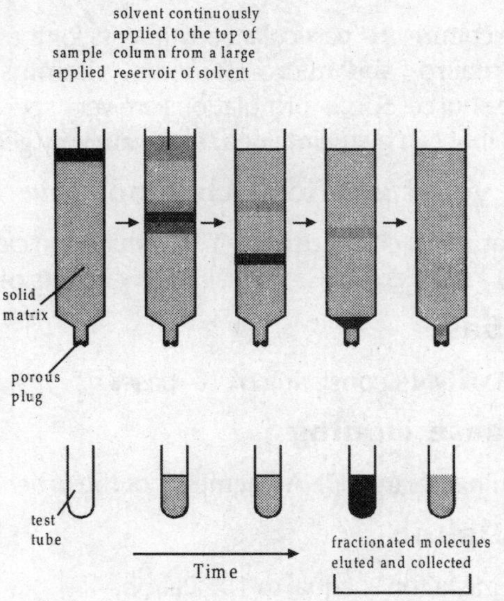

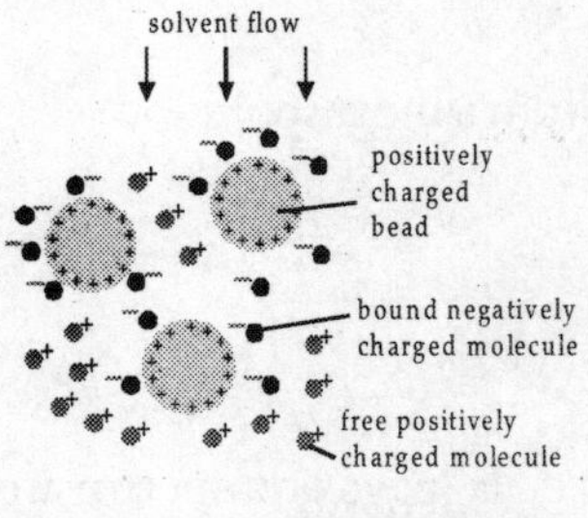

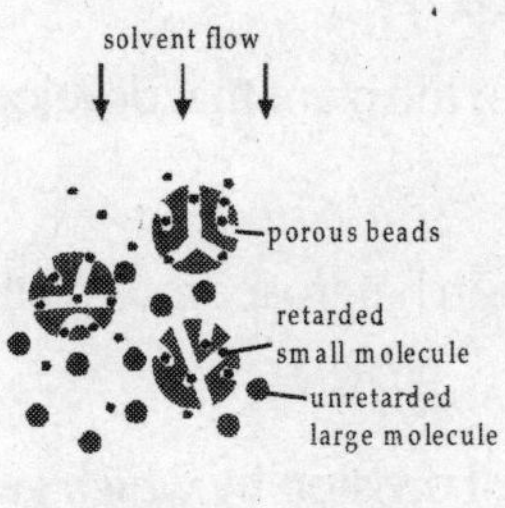

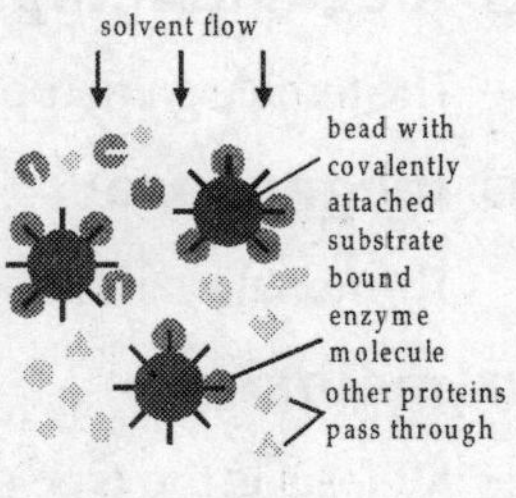

ION-EXCHANGE CHROMATOGRAPHY

Ion-exchange columns are packed with small beads that carry positive or negative charges that retard proteins of the opposite charge. The association between a protein and the matrix depends on the pH and ionic strength of the solution passing down the column. These can be varied in a controlled way to achieve an effective separation.

GEL-FILTRATION CHROMATOGRAPHY

Gel-filtration columns separate proteins according to their size. The matrix consists of tiny porous beads. Protein molecules that are small enough to enter the holes in the beads are delayed and travel more slowly through the column. Proteins that cannot enter the beads are washed out of the column first. Such columns also allow an estimate of protein size.

AFFINITY CHROMATOGRAPHY

Affinity columns contain a matrix covalently coupled to a molecule that interacts specifically with the protein of interest (e.g., an antibody, or an enzyme substrate). Proteins that bind specifically to such a column can finally be released by a pH change or by concentrated salt solutions, and they emerge highly purified.

❐ median

In a set of measurements, the central value above and below which there are an equal number of measurements.

❐ medium

1. In plant tissue culture, a term for the liquid or solid formulation upon which plant cells, tissues or organs develop.
2. In general terms, a substrate for plant growth, such as nutrient solution, soil, sand, etc., e.g. potting medium.

❐ medium formulation

In tissue culture, the particular constituents for the culture medium, commonly comprising macro- and micro-elements, vitamins, plant hormones, and a carbohydrate source. Some formulations are very specific to the kind of explant or plant species that can be maintained; some are very general.

❐ mega yeast artificial chromosome

A yeast artificial chromosome (YAC) which can carry particularly large inserts (up to 1Mbp) - standard YACs typically carry inserts of up to 500kbp.

❐ megabase

A length of DNA consisting of 10^6 bases.

❐ megabase cloning

The cloning of large DNA fragments of the order of 1Mb.

❐ megaDalton

One megaDalton is equal to 10^6 Dalton.

❐ megagametophyte

The female gametophyte; the plant that develops from a megaspore.

❐ megaspore

The female gametophyte in heterosporous plants.

❐ meiosis

Meiosis is the type of cell division by which germ cells (eggs and sperm) are produced. Meiosis involves a reduction in the amount of genetic material.

Meiosis comprises two successive nuclear divisions with only one round of DNA replication. Four stages can be described for each nuclear division.

❐ First division of meiosis

- **Prophase 1:** Each chromosome duplicates and remains closely associated. These are called sister chromatids. Crossing-over can occur during the latter part of this stage.
- **Metaphase 1:** Homologous chromosomes align at the equatorial plate.

- **Anaphase 1**: Homologous pairs separate with sister chromatids remaining together.
- **Telophase 1**: Two daughter cells are formed with each daughter containing only one chromosome of the homologous pair.

❐ Second division of meiosis: Gamete formation

- **Prophase 2**: DNA does not replicate.
- **Metaphase 2**: Chromosomes align at the equatorial plate.
- **Anaphase 2**: Centromeres divide and sister chromatids migrate separately to each pole.
- **Telophase 2**: Cell division is complete. Four haploid daughter cells are obtained.

One parent cell produces four daughter cells. Daughter cells have half the number of chromosomes found in the original parent cell and with crossing over, are genetically different.

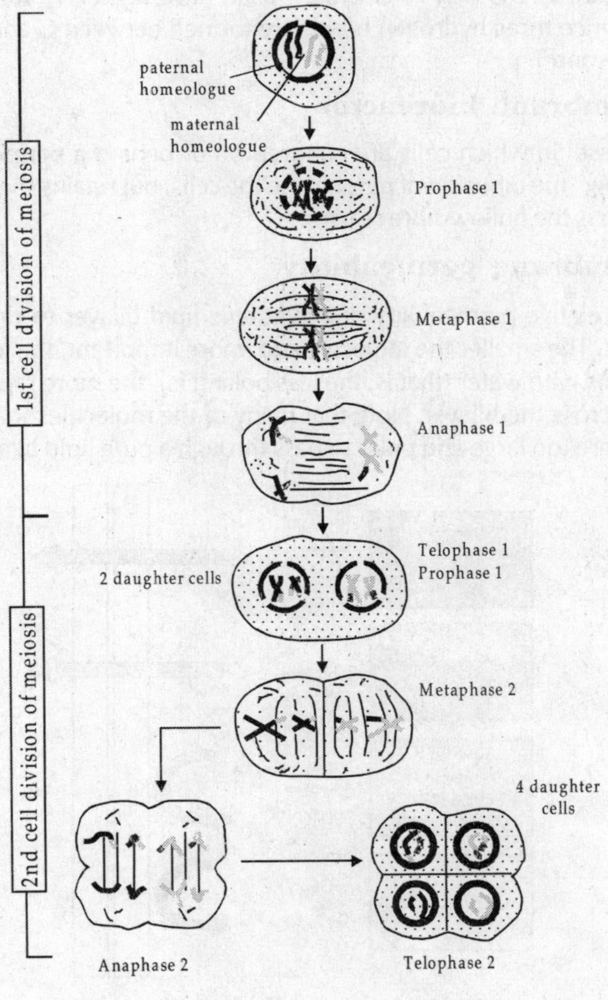

❒ meiotic analysis

The use of patterns of chromosome pairing at meiotic prophase and metaphase to detect relationships between chromosomes, from which can be deduced the relationship between the parents of the organism studied.

❒ meiotic drive

Any mechanism that causes a particular allele or chromosome to be over-represented in a population of gametes.

❒ melanin

Dark pigment, produced by specialized epidermal cells called melanocytes.

❒ melting temperature

The temperature at which a double-stranded DNA molecule denatures into separate single strands. T_m is determined by the length of the molecule and its base composition. DNAs rich in G:C base pairs have higher T_m than A:T rich DNA, because since three hydrogen bonds are formed between G and C, but only two between A and T.

❒ membrane bioreactor

A vessel in which cells are cultured on or behind a permeable membrane, which allows the diffusion of nutrients to the cells, but retains the cells themselves. A variation is the hollow-fibre reactor.

❒ membrane permeability

The relative permeability of a synthetic lipid bilayer to different classes of molecules. The smaller the molecule and, more important, the fewer its favorable interactions with water (that is, the less polar it is), the more rapidly the molecule diffuses across the bilayer. Note that many of the molecules that the cell uses as nutrients are too large and polar to pass through a pure lipid bilayer.

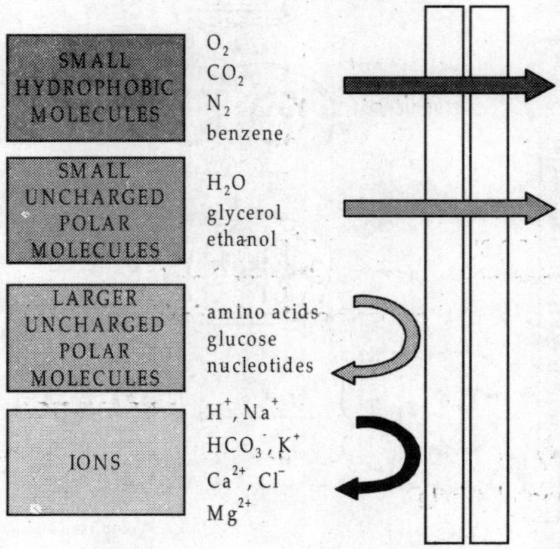

▢ memory cell

Long-lived B cells and T cells that mediate rapid secondary immune responses to a previously encountered antigen.

▢ mendel: experiment 1

This drawing illustrates Mendel's "Experiment 1", in which he demonstrated his concept of heredity in the mating of pea plants.

Mendel suspected that heredity depended on contributions from both parents and that specific characteristics from each parent were passed on, rather than being blended together in the offspring.

A parent homozygous for the spherical seed allele is crossed with a parent homozygous for the wrinkled seed allele. Each parent makes gametes of only one kind, either S or s, and these combine at fertilization to form plants that all have the genotype Ss and the spherical seed phenotype.

When the F1 plants self-pollinate they produce two kinds of eggs, S and s, and the same two types of male sex cells. These combine randomly in four different ways to form F2 plants.

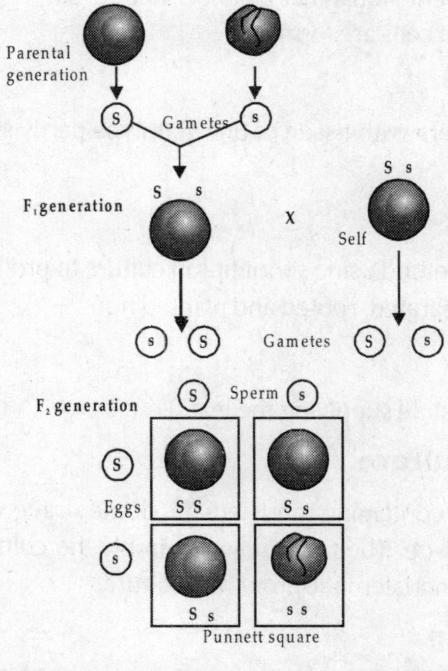

Three of the four possible combinations produce "spherical seed" phenotype, and the fourth produces "wrinkled seed" phenotype, so that the observed ratio is 3:1.

The wrinkled seed phenotypes can only correspond to the "ss" genotype. The spherical seed phenotype corresponds, in one case to the "SS" genotype, and in the two others to the "Ss" genotype. Two of the offsprings are homozygous (SS or ss); the two others are heterozygous (Ss).

The fact that the spherical seed phenotype is observed with the genotypes "SS" or "Ss" can only be explained if the character "spherical seed" is dominant and the character "wrinkled seed" recessive.

The illustration at the previous page, called a Punnett Square, is a handy device for keeping track of the different ways in which gametes combine at fertilization.

❐ mendelian population

A natural, interbreeding unit of sexually reproducing plants or animals sharing a common gene pool.

❐ mendelian segregation

Occurs when alleles are inherited according to Mendel's Laws.

❐ mendel's laws

Two laws summarizing Gregor Mendel's theory of inheritance. The Law of Segregation states that each hereditary characteristic is controlled by two 'factors' (now called alleles), which segregate and pass into separate germ cells. The Law of Independent Assortment states that pairs of 'factors' segregate independently of each other when germ cells are formed.

❐ mericlinal

Refers to a chimera with tissue of one genotype partly surrounded by that of another genotype.

❐ mericloning

A propagation method using shoot tips in culture to proliferate multiple buds, which can then be separated, rooted and planted out.

❐ meristele

The branch of a stele supplying the leaf.

❐ meristem culture

A tissue culture containing meristematic dome tissue without adjacent leaf primordia or stem tissue. The term may also imply the culture of meristemoidal regions of plants, or meristematic growth in culture.

❐ meristem tip

An explant comprising the meristem (meristematic dome) and usually one pair of leaf primordia. Also refers to explants originating from apical meristem tip or lateral or axillary meristem tip.

❐ meristem tip culture

Cultures derived from meristem tip explants. Used widely to achieve virus elimination and axillary shoot proliferation, less commonly for callus production.

❑ meristem

Undifferentiated but determined plant tissue, in which the cells are capable of active division and differentiation into specialized tissues such as shoots and roots.

❑ meristemoid

A localized group of callus cells, characterized by their accumulation of starch, RNA and protein, and giving rise to adventitious shoots or roots.

❑ merozygote

Partial zygote produced by a process of partial genetic exchange, such as transformation in bacteria.

❑ mesoderm

The middle germ layer that forms in the early animal embryo and gives rise to parts such as bone and connective tissue.

❑ mesophile

A micro-organism able to grow in the temperature range 20-50 °C; optimal growth often occurs at about 37 °C.

❑ mesophyll

Leaf parenchyma tissue found between epidermal layers.

❑ messenger RNA

Abbreviation: mRNA.

❑ metabolic cell

A cell that is not dividing.

❑ metabolism

The biochemical processes whereby nutritive material is converted to living matter, or aids in building living matter, or by which complex substances and food are broken down into simple substances.

❑ metabolite

A low-molecular-weight biological compound that is usually synthesized enzymically.

❑ metabolomics

The large-scale study of the full complement of secondary metabolites produced by a given species in all its tissues and growth stages.

❑ metacentric chromosome

A chromosome in which the centromere is located in the middle and, consequently, the chromosome arms are of about equal length.

❐ metal affinity chromatography

A chromatographic technique, in which a compound interacting with a specific metal ion can be captured by immobilizing the relevant ion on the column's solid matrix.

❐ metalloenzyme

An enzyme which requires the presence of a metal in order to be catalytically active.

❐ metallothionein

A protective protein that binds heavy metals such as cadmium and lead.

❐ metaphase

The stage of cell division when chromosomes are aligned at the cell center prior to separation. A "metaphase spread" refers to the view of a cell's chromosomes in the metaphase stage on a slide.

Mitosis

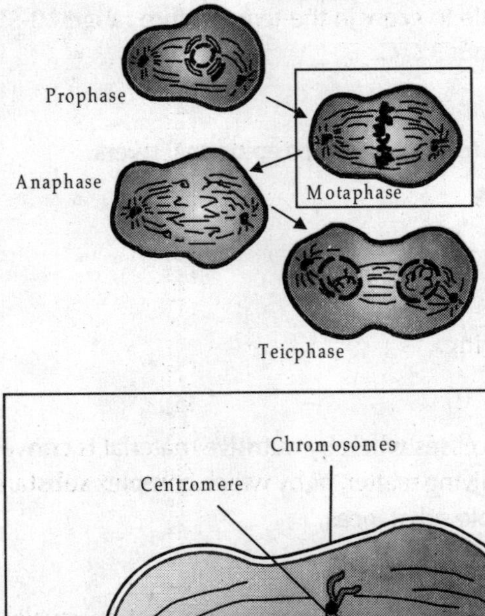

Metaphase

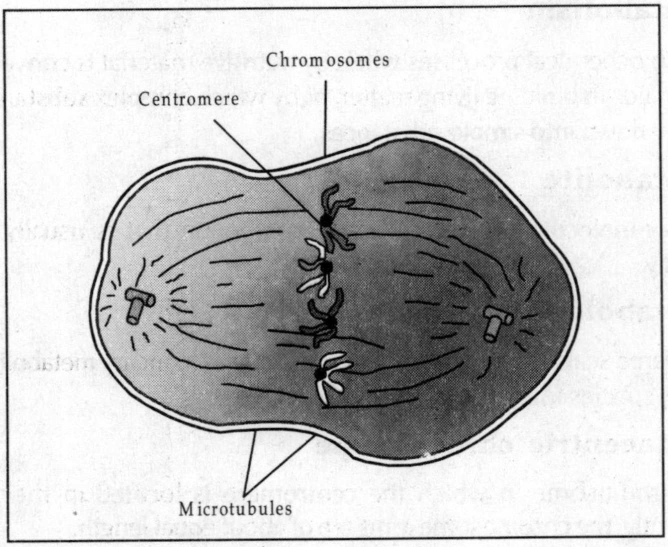

❒ metastasis

The spread of cancer cells to previously unaffected organs.

❒ methylation

The addition of a methyl group (-CH_3) to a molecule, most commonly in the context of DNA where cytosine and, less often, adenine residues can be modified in this way, resulting in a change in transcription.

❒ mhc

Abbreviation for major histocompatibility complex.

❒ microalgal culture

Culture in bioreactors of microalgae (including seaweeds).

❒ micro-array

A large set of cloned DNA molecules immobilized as a compact and orderly pattern of sub-microlitre spots onto a solid matrix (typically a glass slide). Used to analyse patterns of gene expression, presence of markers, or nucleotide sequence. The major advantage of micro-arrays is the extent to which the process of genotyping can be automated, thereby enabling large numbers of individuals to be simultaneously genotyped at many loci. A similar approach may be used with other immobilized components for other purposes.

❒ microarray technology

A new way of studying how large numbers of genes interact with each other and how a cell's regulatory networks control vast batteries of genes simultaneously. The method uses a robot to precisely apply tiny droplets containing functional DNA to glass slides. Researchers then attach fluorescent labels to DNA from the cell they are studying. The labeled probes are allowed to bind to complementary DNA strands on the slides. The slides are put into a scanning microscope that can measure the brightness of each fluorescent dot; brightness reveals how much of a specific DNA fragment is present, an indicator of how active it is.

Microarray is based on a database of over 40,000 fragments of genes called expressed sequence tags (ESTs). Minuscule amounts of hundreds or thousands of these ESTs are arranged on a single microscope slide by a robot. Next the genetic messenger molecules, which signal the production of proteins, from a particular cell are labeled with fluorescent tags and allowed to hybridize, or bind, to the ESTs on the slide whose sequences are complementary to the those of the messengers. After a scanner measures the fluorescence of each sample on the slide, scientists can determine how active the genes represented by the ESTs are in the cell. Strong fluorescence indicates that many of the cell's messengers hybridized to the EST and, therefore, that the gene is very active in the cell. Conversely, no fluorescence indicates that none of the messenger molecules hybridized to the EST and that the gene is inactive in the cell.

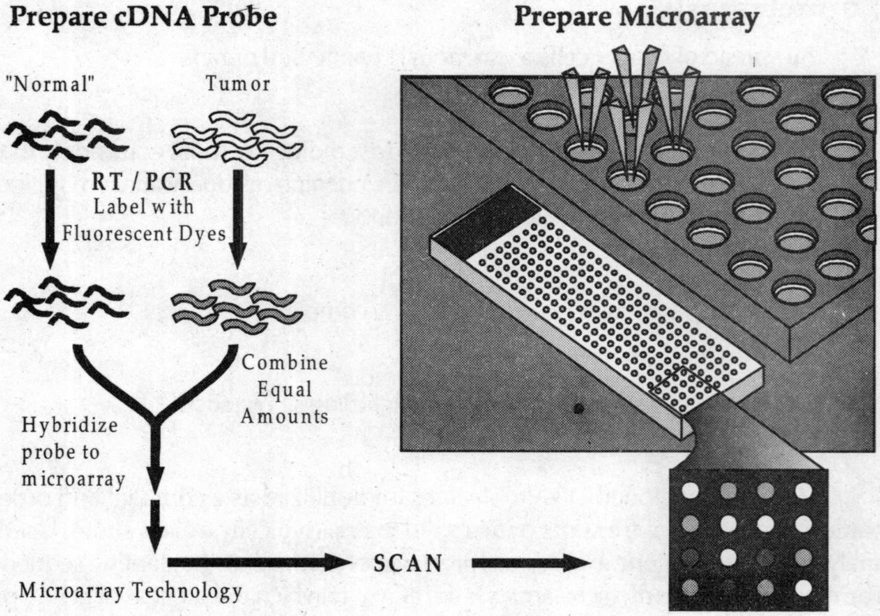

▫ microbial mat

Layered microbial populations, usually growing on the surface of a solid medium or on a membrane.

▫ microbody

A frequently spherical cellular organelle, bound by a single membrane, 20-60 nm in diameter, and containing a variety of enzymes.

▫ micro-carrier

Small particles used as a support material for (particularly mammalian) cells, which are too fragile to be pumped and stirred as bacterial cells are in a large-scale culture.

▫ microdissection

A method for dissecting and cloning from defined subchromosomal regions by microscopic examination and manipulation.

▫ microdroplet array

A technique used to simultaneously evaluate large numbers of media modifications, employing small quantities of medium into which are placed small numbers of cells or protoplasts. These form a monolayer at the droplet meniscus and can easily be examined. *Synonyms*: multiple drop array, hanging droplet technique.

▫ micro-element

A nutritional element required in very small quantities.

❐ micro-encapsulation

A process of enclosing a substance in very small sealed capsules from which material is released by heat, solution or other means.

❐ micro-environment

A small-scale environment in which the conditions (temperature, humidity, pH etc.) are distinct; typically used in connection with the surroundings of a living object.

❐ microfibril

Microscopic fibres visible only at the high magnification of the electron microscope.

❐ micro-injection

The introduction of small amounts of (usually) liquid material (DNA, RNA, enzymes, cytotoxic agents) into a defined tissue or single cell with a fine, microscopic needle.

❐ micro-isolating system

Mechanical separation of single cells or protoplasts thus allowing them to proliferate individually.

❐ micronucleus

A nucleus, distinct from and smaller than the main nucleus, but lying within the same cell. They usually arise following abnormal meiotic or mitotic telophases, where individual chromosomes or chromosome fragments do not reach the pole.

❐ micronutrient

An essential element normally required in cell culture at concentrations under 0.5 millimole/litre.

❐ micro-organism

Organism visible only under magnification.

❐ microplast

Vesicle produced by subdivision and fragmentation of protoplasts or thin-walled cells.

❐ micropropagation

Miniaturized *in vitro* multiplication and/or regeneration of plant material under aseptic and controlled environmental conditions.

❐ micropyle

1. A small opening in the surface of a plant ovule through which the pollen tube passes prior to fertilization.
2. A small pore in some animal cells or tissues.

❑ microsatellite

A segment of DNA characterized by a variable number of copies(typically 5-50) of a sequence of around 5 or fewer bases (called a repeat unit). At any one locus (genomic site), there are usually several different "alleles" in a population, each allele identifiable according to the number of repeat units. This existence of multiple alleles (high level of polymorphism) has enabled microsatellites to be developed as powerful markers in many different species. They are detected by the polymerase chain reaction.

❑ microspore

The immature male gametophyte in seed plants; the haploid male cell that ripens into a pollen grain.

❑ microtuber

Miniature tuber, produced in tissue culture, which is readily regenerable into a normal tuberous plant.

❑ microtubule

Self-assembling components of the cytoskeleton. Microtubules are cylindrical protein polymers, interconnected by cross-bridging proteins, which structurally and dynamically organize functional activities in living cells. They form the spindle during mitosis.

❑ middle lamella

A thin exclusively plant membrane separating two adjacent protoplasts and remaining as a distinct cementing layer between adjacent cell walls.

❑ mid-parent value

The average of the phenotypic measure, with respect to a given trait, of the two parents used to generate the population being analysed.

❑ mineralization

The conversion of organic compounds into inorganic (mineral) ones. For example, the conversion of ethanol into carbon dioxide and water.

❑ minimum effective cell density

The cell density below which reproducible cell growth fails. The minimum density is determined by both the source function of the tissue (species, explant, cell line) and by the culture phase of the initial inoculum.

❑ minimum inoculum size

The critical volume of inoculum required to initiate culture growth, due to the diffusive loss of cell materials into the medium. The subsequent culture growth cycle is dependent on the inoculum size, which is determined by the volume of medium and size of the culture vessel.

❑ mini-prep

A small-scale preparation of plasmid or phage DNA. Used to purify cloned DNA from the DNA of the vector.

❐ minisatellite

A form of variable number tandem repeats in which the repeat unit size ranges from 10-100 bp. Used for DNA fingerprinting following Southern hybridization. Generally concentrated at the ends of chromosomes and in regions with a high frequency of recombination.

❐ minituber

Small tubers (5-15 mm in diameter) formed on shoot cultures or cuttings of tuber-forming crops, such as potato.

❐ mismatch repair

A DNA repair process that corrects mismatched base pairs.

❐ mismatch

The occurrence of a non-complementary pairs of bases in a double helix of DNA, e.g. A:C, G:T.

❐ missense mutation

A mutation that changes a codon for one amino acid into a codon specifying another amino acid.

❐ mist propagation

Application of fine droplets of water to maintain humidity around plantlets or cuttings, which have not yet developed effective roots.

❐ mite

Free-living and parasitic arachnids. Infestation of plant crops reduces yield through the destruction of leaf tissue. Also can infest plant tissue culture work areas, contaminating culture vessels, thereby spreading bacteria and fungi.

❐ mitochondrial DNA

A circular DNA found in mitochondria. In mammals, mtDNA makes up less than 1% of the total DNA, but in plants the amount is variable. It encodes rRNA and tRNA and some mitochondrial proteins (up to 30 in animals).

❐ mitochondrion

Organelle possessing its own DNA which appear in all eukaryotic cells (and never in prokaryotic cells) and produce adenosine triphosphate as an energy source for the cell via oxidative phosphorylation. Mitochondria contain many enzymes of the respiratory cycle, although most of these proteins are nuclear encoded.

❐ mitogen

A substance that can cause cells to initiate mitosis.

❑ mitosis

Overview of the major events in mitosis: In biotechnology, mitosis is the process of chromosome segregation and nuclear division that follows replication of the genetic material in eukaryotic cells. This process assures that each daughter nucleus receives a complete copy of the organism's genome. In most eukaryotes mitosis is accompanied with cell division or cytokinesis, but there are many exceptions, for instance among the fungi. There is another process called meiosis, in which the daughter nuclei receive half the chromosomes of the parent, which is involved in gamete formation and other similar processes.

Cell cycle: Interphase followed by mitosis

Mitosis is divided into several stages, with the remainder of the cell's growth cycle considered interphase. Properly speaking, a typical cell cycle involves a series of stages: G1, the first growth phase; S, where the genetic material is duplicated; G2, the second growth phase; and M, where the nucleus divides through mitosis. Mitosis is divided into prophase, prometaphase, metaphase, anaphase, and telophase.

Diagram showing the changes which occur in the centrosomes and nucleus of a cell in the process of mitotic division. I to III, prophase; IV, metaphase; V and VI, anaphase; VII and VIII, telophase.

The whole procedure is very similar among most eukaryotes, with only minor variations. As prokaryotes lack a nucleus and only have a single chromosome with no centromere, they cannot be properly said to undergo mitosis. Instead they undergo what is known as binary fission.

Phases

Prophase: The genetic material (DNA), which normally exists in the form of chromatin condenses into a highly ordered structure called a chromosome. Since the genetic material has been duplicated, there are two identical copies of each chromosome in the cell. Identical chromosomes (called sister chromosomes) are attached to each other at a DNA element present on every chromosome called the centromere. A protein called cohesin acts as a glue joining two sister chromosomes (a.k.a. sister chromatids) along their lengths. When chromosomes are paired up and attached, each individual chromosome in the pair is called a chromatid, while the whole unit (confusingly) is called a chromosome. Just to be even more confusing, when the chromatids separate, they are no longer called chromatids, but are called chromosomes again. The goal of mitosis is to assure that one copy of each sister chromatid - and only one copy - goes to each daughter cell after cell division.

The other important piece of hardware in mitosis is the centriole. Each centrosome contains two centrioles. The two centrosomes of the cell serve as organizing centers for microtubules. During prophase, the two centrosomes - which replicate independently of mitosis - begin recruiting microtubules (which may be thought of as cellular ropes or poles) and forming mitotic spindles. By increasing the length of the spindle (growing the microtubules), the centrioles push apart to opposite ends of the cell nucleus. It should be noted that many eukaryotes, for instance plants, lack centrioles, although the basic process is still similar.

Prometaphase: Some biology texts do not include this phase, considering it a part of prophase. In this phase, the nuclear membrane dissolves in some eukaryotes, reforming later once mitosis is complete. This is called open mitosis, found in most multicellular forms. Many protists undergo closed mitosis, in which the nuclear membrane persists throughout.

Now kinetochores begin to form at the centromeres. This is a complex protein structure that may be thought of as an 'eyelet' for the microtubule 'rope' - it is the attaching point by which chromosomes may be secured. The kinetochore is an enormously complex structure that is not yet fully understood. The key feature of a kinetochore is that it contains a molecular motor that uses energy from ATP molecules to move a chromatid along a microtubule, like a train on a train track. Two kinetochores form on each chromosome - one for each chromatid.

When the spindle grows to sufficient length, the microtubules begin searching for kinetochores to attach to.

Metaphase: As microtubules find and attach to kinetochores, they begin to line up in the middle of the cell. Proper segregation requires that every kinetochore be attached to a microtubule before separation begins. It is thought that unattached kinetochores control this process by generating a signal - the mitotic spindle checkpoint - that tells the cell to wait before proceeding to anaphase. There are many theories as to how this is accomplished, some of them involving the generation of tension when both microtubules are attached to the kinetochore.

When chromosomes are bivalently attached - when both kinetochores are attached to microtubules emanating from each centriole - they line up in the middle of the spindle, forming what is called the metaphase plate. This does not occur in every organism - in some cases chromosomes move back and forth between the centrioles randomly, only roughly lining up along the midline.

Anaphase: When every kinetochore is attached to a microtubule and the chromosomes have lined up along the middle of the spindle, the cell proceeds to anaphase. This is divided into two phases. First, the proteins that bind the sister chromatids together are cloven, allowing them to separate. They are pulled apart by the microtubules, towards the respective centrioles to which they are attached. Next, the spindle axis elongates, driving the centrioles (and the set of chromosomes to which they are attached) apart to opposite ends of the cell. These two stages are sometimes called 'early' and 'late' anaphase.

At the end of anaphase, the cell has succeeded in separating identical copies of the genetic material into two distinct populations.

Telophase: Daughter chromosomes arrive at poles by this stage. The nuclear membrane reforms around the genetic material and the chromosomes are unfolded back into chromatin. This is often followed by cytokinesis or cleavage, where the cellular membrane pinches off between the two newly separated nuclei, to form two new daughter cells. At this point, each new nucleus contains one copy of each chromosome. Mitosis is complete.

Endomitosis: Endomitosis is a process that occurs when chromosomal replication takes place without nuclear or cellular division, resulting in cells with many copies of the same chromosome. This process may also be referred to as endoreplication.

❐ mixed bud

A bud containing both rudimentary leaves and flowers.

❐ mixoploid

Groups of cells with variable (a mix of euploid, aneuploid, polyploid) chromosome numbers.

❐ mobilization

1. The transfer between bacteria of a non-conjugative plasmid by a conjugative plasmid.
2. The transfer between bacteria of chromosomal genes by a conjugative plasmid.

❐ mobilizing function

The genes on a plasmid that give it the ability to facilitate the transfer of either a non-conjugative or a conjugative plasmid from one bacterium to another.

❐ mode

In a frequency distribution, the class having the greatest frequency.

❐ model

1. A mathematical description of a biological phenomenon.
2. A simplified biological system used to test hypotheses (e.g. *Arabidopsis thaliana* as a model plant).

❐ modification

Enzymatic attachment of one or more chemical groups to a macromolecule, affecting its biological activity or properties. *See*: methylation, glycosylation, phosphorylation.

❐ modifying gene

A gene that affects the expression of some other gene.

❐ moet

Abbreviation for multiple ovulation and embryo transfer.

❐ molecular biology

The study of living processes at the molecular level.

❐ molecular cloning

The biological amplification of a DNA sequence via the mitotic division of a host cell into which it has been transformed or transfected.

❐ molecular genetics

The study of the expression, regulation and inheritance of genes at the level of DNA and its transcription products.

❐ molecular marker

A genetic marker which is assayed at the DNA level.

❐ molecule

The stable union of two or more atoms; some organic molecules contain very large numbers of atoms.

❐ molecules drawn to scale

Small molecules, proteins, and a ribosome drawn to scale. Ribosomes are part of the machinery the cell uses to make proteins: each ribosome is composed of about 90 macromolecules (protein and RNA molecules).

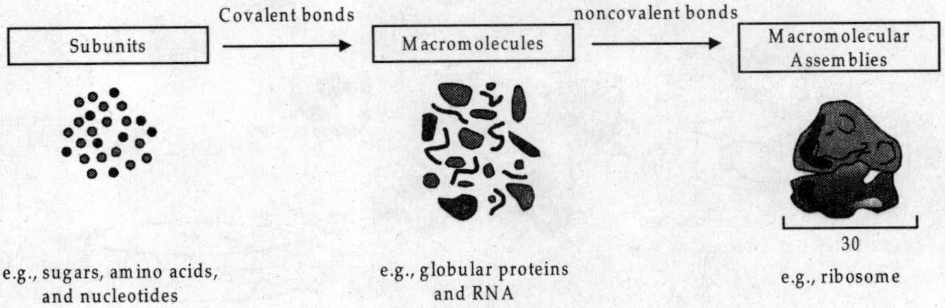

❐ monoclonal antibody

An antibody, produced by a hybridoma, directed against a single antigenic determinant of an antigen.

❐ monoclonal antibody production

Process by which large quantities of antibodies (targeted against a particular antigen X) can be produced.

A mouse is immunized by injection of an antigen X to stimulate the production of antibodies targeted against X. The antibody forming cells are isolated from the mouse's spleen.

Monoclonal antibodies are produced by fusing single antibody-forming cells to tumor cells grown in culture. The resulting cell is called a hybridoma.

Each hybridoma produces relatively large quantities of identical antibody molecules. By allowing the hybridoma to multiply in culture, it is possible to produce a population of cells, each of which produces identical antibody molecules. These antibodies are called "monoclonal antibodies" because they are produced by the identical offspring of a single, cloned antibody producing cell.

Once a monoclonal antibody is made, it can be used as a specific probe to track down and purify the specific protein that induced its formation.

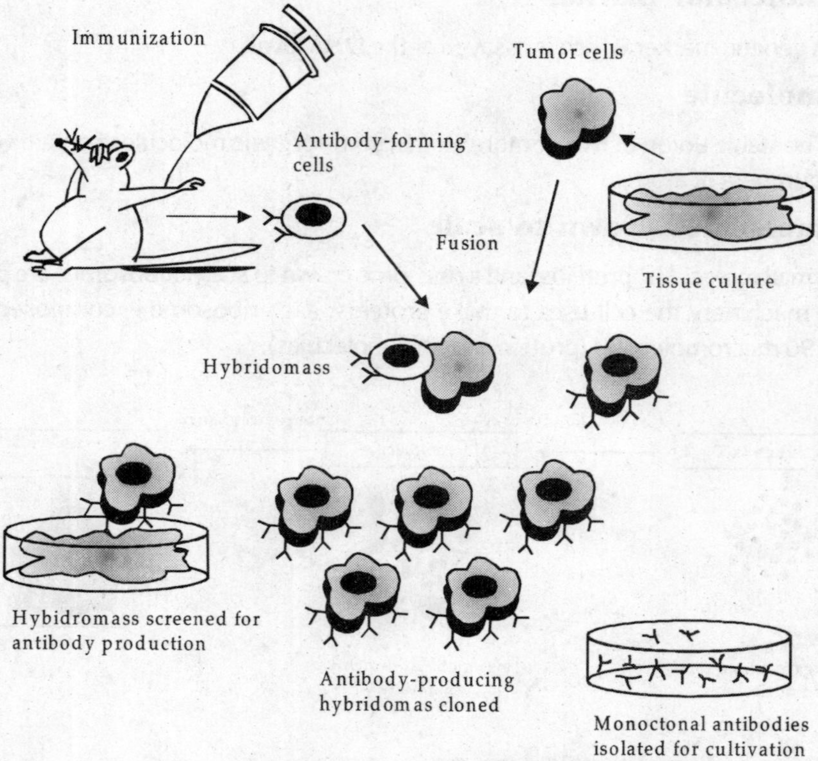

Monoclonal Antibody Production

❐ **monocot**

Abbreviation for monocotyledon.

❐ **monocotyledon**

A flowering plant whose embryo has one cotyledon. Examples are cereals (corn, wheat, rice etc.), banana, and lily.

❐ **monoculture**

The agricultural practice of cultivating a single crop over a whole farm or area.

❐ **monoecious**

A plant species that has separate male and female flowers on the same plant (e.g. maize).

❐ **monogastric animal**

A non-ruminant animal with a simple stomach.

❐ **monogenic**

Trait controlled by a single gene.

☐ monohybrid cross

A cross between parents differing in only one trait or in which only one trait is being considered.

☐ monohybrid

Heterozygous with respect to one gene.

☐ monokine

Generic name for proteins that are released by monocytes to act on other cells involved in the immune response. A sub-class of cytokines.

☐ monolayer

A single layer of cells growing on a surface.

☐ monolignols

The building blocks of lignin that undergo polymerization.

☐ monomer

A small molecule (in the biological sciences typically individual amino acids, nucleotides or monosaccharides) that can combine with identical or similar others to form a larger, more complex molecule called a polymer.

☐ monomorphic

Absence of variation for a marker, gene, chromosome, or genetically determined trait in a population.

☐ monophyletic

A group of organisms that are assumed to have originated from the same ancestor.

☐ monosaccharide

A simple sugar (e.g. glucose, fructose).

☐ monosomic

A form of aneuploidy in which a diploid organism lacks one member of a homologous chromosome pair.

☐ mono-unsaturates

Oils containing mono-unsaturated fatty acids (i.e. where one $-CH_2-CH_2-$ group in the hydrocarbon chain is replaced by $-CH=CH-$).

☐ monozygotic twin

One of a pair of twins derived from a single fertilized egg.

☐ morphogen

A substance that stimulates the development of form or structure in an organism.

❐ morphogenesis

The development, through growth and differentiation, of form and structure in an organism.

❐ morphogenic response

The effect on the developmental history of a plant or its parts exposed to a given set of growth conditions or to a change in the environment.

❐ morphology

Shape, form, external structure or arrangement.

❐ mosaic

An organism or part of an organism that is composed of cells with different origin.

❐ motif

A conserved sequence of nucleotides or amino acids that can be associated with some function of, DNA or a protein respectively.

❐ mrna

Abbreviation for messenger RNA. The RNA molecule resulting from transcription of a protein-encoding gene, following any splicing:

1. The information encoded in the mRNA molecule is translated into a Protein product by the ribosomes.

Template for protein synthesis. Each set of three bases, called codons, specifies a certain protein in the sequence of amino acids that comprise the protein. The sequence of a strand of mRNA is based on the sequence of a complementary strand of DNA.

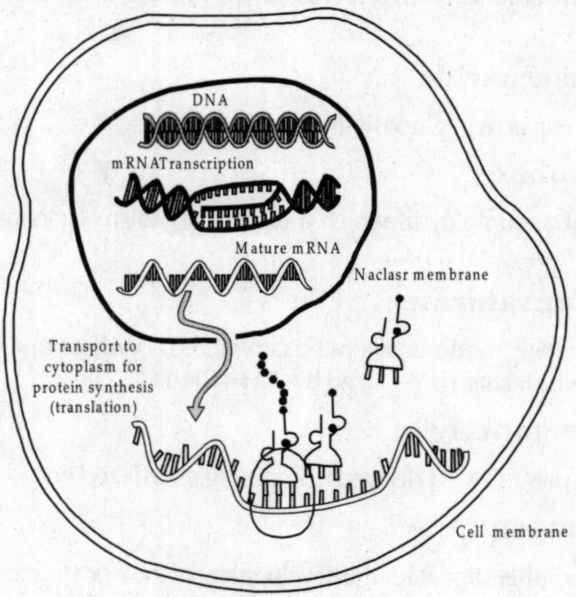

❐ mru

Abbreviation for minimum recognition units.

❐ mtdna

Abbreviation for mitochondrial DNA.

❐ multi-copy

Describing plasmids which replicate to produce many copies per host bacterial cell.

❐ multigene family

A set of genes (not necessarily mapping to the same genomic location) that are related in nucleotide sequence and/or that produce polypeptides with similar amino acid sequences. Sequence similarity does not always result in functional similarity.

❐ multigenic

Trait controlled by several genes, as opposed to monogenic.

❐ multi-locus probe

A DNA sequence that hybridizes to a number of different genomic sites.

❐ multimer

A protein made up of more than one polypeptide chain.

❐ multiple alleles

The existence of more than two alleles at a locus in a population.

❐ multiple arbitrary amplicon profiling

A collective term for a number of related polymerase chain reaction techniques, all of which use arbitrary primers, and which generate a number of distinct amplification products.

❐ multiple ovulation and embryo transfer

A technology by which a single female that usually produces only one or two offspring can produce a litter of offspring. Involves stimulation of a female to shed large numbers of ova; natural mating or artificial insemination; collection of fertilized ova (either surgically, or non-surgically through the cervix); and transfer (usually non-surgical, through the cervix) of these fertilized ova to recipient females.

❐ multiplex

1. The simultaneous amplification of a number of amplicons in a single polymerase chain reaction, achieved by including more than one set of primers in the reaction mix.
2. The inheritance pattern of alleles in autopolyploids.

❐ multivalent vaccine

A vaccine designed to elicit an immune response either to more than one infectious agent or to several different antigenic determinants of a single agent.

❐ muscle contraction: sliding filaments

A. The myosin and actin filaments of a sarcomere overlap with the same relative polarity on either side of the midline. Recall that actin filaments are anchored by their plus ends to the Z disc and that myosin filaments are bipolar.

B. During contraction, the actin and myosin filaments slide past each other without shortening. The sliding motion is driven by the myosin heads walking toward the plus end of the adjacent actin filament.

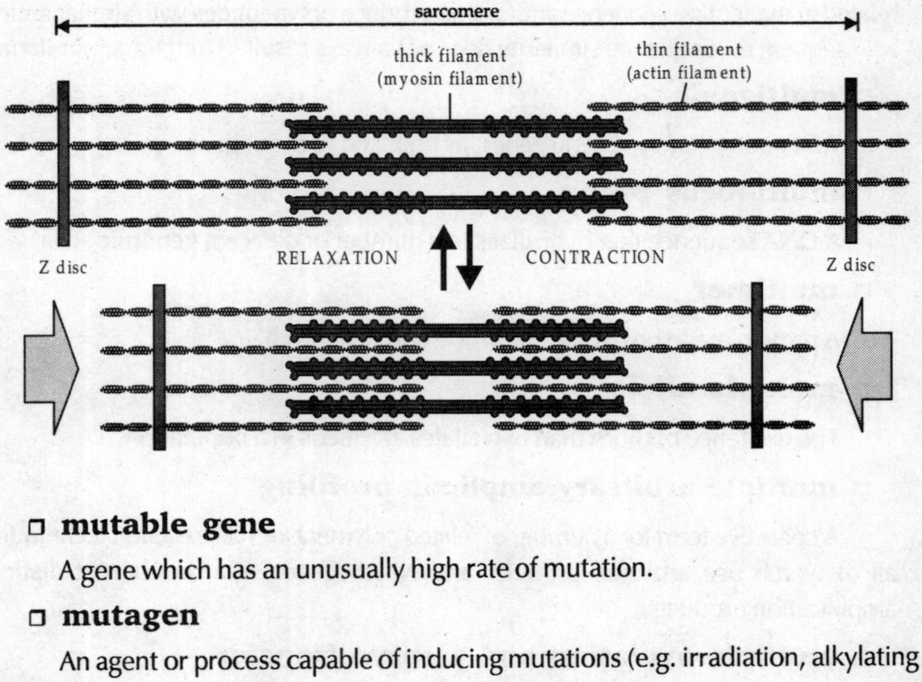

❐ mutable gene

A gene which has an unusually high rate of mutation.

❐ mutagen

An agent or process capable of inducing mutations (e.g. irradiation, alkylating agents).

❐ mutagenesis

Induction of heritable change(s) in the genetic constitution of a cell through alterations to its DNA.

❐ mutant

An organism or an allele bearing a mutation. Usually applied when a characteristic change in phenotype can be recognized.

❐ mutation

Any change in the genome with respect to a defined wild type. Can occur at the level of ploidy, karyotype, or nucleotide sequence. Most of the latter mutations are silent (i.e. cannot be associated with any change in phenotype), either because the DNA sequence affected is in the non-coding part of the genome, or because the specific change does not alter the function of a coding sequence. *See*: back mutation, single nucleotide polymorphism.

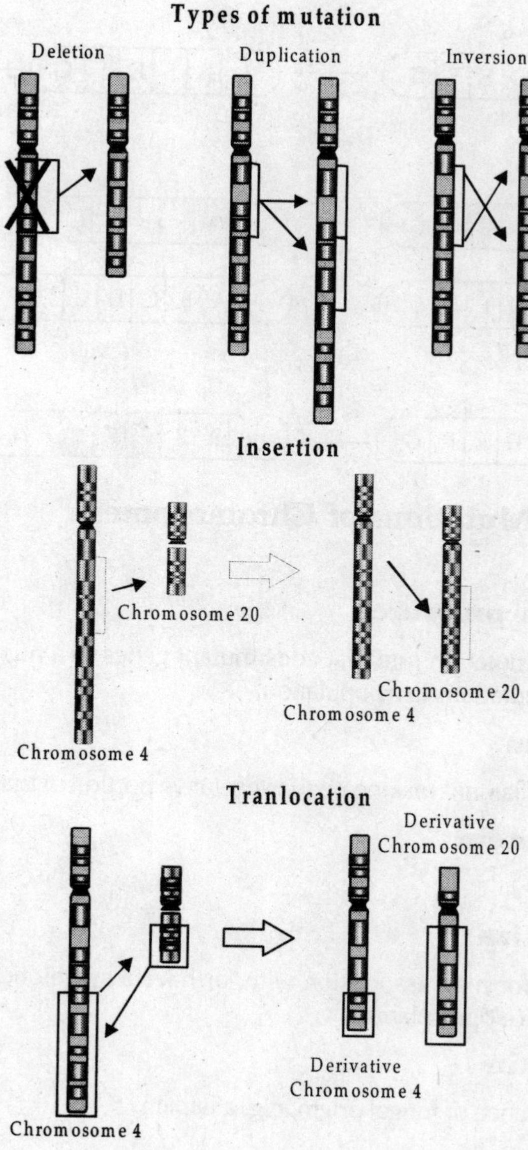

▫ mutation of chromosomes

A permanent structural alteration in DNA. In most cases, such DNA changes either have no effect or cause harm, but occasionally a mutation can improve an organism's chance of surviving and passing the beneficial change on to its descendants.

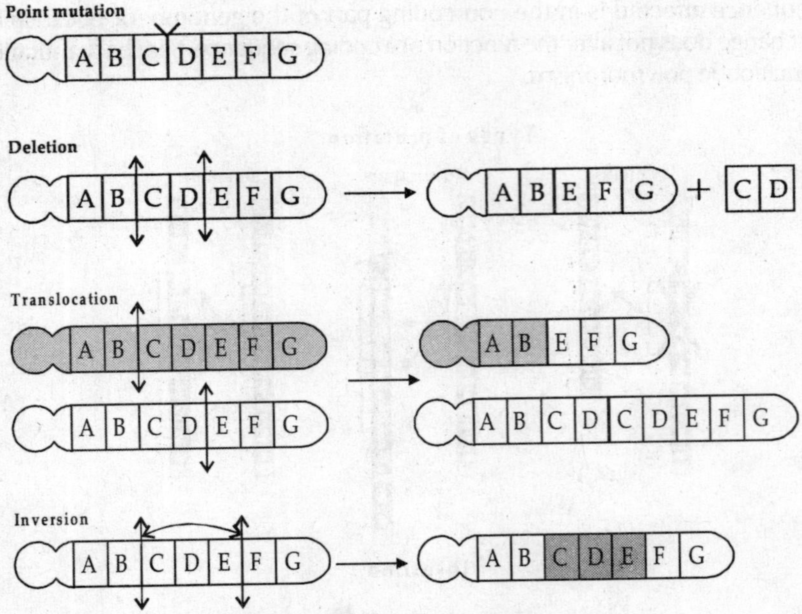

Mutations of Chromosomes

❏ mutation pressure

A constant mutation rate that adds mutant genes to a population; repeated occurrences of mutations in a population.

❏ mycelium

Threadlike filament making up the vegetative portion of thallus fungi.

❏ mycoprotein

Fungal protein.

❏ mycorrhiza

Fungi that form an association with, or have a symbiotic relationship with roots of more developed plants.

❏ mycotoxin

Toxic substance of fungal origin, e.g. aflatoxin.

❏ myeloma

A plasma cell cancer.

❏ myosin and actin model

The regulation of contractility in striated muscle by calcium is effected via tropomyosin on the actin filament. The tropomyosin molecule is a coiled-coil 42 nm in length, which covers seven monomers on one strand of the actin double helix. On each strand these units are disposed end-to-end to form what appears to be a continuous chain. The steric blocking model of Haselgrove and Huxley proposes that myosin binding sites are blocked by tropomyosin·troponin (TmTn) in the absence of calcium, and that tropomyosin moves to a position which allows myosin binding when calcium is bound to troponin-C.

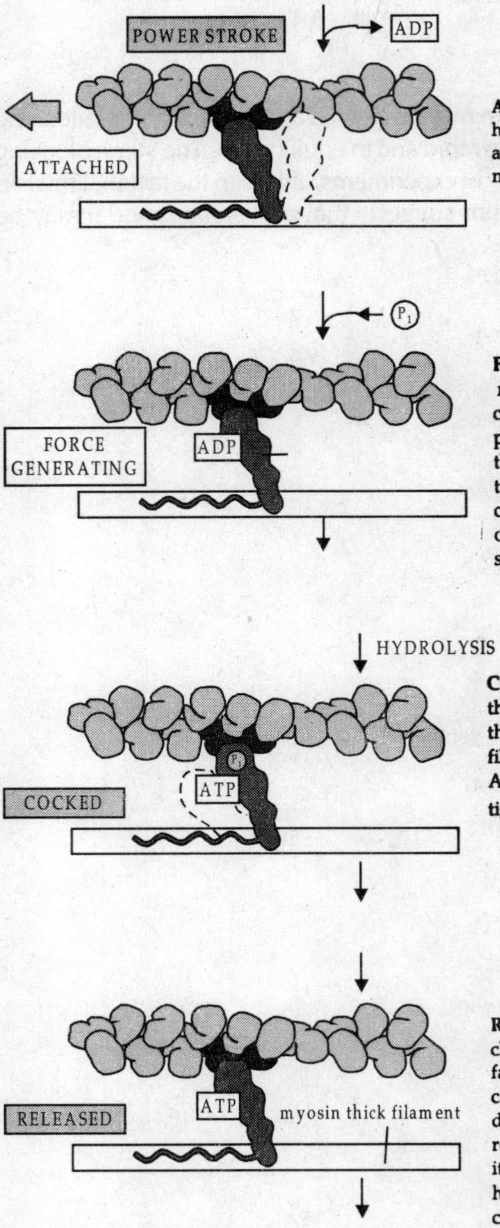

ATTACHED -At the end of the cycle, the myosin head is again locked tightly to the actin filament in a rigor configuration. Note that the head has moved to a new position on the actin filament.

FORCE-GENERATING -The weak binding of the myosin head to a new site on the actin filament causes release of the inorganic phosphate produced by ATP hydrolysis, concomitantly with the tight binding of the head to actin. This release triggers the power stroke-the force-generating change in shape during which the head regains its original conformation. In the course of the power stroke, the head loses

COCKED - The cleft closes like a clam shell around the ATP molecule, triggering a large shape change that causes the head to be displaced along the filament by a distance of about 5 nm. Hydrolysis of ATP occurs, but the ADP and P_1 produced remain tightly bound to the protein.

RELEASED- A molecule of ATP binds to the large cleft on the "back" of the head (that is, on the side farthest from the actin filament) and immediately causes a slight change in the conformation of the domains that make up the actin-binding site. This reduces the affinity of the head for actin and allows it to move along the filament. (The space drawn here between the head and actin emphasizes this change, although in reality the head probably remains very close to the acting.)

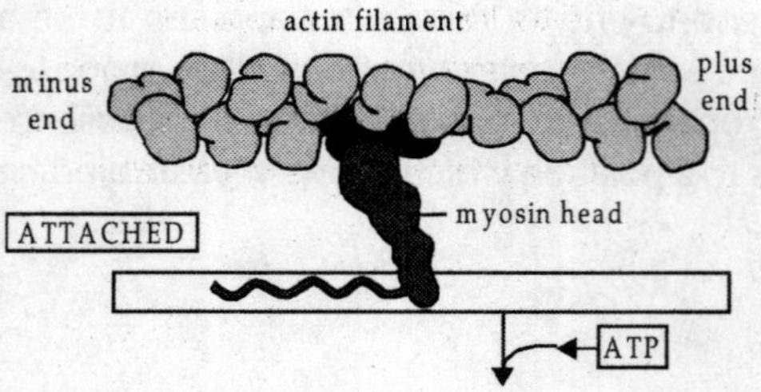

ATTACHED- At the start of te cycle shown in this figure, a myosin head lacking a bound nucleotide is locked tightly onto an actin filament in a rigor configuration (so named because it is responsible for rigor mortis, the rigidity of death. In an actively contracting muscle this state is very short-lived, being rapidly terminated by the binding of a molecule of ATP.

The two positions of tropomyosin may be viewed as two states of an allosteric system if transitions between them are rapid and in equilibrium. The steric blocking model is broadly confirmed by a variety of experiments, although the fact that myosin binds weakly in the absence of calcium suggests that a two-state model may be oversimplified.

N

❐ naked bud

A bud not protected by bud scales.

❐ narrow-host-range plasmid

A plasmid that can replicate in one, or at most a few, different bacterial species.

❐ narrow-sense heritability

The proportion of the phenotypic variance that is due to variation in breeding values; the proportion of the phenotypic variance that is due to additive genetic variance.

❐ native protein

The naturally occurring form of a protein.

❐ natural selection

The differential survival and reproduction of organisms because of differences in characteristics that affect their ability to utilize environmental resources.

❐ necrosis

Death of tissue evidenced by discolouration, dehydration and loss of organization.

❐ negative autogenous regulation

Inhibition of the expression of a gene or set of co-ordinately regulated genes by the product of the gene or the product of one of the genes.

❐ negative control system

A mechanism by which a regulatory protein is required to turn off gene expression.

❐ negative selection

Selection against individuals possessing a certain character.

❐ nematode

Slender, unsegmented worms, often parasitic. Also known as eelworm, especially when phytoparasitic.

❑ neomycin phosphotransferase II

(Abbreviation: npt-II). An enzyme which detoxifies the antibiotic neomycin, used as a marker gene to select for successfully transformed cells in plant transgenesis.

❑ neoplasm

Localized cell multiplication, forming a tumour. Generally the result of genetic transformation. Neoplasmic cells differ in structure and function from the original cell type.

❑ neoteny

The retention of juvenile body characters in the adult state, or the occurrence of adult characters in the juvenile state.

❑ net photosynthesis

Photosynthetic activity minus respiratory activity, measured by the net absorption of carbon dioxide.

❑ neutral mutation

A mutation that changes the nucleotide sequence of a gene, but has no observable effect on the fitness of the organism.

❑ neutral theory

The theory that much of evolution has been primarily due to random drift of neutral mutations.

❑ neutrophil

A type of leukocyte involved in the early inflammatory response.

❑ nft

Abbreviation for nutrient film technique.

❑ nick

To break (*or* a break in) a phosphodiester bond in one of the strands of a double-stranded DNA molecule.

❑ nick translation

A procedure for labelling DNA by treating a fragment with DNAse to produce single-stranded nicks followed by excision of a nucleotide and repair of the gaps with radiolabelled nucleotide.

❑ nicked circle

During the extraction of plasmid DNA from a bacterial cell, one strand of the DNA often becomes nicked. This relaxes the torsional strain which normally ensures a supercoiled structure.

□ *nif* gene cluster

Group of bacterial genes responsible for the biological fixation of atmospheric nitrogen.

□ nitrate

The form of nitrogen that can be used directly by plants; a major component of inorganic fertilizers.

□ nitrification

The natural process in which nitrogen in plant and animal wastes is oxidized, first to nitrites and then to nitrates, through the action of soil-borne microbes.

□ nitrocellulose

A derivative of cellulose, which has the property of binding to many biological macromolecules, in particular DNA, RNA and protein. Filters made from nitrocellulose are commonly used in Southern and northern blotting experiments.

□ nitrogen assimilation

The incorporation of nitrogen into the cells of living organisms.

□ nitrogen fixation

The conversion of atmospheric nitrogen gas to oxidized forms that can be assimilated by plants, particularly by blue-green algae and some genera of bacteria (e.g. *Rhizobium* spp.; *Azotobacter* spp.). An important source of nitrogen in unfertilized soils.

□ nitrogenous base

The purines (adenine and guanine) and pyrimidines (cytosine, and thymine or uracil) that are present in DNA and RNA.

□ no

Abbreviation for nucleolar organizer.

□ *nod* box

A DNA sequence that controls the transcriptional regulation of *Rhizobium* nodulation genes.

□ nodal culture

The culture of a lateral bud and a section of adjacent stem tissue.

□ node

Slightly swollen structure on the stem, where leaves and buds arise and where branches originate. Stems have nodes but roots do not.

□ nodular

Term commonly used to describe the pebbly (rough) texture of a callus.

❐ nodulation

The formation of nodules on the roots of legumes following colonization by symbiotic nitrogen-fixing bacteria.

❐ nodule

Swollen globular structures formed on the roots of legumes, containing the nitrogen-fixing bacteria.

❐ non repetitive dna/rna

A nucleotide sequence which does not include a significant proportion of repetitive sequences of nucleotides.

❐ non-additive genetic variation

The proportion of the total genetic variation in a population that does not respond to simple mass selection and that causes specific pairwise crosses to depart from performance values predicted by the breeding values of the parents.

❐ non-autonomous

A term referring to biological units that cannot function by themselves; such units require the assistance of another unit, or "helper".

❐ non-disjunction

Failure of separation of homologous chromosomes or chromatids in mitosis or meiosis, resulting in too many chromosomes in some daughter cells and too few in others.

❐ non-histone chromosomal protein

In chromosomes, all of the proteins except the histones.

❐ nonsense mutation

A mutation which converts an amino-acid-specifying codon into a stop codon, e.g. a single base change from UAU to UAG generates the premature termination of the polypeptide chain at the position where a tyrosine was incorporated in the wild type.

❐ non-target organism

An organism which is affected by a treatment (e.g. pesticide application) for which it was not the intended recipient.

❐ non-template strand

The non-transcribed strand of DNA.

❐ nor

Abbreviation for nucleolar organizer region.

❐ northern blot

A technique analogous to Southern blotting, but involving the transfer of RNA from an agarose gel to a membrane, prior to probing.

❐ nucellar embryo

An embryo which has developed vegetatively from somatic tissue surrounding the embryo sac, rather than by fertilization of the egg cell.

❐ nucellus

Tissue forming the major part of the young ovule in which the embryo sac develops; Also known as megasporangium.

❐ nuclear transfer

A technology by which novel animals are generated by cloning a single diploid somatic cell. It involves inserting a single diploid cell from a culture of cells into an enucleated ovum. The resultant diploid ovum develops into an embryo that is placed in a recipient female, which gives birth to the cloned animal in the normal manner. Note that the term is somewhat of a misnomer, since it is a whole cell that is transferred, not just the nucleus.

❐ nuclease

A class of largely bacterial enzymes that degrade DNA or RNA molecules by catalysing the cleavage of the phosphodiester bonds that link adjacent nucleotides. For deoxyribonuclease (DNAse) the substrate is DNA, for ribonuclease (RNAse) the substrate is RNA, and for S1 nuclease, the substrate is single-stranded DNA or RNA. Endonucleases cleave at internal sites in the substrate molecule, while exonucleases progressively cleave from the end of the substrate molecule. Nucleases have varying degrees of base-sequence specificity, the most specific being the restriction endonucleases.

❐ nucleic acid

A macromolecule consisting of polymerized nucleotides. Two forms are found, DNA and RNA. Nucleic acids may be linear or circularized, and single- or double-stranded.

❐ nucleic acid hybridization

The two strands of a DNA molecule are denatured by heating to about 100°C = 212°F (a to b). At this temperature, the complementary base pairs that hold the double helix strands together are disrupted and the helix rapidly dissociates into two single strands.

The DNA denaturation is reversible by keeping the two single stands of DNA for a prolonged period at 65°C = 149°F (b to a). This process is called DNA renaturation or hybridization.

Similar hybridization reactions can occur between any single stranded nucleic acid chain: DNA/DNA, RNA/RNA, DNA/RNA. If an RNA transcript is introduced during the renaturation process, the RNA competes with the coding DNA strand and forms double-stranded DNA/RNA hybrid molecule (c to d).

These hybridization reactions can be used to detect and characterize nucleotide sequences using a particular nucleotide sequence as a probe.

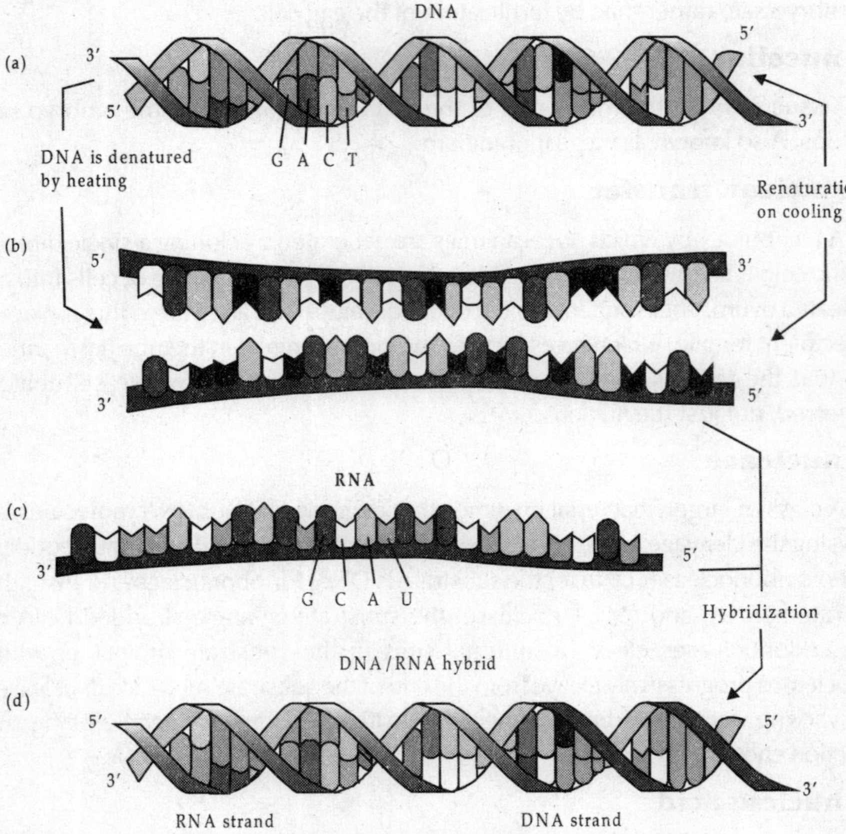

Nucleic Acid Hybridization

❑ nuclein

The term used by Friedrich Miescher to describe the nuclear material he discovered in 1869, which today is known as DNA.

❑ nucleo-cytoplasmic ratio

In a cell, the ratio of nuclear to cytoplasmic volume. This ratio is high in meristematic cells and low in differentiated cells.

❑ nucleolar organizer region

A chromosomal segment containing a large array of genes that encode ribosomal RNA; located at the secondary constriction of specific chromosomes.

❑ nucleolus

An RNA-rich nuclear organelle in the nucleus of eukaryotic cells, produced by a nucleolar organizer. It represents the storage place for ribosomes and ribosome precursors. The nucleolus consists primarily of ribosomal precursor RNA, ribosomal RNA, their associated proteins, and some, perhaps all, of the enzymatic equipment (RNA polymerase, RNA methylase, RNA cleavage enzymes) required for synthesis, conversion and assembly of ribosomes. Subsequently the ribosomes are transported to the cytoplasm.

❐ nucleoplasm

The non-staining or slightly chromophilic, liquid or semi-liquid, ground substance of the interphase nucleus and which fills the nuclear space around the chromosomes and the nucleoli. Little is known of the chemical composition of this ground substance, which is not easily defined. Sometimes called "karyoplasm" when it is gel-like, and "karyolymph" when it is a colloidal fluid.

❐ nucleoprotein

Conjugated protein composed of nucleic acid and protein; the material of which the chromosomes are made.

❐ nucleoside

A base (purine or pyrimidine) that is covalently linked to a 5-carbon (pentose) sugar. When the sugar is ribose, the nucleoside is a ribonucleoside; when it is 2-deoxyribose, the nucleoside is a deoxyribonucleoside. *See*: nucleotide.

❐ nucleoside analogue

A synthetic molecule that resembles a naturally occurring nucleoside.

❐ nucleosome

Nucleosome: Subunit of chromatin composed of a short length of DNA wrapped around a core of histone proteins.

The human genome contains about 3 billion nucleotide pairs organized as 23 chromosomes pairs. If uncoiled, the DNA contained by each of those chromosomes would measure between 1.7 and 8.5 cm (0.67 to 3.35 inches) long. This is too long to fit into a cell. Moreover, if chromosomes were composed of extended DNA, it is difficult to imagine how the DNA could be replicated and segregated into two daughter cells without breaking down.

In fact chromosomal DNA is packaged into a compact structure with the help of specialized proteins called histones. The complex DNA plus histones in eucaryotic cells is called chromatin.

The fundamental packing unit is known as a nucleosome. Each nucleosome is about 11nm in diameter. The DNA double helix wraps around a central core of eight histone protein molecules (an octamer) to form a single nucleosome. A second histone (H1 in the illustration) fastens the DNA to the nucleosome core. The total mass of this complex is about 100,000 daltons.

Nucleosomes are usually packed together, with the aid of a histone (H1,) to form a 30nm large fiber. As a 30nm fiber, the typical human chromosome would be about 0.1cm in length and would span the nucleus 100 times. This suggests higher orders of packaging, to give a chromosome the compact structure seen in a typical karyotype (metaphase) cell.

Spherical sub-units of eukaryotic chromatin that are composed of a core particle consisting of an octamer of histones (two molecules each of histones H_{2a}, H_{2b}, H_3 and H_4) and 146 bp of DNA.

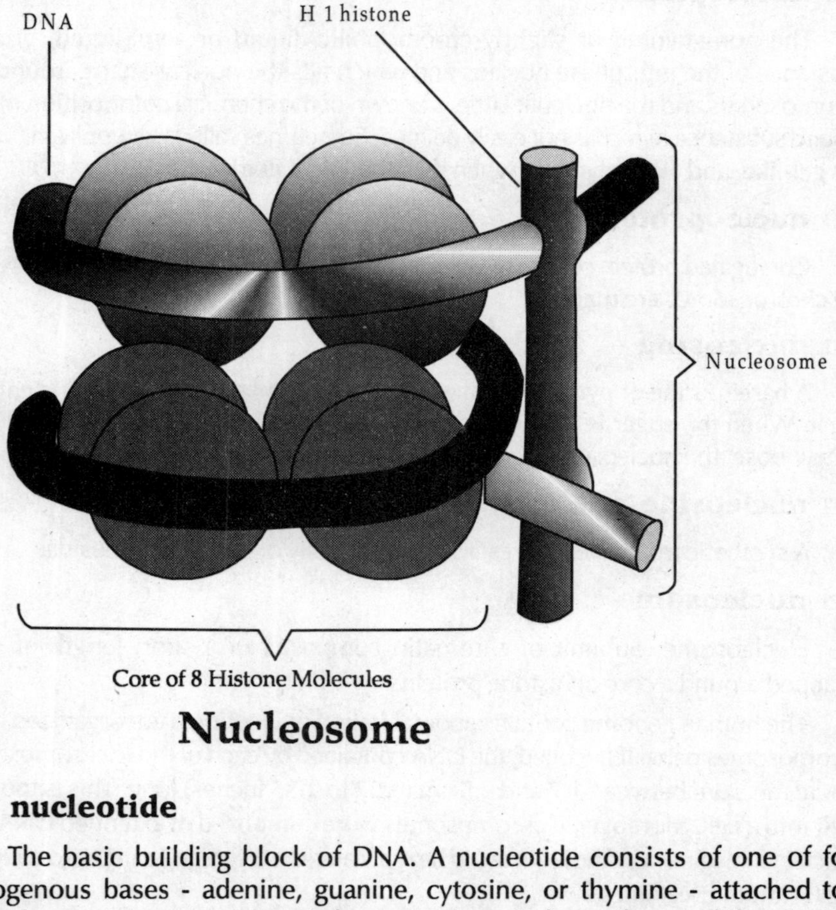

Nucleosome

☐ nucleotide

The basic building block of DNA. A nucleotide consists of one of four nitrogenous bases - adenine, guanine, cytosine, or thymine - attached to a phosphate-sugar group. In DNA the sugar group is deoxyribose, while in RNA (a DNA-related molecule that helps to translate genetic information into proteins), the sugar group is ribose, and the base uracil substitutes for thymine. Each group of three nucleotides in a gene is known as a codon. A nucleic acid is a long chain of nucleotides joined together, and therefore is sometimes referred to as a "polynucleotide."

A nucleoside with one or more phosphate groups linked at the 3'- or 5'-hydroxyl of a pentose sugar. When the sugar is ribose, the nucleotide is a ribonucleotide; when it is 2-deoxyribose, the nucleotide is a deoxyribonucleotide. RNA and DNA are polymers of, respectively, ribonucleoside 5'-monophosphates and deoxyribonucleoside 5'-monophosphates. Nucleotides containing the bases adenine, guanine and cytosine (A, G, C) occur in both DNA and RNA; thymine (T) occurs only in DNA, and uracil

(U) only in RNA. Ribonucleoside mono-, di-, and triphosphates for which a specific base is not assigned are abbreviated NMP, NDP, and NTP, while deoxyribonucleoside mono-, di-, and tri-phosphates are abbreviated dNMP, dNDP, and dNTP. Otherwise, the "N" is replaced by the base letter abbreviation.

Deoxyribonucleic Acid (DNA)

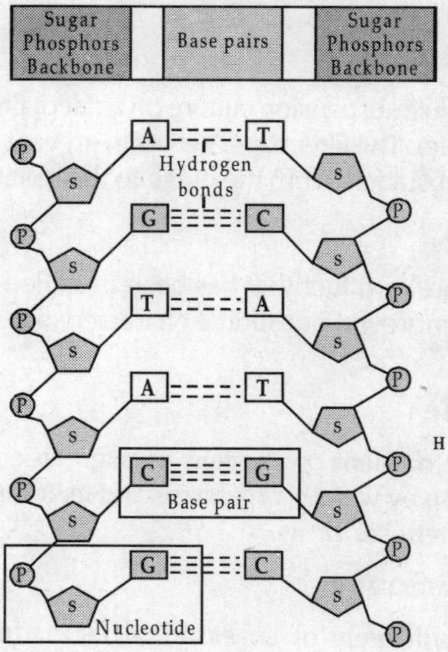

▫ nucleus

A dense protoplasmic membrane-bound region of a eukaryotic cell that contains the chromosomes, separated from the cytoplasm by a membrane; present in all eukaryotic cells except mature sieve-tube elements and red blood cells.

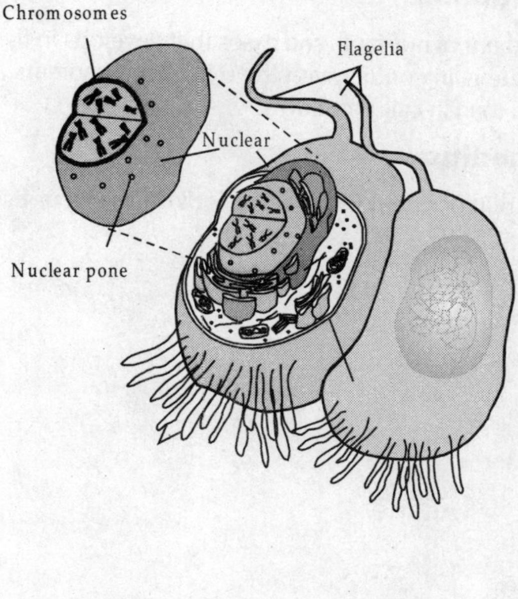

❒ null allele

An allele which produces no functional gene product.

❒ nullisomy

An otherwise diploid cell or organism lacking both members of a homologous chromosome pair.

❒ nurse culture

Culturing cells from a suspension culture on a raft of filter paper above a callus tissue piece (nurse tissue). The filter paper serves to prevent tissue union but allows the flow of essential substances from the nurse to the isolated cells.

❒ nutriceutical

A conventional food product that has been modified (potentially by genetic engineering) to have improved nutritional characteristics and/or pharmaceutical properties.

❒ nutrient cycle

The passage of a nutrient or element through an ecosystem, including its assimilation and release by various organisms and its transformation into various organic or inorganic chemical forms.

❒ nutrient deficiency

Absence or insufficiency of an essential factor for normal growth and development.

❒ nutrient film technique

A hydroponic technique used to grow plants. NFT delivers a thin film of water or nutrient solution either continuously or through on-off cycles.

❒ nutrient gradient

A diffusion gradient of nutrients and gases that develops in tissues where only a portion of the tissue is in contact with the medium. Gradients are less likely to form in liquid media than in callus cultures.

❒ nutrient medium

A solid, semi-solid or liquid formulation for *in vitro* cell or tissue growth.

O

❐ octoploid

An organism, or a tissue whose cells contain eight haploid sets of chromosomes.

❐ oestrogen

The generic term for a group of female sex hormones which control the development of sexual characteristics and control oestrus. *AlteRNAtive spelling*: estrogen.

❐ oestrous cycle

The cycle of reproductive activity shown by most sexually mature non-pregnant female mammals. *AlteRNAtive spelling*: estrous.

❐ oestrus

In female mammals, the period of sexual excitement and acceptance of the male. *Synonyms*: rut, heat. *AlteRNAtive spelling*: estrus.

❐ offset

Young plant produced at the base of a mature plant.

❐ offshoot

Short, usually horizontal, stem produced near the crown of a plant.

❐ offspring

New individuals resulting from sexual or asexual reproduction. *Synonym*: progeny.

❐ okazaki fragment

Replicated DNA fragments formed on the lagging strand in DNA synthesis from a double-stranded template. The fragments are subsequently joined together by DNA ligase.

❐ ola

Abbreviation for oligonucleotide ligation assay.

❐ oligomer

A molecule formed by the covalent joining of a small (undefined) number of monomers.

❐ oligonucleotide

A nucleotide oligomer. Often synthesized for use as primers for *in vitro* DNA synthesis.

❐ oligonucleotide ligation assay

A diagnostic technique for determining the presence or absence of a single nucleotide polymorphism within a target DNA sequence, often indicating whether the gene is wild type (normal) or mutant (usually defective).

❐ oligosaccharide

Carbohydrate consisting of several linked monosaccharide units.

❐ oncogene

A gene that causes cells to grow in an uncontrolled (i.e. tumourous) manner. Oncogenes are mutant forms of normal functional genes (called proto-oncogenes) that have a role in regulating cell proliferation. *See*: cellular oncogene; dominant oncogene; immortalizing oncogene; recessive oncogene, p53 gene.

❐ oncogenesis

The progression of cytological, genetic and cellular changes that culminate in the development of a tumour.

❐ onco-mouse

A mouse that has been genetically modified to incorporate an oncogene, which acts as an animal model for studies of human cancer.

❐ ontogeny

Developmental life history of an organism.

❐ oocyte

The egg mother cell; it undergoes two meiotic divisions (oogenesis) to form the egg cell. The primary oocyte refers to the state before completion of the first meiotic division; the secondary oocyte after completion of the first meiotic division.

❐ oogenesis

The formation and growth of the egg in an animal ovary.

❐ oogonium

1. A germ cell of the female animal, that gives rise to oocytes by mitotic division.
2. The female sex organ of algae and fungi.

❐ oosphere

The non-motile female gamete in plants and some algae.

❐ oospore

A spore developing from a zygote of certain algae and fungi, following the fusion of heterogametes.

❐ open continuous culture

A continuous culture system, in which inflow of fresh medium is balanced by outflow of a corresponding volume of spent medium plus cells. In the steady state, the rate of cell wash-out equals the rate of formation of new cells in the system. *See*: continuous culture; batch culture; closed continuous culture.

❐ open pollination

Pollination by wind, insects or other natural mechanisms.

❐ open reading frame

A sequence of nucleotides in a DNA molecule that has the potential to encode a peptide or protein: comprisesa start triplet (ATG), followed by a series of triplets (each of which encodes an amino acid), and ending with a stop codon (TAA, TAG or TGA). The term is generally applied to sequences of DNA fragments, for which no function has yet been determined. The number of ORFs provides an estimate of the number of genes transcribed from the DNA sequence.

❐ operational definition

An operation or procedure that can be carried out to define or delimit something.

❐ operator

The region of DNA that is upstream from a gene or genes and to which one or more regulatory proteins (repressor or activator) bind to control the expression of the gene(s).

❐ operon

A functionally integrated genetic unit for the control of gene expression in bacteria. It consists of one or more genes that encode one or more polypeptide(s) and the adjacent site (promoter and operator) that controls their expression by regulating the transcription of the structural genes.

❐ opine

The condensation product of an amino acid with either a keto-acid or a sugar, produced by the plant host as a result of *Agrobacterium* infection, and used exclusively by the *Agrobacterium* as a carbon source for growth and reproduction within the plant.

❐ opu

Abbreviation for ovum pickup.

❐ orf

Abbreviation for open reading frame.

❐ organ

A tissue or group of tissues that constitute a morphologically and functionally distinct part of an organism.

❐ organ culture

The aseptic culture of complete living organs of animals and plants outside the body in a suitable culture medium. Animal organs must be small enough to allow the nutrients in the culture medium to penetrate all the cells.

❐ organellar gene

Genes located on organelles outside the nucleus.

❐ organelle

A membrane-bounded specialized region within a cell, such as the mitochondrion or dictyosome, that carries out a specialized function in the life of a cell.

❐ organic complex

A chemically undefined compound added to nutrient media to stimulate growth, e.g. coconut milk; malt extract; yeast extract; casein hydrolysate.

❐ organism

An individual living system, such as animal, plant or micro-organism, that is capable of reproduction, growth and maintenance.

❐ organized growth

The development under tissue culture conditions of organized explants (meristem tips or shoot tips, floral buds or organ primordia).

❐ organized tissue

Composed of normally differentiated cells.

❐ organogenesis

The initiation of adventitious or *de novo* shoots or roots from callus, meristem or suspension cultures.

❐ organoid

An organ-like structure produced in culture.

❐ organoleptic

Having an effect on one of the organs of sense, such as taste or smell.

❐ origin of replication

The nucleotide position on a DNA sequence from which DNA synthesis (replication) is initiated.

❐ orphan gene

A gene or DNA sequence whose function is not known.

❐ orphan receptor

A receptor for which a cellular function or ligand has yet to be identified.

❐ ortet

The plant from which a clone is obtained.

❐ orthologous

Homologous genes/gene products that have evolved divergently *between* species; many rice genes have orthologues in other cereal genomes, because of the common ancestry of cereal species.

❐ osmosis

Diffusion of water across a semi-permeable membrane from zones of low solute concentration to zones of higher solute concentration.

❐ osmotic potential

Change in the energy state of solvent brought about by dissolving a substance in the solvent - water in the biological sciences. The potential of aqueous solutions is always negative compared to pure water. Solvent flows from higher to lower osmotic potential solutions by diffusion or osmosis.

❐ osmotic swelling

Three ways in which cells avoid osmotic swelling. (A) The animal cell keeps the intracellular solute concentration low by pumping out ions. (B) The plant cell is saved from swelling and bursting by its tough wall. (C) The protozoan avoids swelling by periodically ejecting the water that moves into the cell.

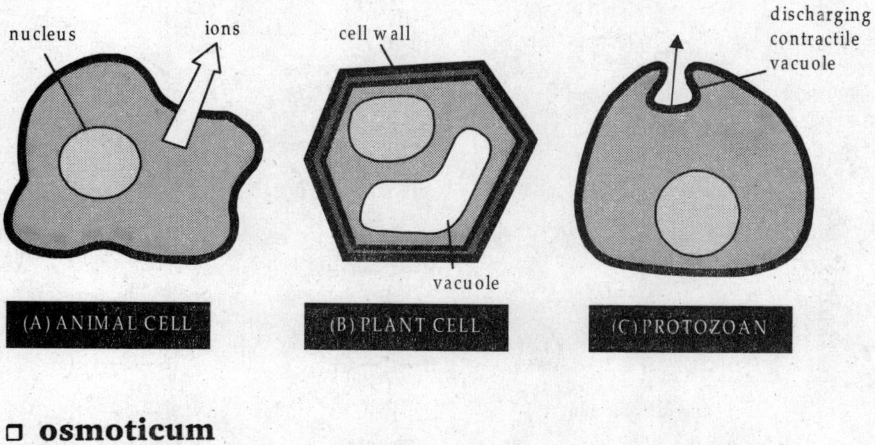

❐ osmoticum

A chemical agent (e.g. polyethylene glycol, mannitol, glucose or sucrose) employed to maintain the osmotic potential of a nutrient medium equivalent to that of the cultured cells, i.e. the medium and the cells are isotonic. Because of this osmotic equilibrium, cells are not damaged *in vitro*.

❒ outbreeding

A mating system characterized by the inter-breeding of genetically unrelated or dissimilar individuals. Since genetic diversity tends to be enhanced, and vigour or fitness of individuals increased by this process, it is often used to counter the detrimental effects of continuous inbreeding.

❒ outflow

The volume of growing cells that is removed from a bioreactor during a continuous fermentation process.

❒ outline of sugar types found in cells part 1

An outline of some of the types of sugars commonly found in cells

Monosaccharides usually have the general formula $(CH_2O)_n$, where n can be 3, 4, 5, 6, 7, or 8, and have two or more hydroxyl groups. They either contain an aldehyde groups are called aldoses or a ketone group and are called ketoses.

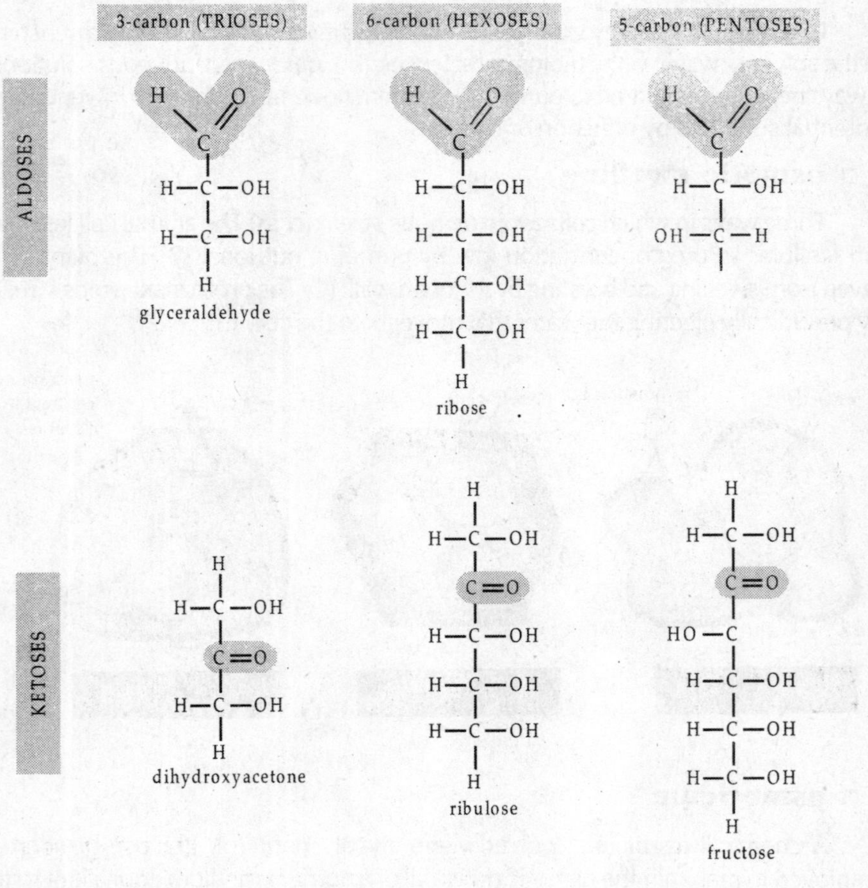

RING FORMATION

In aqueous solution, the aldehyde or ketone group of a sugar molecule tends to react with a hydroxyl group of the same molecule, thereby closing the molecule into a ring.

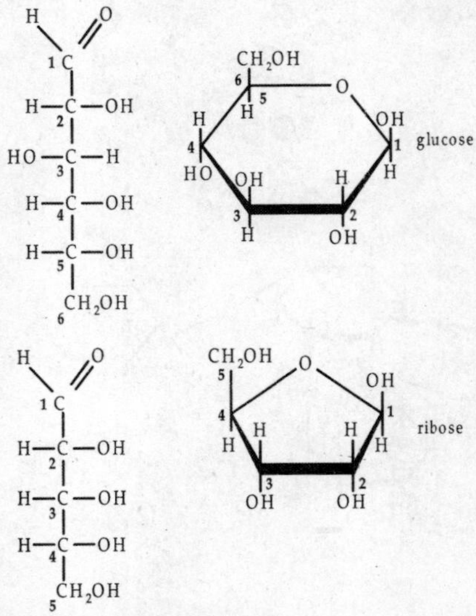

Note that each carbon atom has a number.

ISOMERS

Many monosaccharides differ only in the spatial arrangement of atoms—that is, they are **isomers.** For example, glucose, galactose, and mannose have the same formula ($C_6H_{12}O_6$) but differ in the arrangement of groups around one or two carbon atoms.

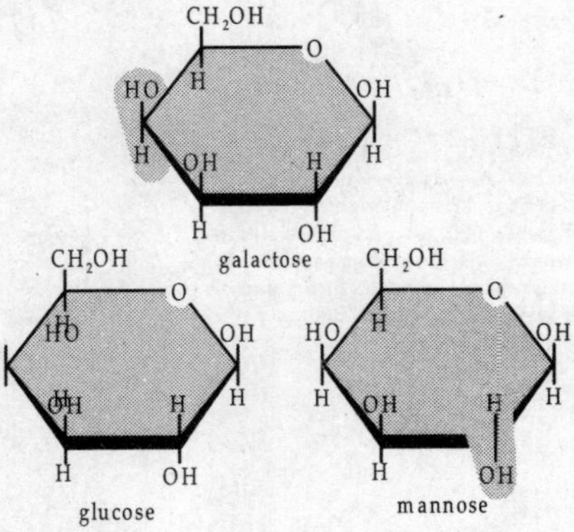

These small differences make only minor changes in the chemical properties of the sugars. But they are recognized by enzymes and other proteins and therefore can have important biological effects.

□ outline of sugar types found in cells part 2

An outline of some of the types of sugars commonly found in cells.

α AND β LINKS

The hydroxyl group on the carbon that carries the aldehyde or ketone can rapidly change from one position to the other. These two positions are called α and β.

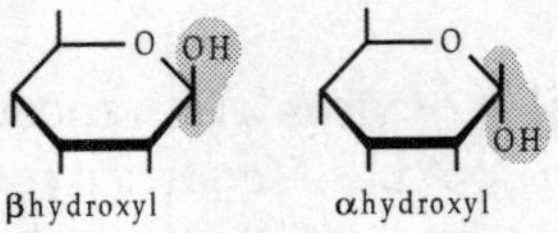

As soon as one sugar is linked to another, the α or β form is [...]

SUGAR DERIVATIVES

The hydroxyl groups of a simple monosaccharide can be replaced by other groups. For example,

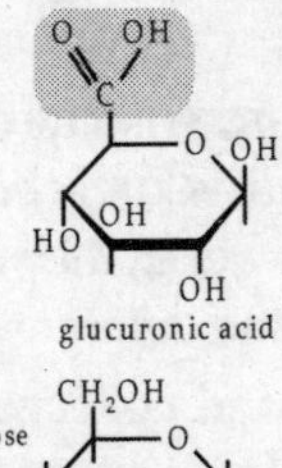

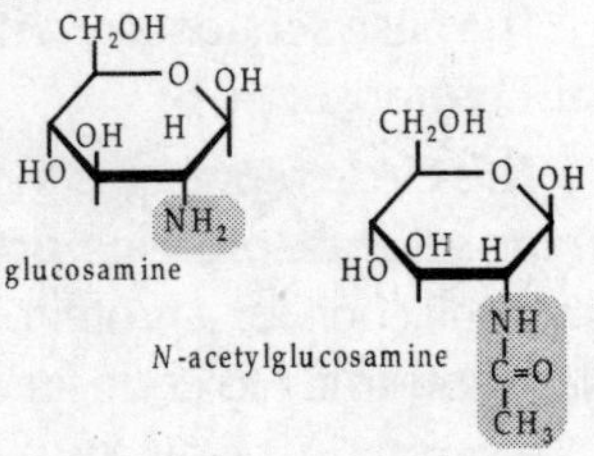

DISACCHARIDES

The carbon that carries the aldehyde or the ketone can react with any hydroxyl group on a second sugar molecule to form a **disaccharide.** Three common disaccharides are

maltose (glucose + glucose)
lactose (galactose + glucose)
sucrose (glucose + fructose)

The reaction forming sucrose is shown here.

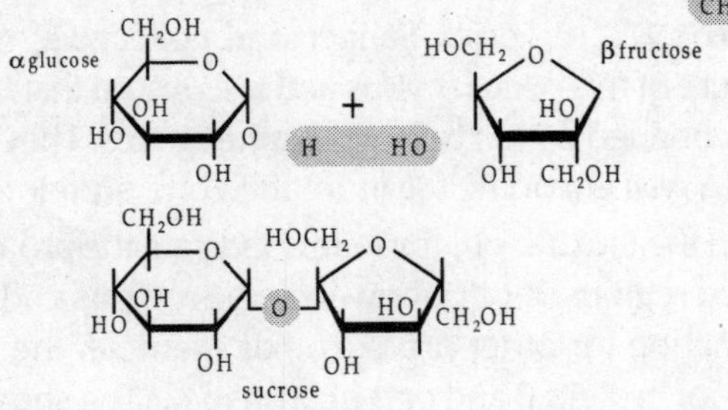

OLIGOSACCHARIDES AND POLYSACCHARIDES

Large linear and branched molecules can be made from simple repeating units. Short chains are called **oligosaccharides**, while long chains are called **polysaccharides**. Glycogen, for example, is a polysaccharide made entirely of glucose units joined together.

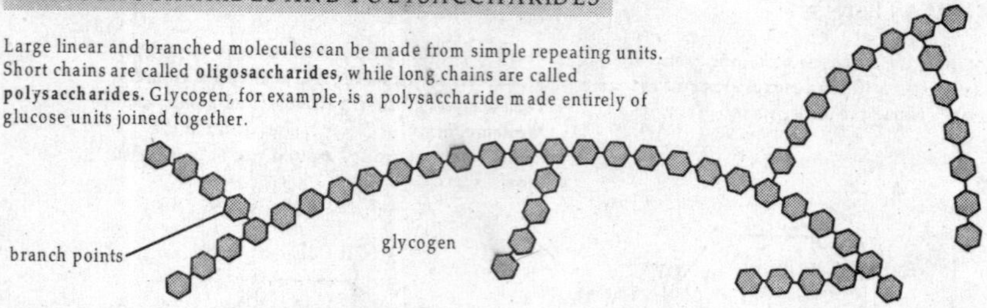

COMPLEX OLIGOSACCHARIDES

In many cases a sugar sequence is nonrepetitive. Many different molecules are possible. Such complex oligosaccharides are usually linked to proteins or to lipids, as is this oligosaccharide, which is part of a cell-surface molecule that defines a particular blood group.

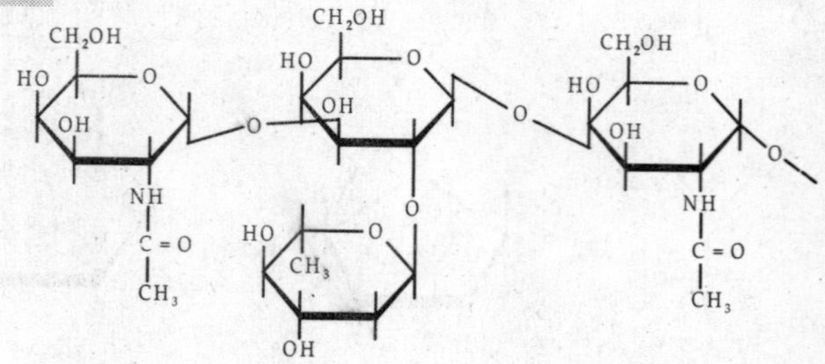

❑ ovary

1. Enlarged basal portion of the pistil of a plant flower that contains the ovules.
2. The reproduction organ in female animals in which eggs are produced.

❑ overdominance

Where heterozygote performance is superior to that of either parental genotype.

❑ overlapping protein codes

The base sequences, which code for proteins, are shown to overlap on a very small chromosome.

This circle represents the chromosome of the **[phi]X174 virus** which encodes for nine separate proteins. Scientists knew how small a DNA base sequence could be and still code for a protein. However, they were puzzled by this little virus whose DNA is too small to code for all the different proteins it produces.

In 1977, Fredrick Sanger and colleagues mapped out the complete DNA structure of this peculiar virus and discovered that the codings for two of the proteins are embedded within two of the other genes. They discovered that the same stretch of DNA was encoding for more than one protein!

In the illustration, the letters indicate the proteins coded for by the genes in the colored regions. Notice how in some regions coding for one protein also includes information for other proteins. For example, the region coding for protein A also codes for protein B and part of protein K. (The short black areas indicate parts of the chromosome that do not code for proteins).

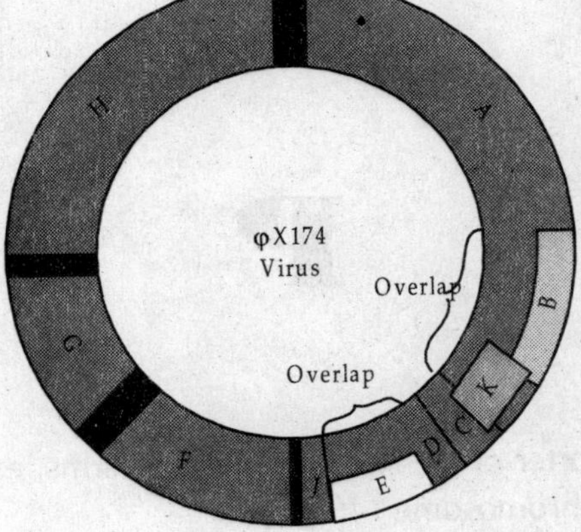

Overlapping Protein Codes

❑ Overlapping reading frame

Start codons in different reading frames generate different polypeptides from the same DNA sequence.

❑ ovulation

The release of mammalian egg(s) from the ovary.

❑ ovule

The part of the seed plant's reproductive organs that comprises the nucellus, the embryo sac and the integuments.

❑ ovum pickup

The non-surgical collection of ova from a female.

❑ oxidative phosphorylation

The enzymatic addition of a phosphate to ADP to make ATP, coupled to electron transport from a substrate to molecular oxygen. A critical reaction for the generation of cellular energy.

❑ oxygen-electrode-based sensor

Sensor in which an oxygen electrode, which measures the amount of oxygen in a solution, is coated with a biological material such as an enzyme which generates or absorbs oxygen when the appropriate substrate is present. When the biological reaction takes place, the amount of oxygen close to the electrode changes and the signal from the electrode changes, thus signalling the concentration of substrate.

P

❐ p

Denotes the shorter of the two chromosome arms, e.g. human 14p is the short arm of human chromosome 14.

❐ p element

A *Drosophila* transposon.

❐ P_1, P_2

Generational symbols for the two parents of a given individual.

❐ p53 gene

A human tumour-suppressor transcription factor gene, damage or mutation to which is believed to be responsible for up to 60% of all human cancer tumours. If, in spite of the presence of p53 protein, a cell begins to divide uncontrollably following damage to its DNA, the p53 gene acts to prevent tumours by triggering apoptosis.

❐ pachynema

A mid-prophase stage in the first meiotic division, between zygonema and diplonema. Chromosomes appear as long, paired threads. Occasionally all four chromatids can be recognized.

❐ packaging cell line

A cell line designed to produce viral particles that do not contain nucleic acid. After transfection of these cells with a full-size viral genome, fully infective viral particles are assembled and released.

❐ packed cell volume

The proportion of a cell culture volume that is occupied by cells. Cell volume is determined by sedimenting using low speed centrifugation.

❐ page

Abbreviation for polyacrylamide gel electrophoresis.

▫ pairing

The pairing of homologous chromosomes during the prophase of the first meiotic division. Pairing is the first prerequisite before crossing over and recombination can occur.

▫ pair-rule gene

A gene that influences the formation of body segments in *Drosophila*.

▫ palaeontology

The study of the fossil record of past geological periods and of the phylogenetic relationships between extinct and contemporary plant and animal species.

▫ palindrome

A segment of double-stranded DNA, in which the order of bases, read 5'?3' in one strand, is the same as that in the complementary antiparallel strand, also read 5'?3'. If the sequence is written in the normal convention, on two lines with paired bases shown one above the other, the base order on one strand runs in the opposite direction to that on the complementary strand. They are often found at the ends of transposable elements, and recognition sites for type II restriction endonucleases are also palindromes.

▫ palisade parenchyma

Elongated cells found just beneath the upper epidermis of leaves, typically rich in chloroplasts.

▫ pamp

Ampicillin-resistant plasmid.

▫ panicle

An inflorescence, the main axis of which is branched; the branches bear loose racemose flower clusters. Rice is a prominent crop plant with a panicle inflorescence.

▫ panicle culture

Aseptic culture of immature panicle explants to induce microspore germination and development.

▫ panmictic population

A population in which mating occurs at random.

▫ panmixis

Random mating in a population.

▫ par

Abbreviation for photosynthetically active radiation.

❐ *par* gene

One of a class of genes required for faithful plasmid segregation at cell division. Initially, par loci were identified on plasmids, but have also been found on bacterial chromosomes.

❐ paracentric inversion

A chromosomal rearrangement in which a segment of chromosome, excluding the centromere, is rotated.

❐ paraffin wax

A translucent, white, solid hydrocarbon with a low melting point. One use is as an embedding medium to support tissue for sectioning for light microscopy observation.

❐ parafilm

A stretchable film based on paraffin wax; used to seal tubes and Petri dishes. Parafilmä is a proprietary name which is incorrectly applied colloquially to similar products.

❐ parahormone

A substance with hormone-like properties that is not a secretory product (e.g. ethylene; carbon dioxide).

❐ parallel evolution

The development of different organisms along similar evolutionary paths due to similar selection pressures acting on them.

❐ paralogous

Homologous genes/gene products that have duplicated and evolved divergently within a species. e.g. beta- and gamma-globulin genes

❐ parameter

A value or measurement that varies with circumstances, and is used as a reference to quantify a situation or a process.

❐ parasexual cycle

A sexual cycle involving changes in chromosome number but differing in time and place from the usual sexual cycle; occurring in those fungi in which the normal cycle is suppressed or apparently absent.

❐ parasite

An organism deriving its food from the living body of another organism.

❐ parasitism

The close association of two or more dissimilar organisms, where the association is harmful to the host, but beneficial to the parasite.

❑ parasporal crystal

Tightly packaged insect pro-toxin molecules that are produced by strains of *Bacillus thuringiensis* during the formation of resting spores.

❑ paratope

Synonym for antibody binding site.

❑ parenchyma

1. A plant tissue consisting of spherical, undifferentiated cells, frequently with air spaces between them.
2. Loose connective tissue formed by large cells.

❑ parthenocarpy

The development of fruit without fertilization.

❑ parthenogenesis

Production of an embryo from an unfertilized egg.

❑ partial digest

Incomplete reaction of a restriction enzyme with DNA, such that only a proportion of the target sites are cleaved. Partial digests are often performed to give an overlapping collection of DNA fragments for use in the construction of a gene bank.

❑ particle radiation

High energy nuclear emissions, used as physical agents of mutagenesis. Three major types typically used: alpha () (positively charged), beta () (negatively charged), or neutrons (uncharged).

❑ parts per million

Unit of concentration: 1 ppm = 1 mg dissolved matter per litre of solution.

❑ parturition

The process of giving birth.

❑ passage number

The number of times cells in culture have been sub-cultured.

❑ passage

The transfer or transplantation of cells from one culture medium to another.

❑ passage time

Time interval between successive sub-cultures.

❑ passive and active transport

If uncharged solutes are small enough, they can move down their concentration gradients directly across the lipid bilayer itself by simple diffusion. Examples of such solutes are ethanol, carbon dioxide, and oxygen. Most solutes, however, can cross the membrane only if there is a membrane transport protein (a carrier protein or a channel protein) to transfer them. As indicated, passive transport, in the same direction as a concentration gradient, occurs spontaneously, whereas transport against a concentration gradient (active transport) requires an input of energy. Only carrier proteins can carry out active transport, but both carrier proteins and channel proteins can carry out passive transport.

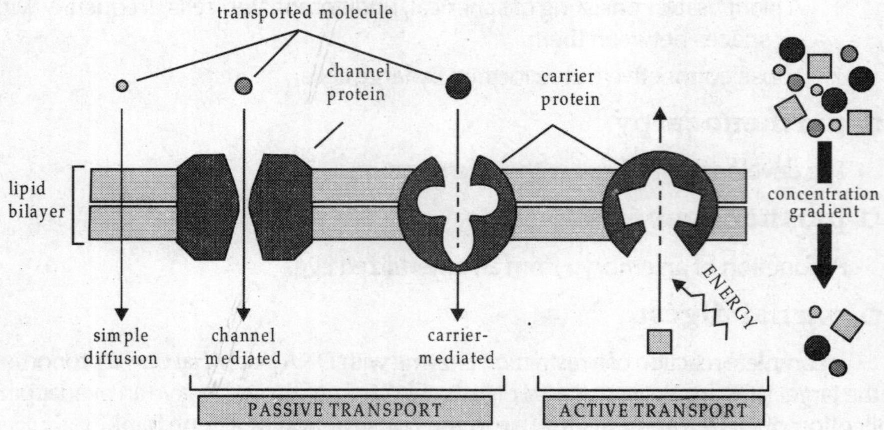

❑ passive immunity

1. Natural acquisition of antibodies by the foetus or neonate (newborn) from the mother.
2. The artificial introduction of specific antibodies by the injection of serum from an immune animal. In both cases, temporary protection is conferred on the recipient.

❑ *pat* gene

A gene obtained from *Streptomyces* sp. encoding resistance to glufosinate-ammonium containing herbicides. (These inhibit plant synthesis of glutamine). Used widely as a means of transgenically inducing herbicide resistance in crop plants.

❑ patent

A legal permission to hold exclusive right - for a defined period of time - to manufacture, use or sell an invention.

❑ paternal

Pertaining to the father.

❑ pathogen

A disease-causing organism (generally microbial: bacteria, fungi, viruses; but can extend to other organisms: e.g. nematodes etc.).

❐ pathogenesis related protein

One of a group of proteins that are characteristically highly expressed as part of a plant's response to pathogen attack. Many of these proteins are similarly expressed following infection with a broad range of pathogens, indicating their role in the hypersensitive response.

❐ pathogen-free

Uncontaminated with pathogen.

❐ pathotoxin

Substance secreted by certain pathogens, in order to attack the host tissue. Some pathotoxins are also toxic to non-hosts, especially animals or humans.

❐ pathovar

Strain of a plant-attacking bacterium or fungus that can be differentiated from others by their interaction with specific host cultivars.

❐ pbr

Abbreviation for plant breeders' rights.

❐ pbr322

One of the first plasmids used for cloning DNA in *E. coli*.

❐ pcr

Abbreviation for polymerase chain reaction.

❐ pcr in forensic science

The DNA sequences that create the variability used in this analysis contain runs of short, repeated sequences, such as GTGTGT . . . , which are found in various positions (loci) in the human genome. The number of repeats in each run is highly variable in the population, ranging from 4 to 40 in different individuals. A run of repeated nucleotides of this type is commonly referred to as a VNTR (variable number of tandem repeat) sequence. Because of the variability in these sequences, each individual will usually inherit a different variant of each VNTR locus from their mother and from their father; two unrelated individuals will therefore not usually contain the same pair of sequences. A PCR reaction using primers that bracket the locus produces a pair of bands of amplified DNA from each individual, one band representing the maternal variant and the other representing the paternal variant. The length of the amplified DNA, and thus its position after electrophoresis, will depend on the exact number of repeats at the locus. (B) The DNA bands obtained from a set of four different PCR reactions, each of which amplifies the DNA from a different VNTR locus. In the schematic example shown here, the same three VNTR loci are analyzed from three suspects (individuals A, B, and C), giving six bands for each person. You can see that although some individuals have several bands in common, the overall pattern is quite distinctive for each. The band pattern can serve as a "fingerprint" to identify an individual nearly uniquely. The fourth lane (F) contains the products of the same PCR reactions carried out on a forensic sample. The starting material for such a PCR reaction can be a single hair that was left at the scene of the crime. From the example, individuals A and C can be eliminated from enquiries, while B remains a clear suspect.

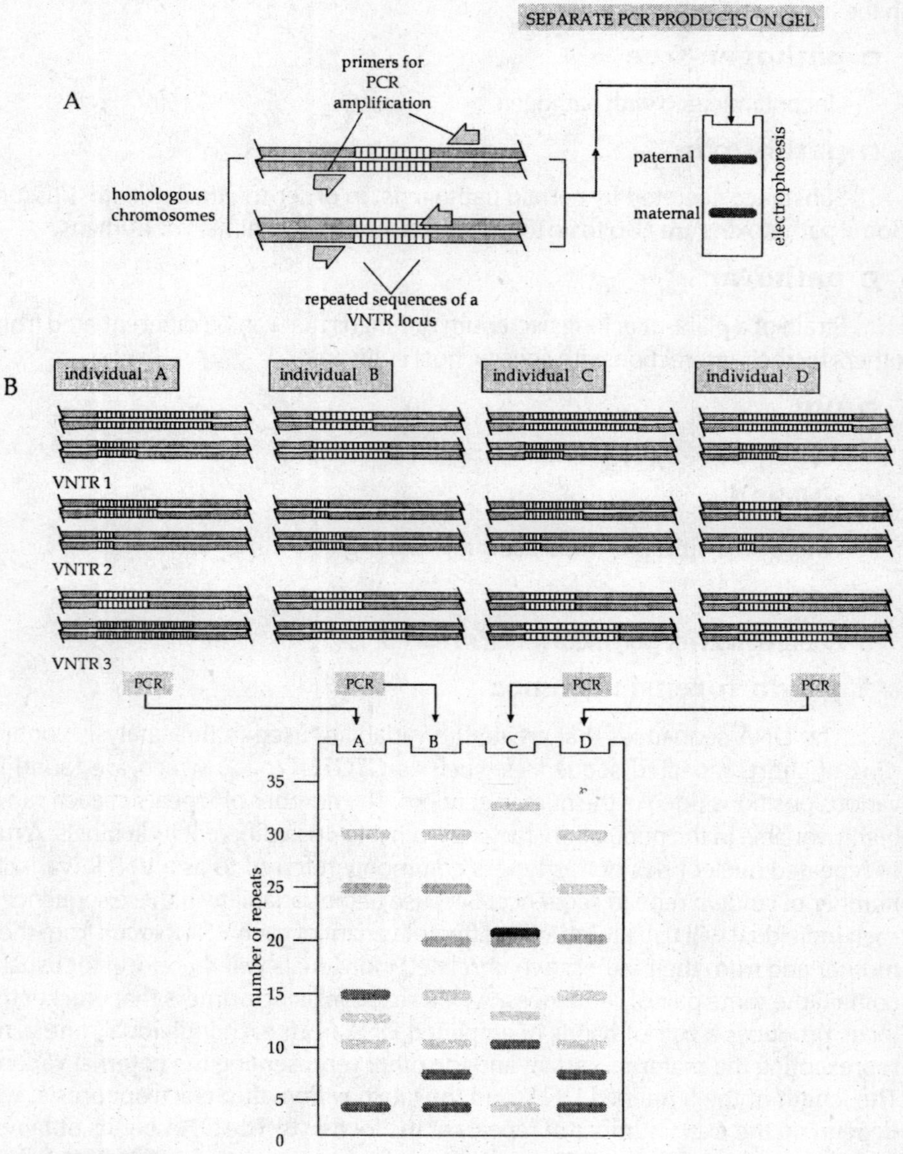

❑ pcr-rflp

Alternative term for cleaved amplified polymorphic sequence.

❑ pcv

Abbreviation for packed cell volume.

❐ pectin

A group of naturally occurring complex polysaccharides, containing galacturonic acid, found in plant cell walls, where their function is to cement cells together. Used as a thickening agent in solid culture media and as a food additive.

❐ pectinase

Enzyme catalysing the hydrolysis of pectin. Used in conjunction with cellulase to solubilize plant cell walls.

❐ pedicel

Stalk or stem of the individual flowers of an inflorescence.

❐ pedigree

A diagram of the heredity of a particular trait through many generations of a family. The ancestry of an individual.

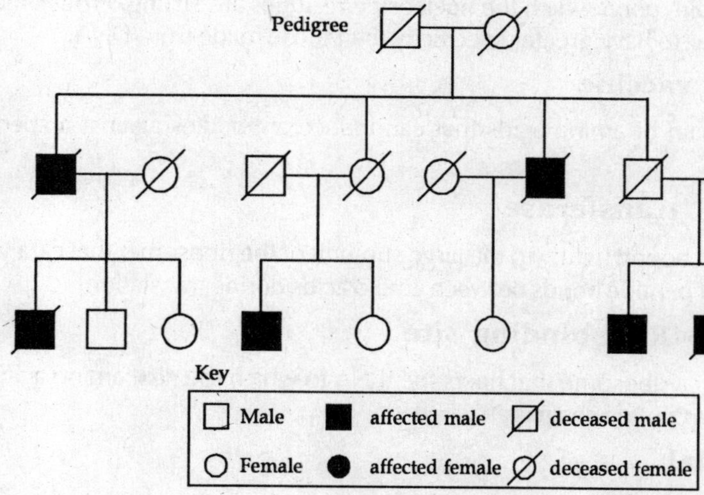

❐ peduncle

Stalk or stem of a flower that is born singly; the main stem of an inflorescence.

❐ peg

Abbreviation for polyethylene glycol.

❐ penetrance

The proportion of individuals in a population that express the phenotype expected from their genotype with respect to a specific gene. Measures the extent to which phenotype predicts genotype.

❐ peptidase

An enzyme that catalyzes the hydrolysis of a peptide bond.

peptide

A sequence of amino acids linked by peptide bonds; a breakdown or build-up unit in protein metabolism. Typically used to describe low molecular weight species.

peptide bond

The chemical bond holding amino acid residues together in peptides and proteins. The (CO-NH) bond is formed by the condensation, with loss of a water molecule, between the carboxyl (-COOH) group of one amino acid and the amino (-NH_2) of the next amino acid.

peptide expression library

A collection of peptide molecules, produced by recombinant cells, in which the amino acid sequences are varied.

peptide nucleic acid

A synthetic oligonucleotide analogue, in which the sugar backbone is replaced by a peptide chain, upon which the nucleoside residues are strung. Probes made from PNA appear to have greater specificity than those made from DNA.

peptide vaccine

A short chain of amino acids that can induce antibodies against a specific infectious agent.

peptidyl transferase

An enzyme bound tightly to the large sub-unit of the ribosome, that catalyses the formation of peptide bonds between amino acids during translation.

peptidyl-tRNA binding site

The site on a ribosome that hosts the tRNA to which the next amino acid for the growing polypeptide chain is attached.

perennial

A plant that flowers continuously for several years.

pericentric inversion

A chromosomal rearrangement in which a segment of chromosome, including the centromere, is rotated.

periclinal chimera

1. Genotypically or cytoplasmically different tissues arranged in concentric layers.
2. A chimera in which one or more layers of tissue derived from one graft member enclose the central tissue derived from the other member of the graft.

periclinal

The orientation of cell wall or plane of cell division parallel to the reference surface.

❑ pericycle

Region of the plant bounded externally by the endodermis and internally by the phloem. Most roots originate from the pericycle.

❑ periplasm

The space between the cell (cytoplasmic) membrane of a bacterium or fungus and the outer membrane or cell wall.

❑ permanent wilting point

The moisture content of a soil below which plants wilt to such an extent that they fail to recover even when fully watered.

❑ permeable

Referring to a membrane, cell or cell system through which small molecules can diffuse.

❑ persistence

Ability of an organism to remain in a particular setting for a period of time after it is introduced.

❑ persistent

Chemicals with a long inactivation or degradation time, such as some pesticides. Persistent substances can become dangerously concentrated in the tissues of organisms at the top end of a food chain.

❑ perv

Abbreviation for porcine endogenous retrovirus.

❑ pesticide

A toxic chemical product that kills harmful organisms (e.g. insecticides, fungicide, weedicides, rodenticides).

❑ petal

One of the parts of the flower that make up the corolla.

❑ petiole

Stalk of leaf.

❑ petri dish

Flat round glass or plastic dish with a matching lid, used for small-scale culturing of organisms, germinating seeds etc. Also referred to as plates, hence to plate a culture.

❑ pfge

Abbreviation for pulsed-field gel electrophoresis.

❑ pg

Abbreviation for polygalacturonase.

❑ ph

Logarithmic measure of acidity/alkalinity of a solution. A pH of 7 is neutral (e.g. pure water), whereas below 7 is acid and above 7 is alkaline.

❑ phage

Abbreviation for bacteriophage.

❑ phagemids

Cloning vectors that contain components derived from both phage and plasmid DNA.

❑ phagocytes

Immune system cells that ingest and destroy viruses, bacteria, fungi and other foreign substances or cells.

❑ phagocytosis

The process by which foreign particles invading the body are engulfed and broken down by phagocytes.

❑ pharmacokinetics

The quantitative measurement of how drugs move around the body, and the processes which control their absorption, distribution, metabolism, and excretion.

❑ phase change

The developmental change from one maturation state to another.

❑ phase state

The coupling or repulsion of two linked genes.

❑ phb

Abbreviation for polyhydroxybutyrate.

❑ pH-electrode-based sensor

Sensor in which a standard pH electrode is coated with a biological material. Many biological processes raise or lower pH, and the changes can be detected by the pH electrode.

❑ phenocopy

An environmentally induced, non hereditary variation in an organism, resembling a genetically determined trait.

❑ phenolic oxidation

Common aspect of the wound response in plants. Phenolic oxidation is often indicated by blackening of tissue and it may be a precursor to growth inhibition or, in severe cases, to tissue necrosis and death.

❐ phenolics

Compounds with hydroxyl group(s) attached to the benzene ring, forming esters, ethers and salts. Phenolic substances produced from newly explanted tissues are liable to oxidise, and as a result form coloured compounds visible in nutrient media.

❐ phenotype

The visible appearance of an individual (with respect to one or more traits) which reflects the reaction of a given genotype with a given environment.

❐ pheromone

A hormone-like substance that is secreted by an organism into the environment as a specific signal to another organism, usually of the same species.

❐ phloem

Specialized vascular plant tissue for the transport of assimilates (generally sugars) from the point of synthesis (in the leaf) to other parts of the plant. It consists of sieve tubes, companion cells, phloem parenchyma and fibres.

❐ phosphatase

A class of enzymes that catalyze the hydrolysis of esters of phosphoric acid, removing a phosphate group from an organic compound.

❐ phosphodiester (phospho-diester) bond

A bond in which a phosphate group joins adjacent carbons through ester linkages. A condensation reaction between adjacent nucleotides results in a phosphodiester bond between 3' and 5' carbons in DNA and RNA.

❐ phospholipase a2

An enzyme which degrades type A2 phospholipids.

❐ phospholipid

A class of lipid molecules in which glycerol is linked to a phosphate group and two fatty acyl groups. Contains both polar and non-polar regions. A major component of biological membranes.

❐ phosphorolysis

The cleavage of a bond by orthophosphate; analogous to hydrolysis referring to cleavage by water.

❐ phosphorylation

The addition of a phosphate group to a compound.

❐ photo-bioreactor

Bioreactor dependent on sunlight, which is taken up by its content of plant material, usually algae.

❑ photoperiod

The length of daylight or period of daily illumination provided for growth.

❑ photoperiodism

The photoperiod required by a plant to switch from the vegetative to the reproductive stage.

❑ photophosphorylation

The formation of ATP from ADP and inorganic phosphate using light energy harvested by photosynthesis.

❑ photoreactivation

A light dependent DNA repair process.

❑ photosynthate

The carbohydrates and other compounds produced in photosynthesis.

❑ photosynthesis

A chemical process by which green plants synthesize organic compounds from carbon dioxide and water in the presence of sunlight.

❑ photosynthesis and respiration

Photosynthesis and respiration as complementary processes in the living world. Photosynthesis uses the energy of sunlight to produce sugars and other organic molecules. These molecules in turn serve as food for other organisms. Many of these organisms carry out respiration, a process that uses O_2 to form CO_2 from the same carbon atoms that had been taken up as CO_2 and converted into sugars by photosynthesis. In the process, the organisms that respire obtain the chemical bond energy that they need to survive. The first cells on the earth are thought to have been capable of neither photosynthesis nor respiration. However, photosynthesis must have preceded respiration on the earth, since there is strong evidence that billions of years of photosynthesis were required before O_2 had been released in sufficient quantity to create an atmosphere rich in this gas. (The earth's atmosphere presently contains 20% O_2.)

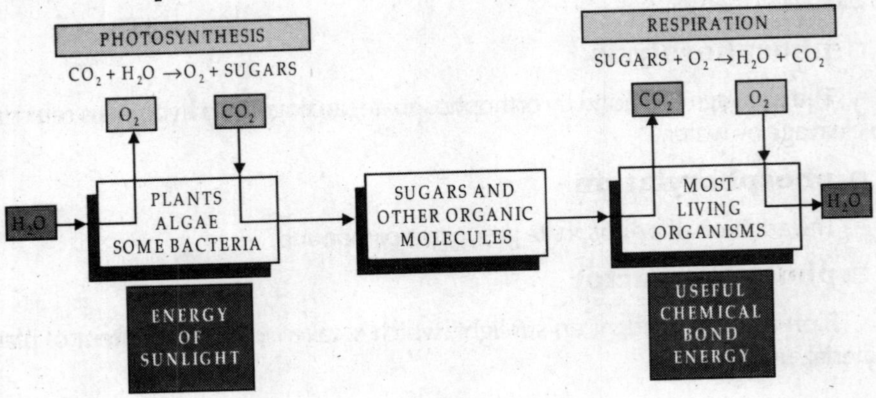

□ **photosynthetic**

Able to use sunlight energy to convert atmospheric carbon dioxide into organic compounds. Nearly all plants, most algae and some bacteria are photosynthetic.

□ **photosynthetic efficiency**

Efficiency of converting light energy into organic compounds.

□ **photosynthetic photon flux**

A measure of the intensity of light utilized by plants for photosynthetic activity.

□ **photosynthetically active radiation**

The part of the radiant energy that is capturable by natural photosynthetic systems (approximately equivalent to the natural light spectrum of wavelengths 400-700nm).

□ **phototropism**

The tendency of plants to direct shoot growth towards the source of light.

□ **phylogeny**

The deduced evolutionary history of related organisms.

□ **physical map**

An indication of the separation, in bp, between pairs of linked loci.

□ **phyto-**

To do with plants.

□ **phytochemical**

Molecules characteristically found in plants.

□ **phytochrome**

A pigment, found in the cytoplasm of green plants, which can exist in two forms Pr (biologically inactive) and Pfr (biologically active). Pfr is converted into Pr by exposure to light of wavelength 730 nm. Involved in the timing of many plant processes, e.g. dormancy, leaf formation, flowering and germination.

□ **phytohormone**

A substance that stimulates growth or other processes in plants. Major species are auxins, abscisic acid, cytokinins, gibberellins and ethylene.

□ **phytoparasite**

Parasite on plants.

□ **phytopathogen**

A plant pathogen.

▫ phytoremediation

The use of plants actively to remove contaminants or pollutants from either soils (e.g. polluted fields) or water resources (e.g. polluted lakes). An example is the exploitation of the Brazil water hyacinth (*Eichhornia crassipes*) to accumulate in its tissues toxic metals such as lead, arsenic, cadmium, mercury, nickel, and copper.

▫ phytosanitary

Plant health, including quarantine.

▫ phytostat

Apparatus designed for the semi-continuous chemostatic culture of plant cells.

▫ phytosterol

One of a group of biologically active phytochemicals present in the seeds of certain plants. Evidence suggests that human consumption of certain phytosterols, such as â-sitosterol, can help to lower total serum cholesterol and low-density lipoproteins levels, thereby reducing the risk of coronary heart disease.

▫ pigment

Compounds that are coloured by the light they absorb. Light absorption is exploited by plants both as a means of energy capture (*see*: photosynthesis) and as a signalling mechanism.

▫ pinocytosis

The engulfing of a minute droplet of liquid by a living cell.

▫ pipette

Widely used device for accurate dispensing of small volumes of liquids.

▫ pistil

Central organ of the flower, typically consisting of ovary, style and stigma. Usually referred to as the female part of a perfect flower.

▫ plant breeders' rights

Legal protection of a new plant variety granted to the breeder or his successor in title. The effect of PBR is that prior authorization is required before the material can be used for commercial purposes.

▫ plant cell culture

In vitro growth of plant cells.

▫ plant cell immobilization

Entrapment of plant cells in gel matrices so that they are protected from physical damage. The cells are suspended in liquified droplets which are then allowed to harden. Commonly used matrices are alginates, agar or polyacrylamide.

◻ plant cells and tissues part 1

The cell types and tissues from which higher plants are constructed.

THE PLANT

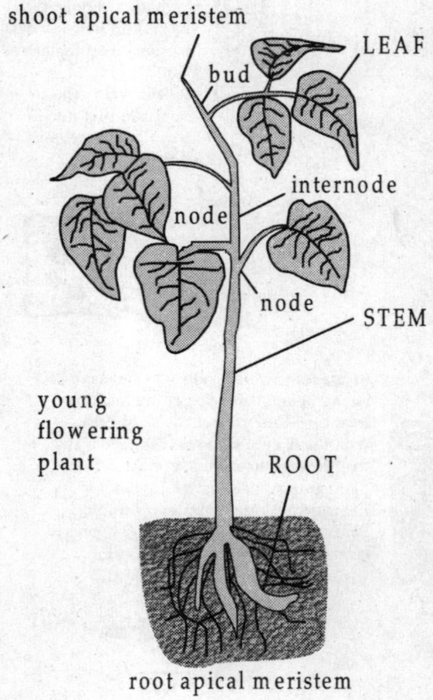

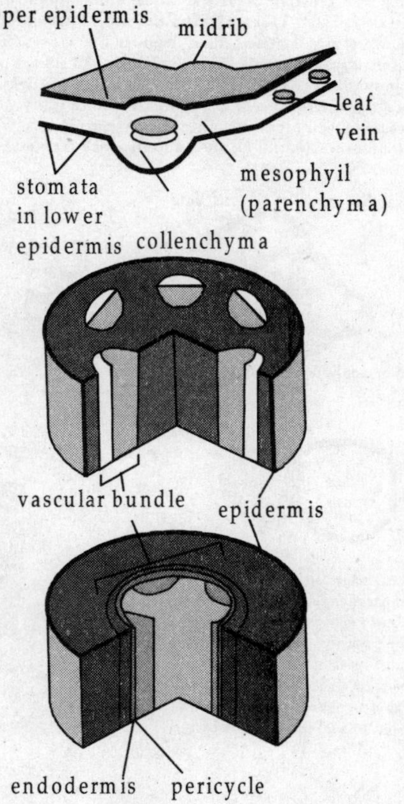

The young flowering plant shown on the left is constructed from three main types of organs: leaves, stems, and roots .
Each plant organ in turn is made from tissue systems:ground ( ),dermal ( ), and vascular ( ).

All three tissue systems derive ultimately from the cell proliferative activity of the shoot or root apical meristems, and each contains a relatively small number of specialized cell types. These three common tissue systems, and the cells that comprise them, are described in this panel.

THE THREE TISSUE SYSTEMS

Cell division, growth, and differentiation give rise to tissue systems with specialized functions.

DERMAL TISSUE: This is the plant's protective outer covering in contact with the environment. It facilitates water and ion uptake in roots and regulates gas exchange in leaves and stems.

VASCULAR TISSUE: Together the phloem and the xylem form a continuous vascular system throughout the plant. This tissue conducts water and solutes between organs and also provides mechanical support.

GROUND TISSUE: This packing and supportive tissue accounts for much of the bulk of the young plant. It also functions in food manufacture and storage.

GROUND TISSUE The ground tissue system contains three main cell types called parenchyma, collenchyma, and sclerenchyma.

Parenchyma cells are found in all plant tissue systems. They are living cells, generally capable of further division, and have a thin primary cell wall. These cells have a variety of functions. The apical and lateral meristematic cells of shoots and roots provide the new cells required for growth. Food production and storage occur in the photosynthetic cells of the leaf and stem (called mesophyll cells); storage parenchyma cells form the bulk of most fruits and vegetables. Because of their proliferative capacity, parenchyma cells also serve as sources of new cells for wound healing and regeneration.

Collenchyma are living cells similar to parenchyma cells except that they have much thicker cell walls and are usually elongated and packed into long ropelike fibers. They are capable of stretching and provide mechanical support in the ground tissue system of the elongating regions of the plant.

Collenchyma cells are especially common in subepidermal regions of stems.

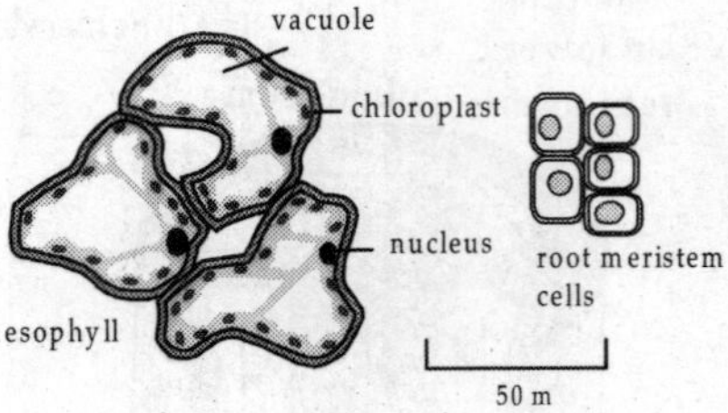

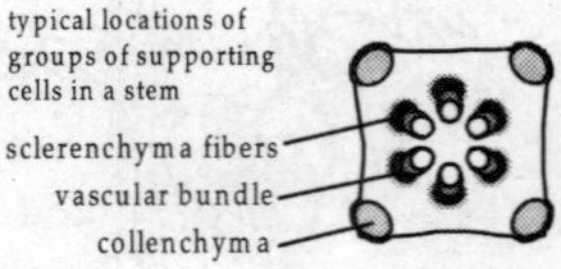

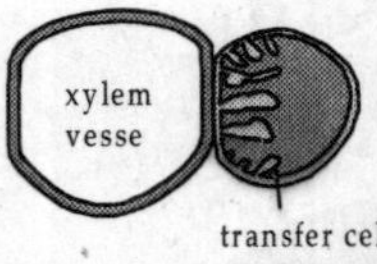

A transfer cell, a specialized form of parenchyma cell, is readily identified by elaborate ingrowths of the primary cell wall. The increase in the area of the plasma membrane beneath these walls facilitates the rapid transport of solutes to and from cells of the vascular system.

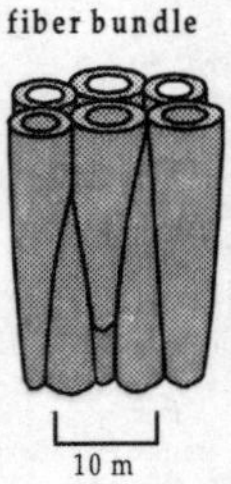

Sclerenchyma, like collenchyma, have strengthening and supporting functions. However, they are usually dead cells with thick, lignified secondary cell walls that prevent them from stretching as the plant grows. Two common types are fibers, which often form long bundles, and sclereids, which are shorter branched cells found in seed coats and fruit.

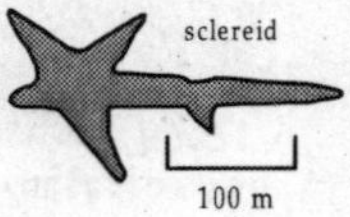

▫ plant cells and tissues part 2

The cell types and tissues from which higher plants are constructed.

DERMAL TISSUE

The **epidermis** is the primary outer protective covering of the plant body. Cells of the epidermis are also modified to form stomata and hairs of various kinds.

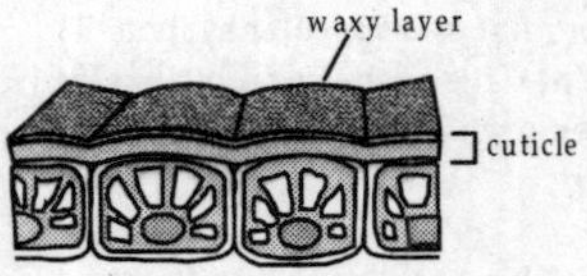

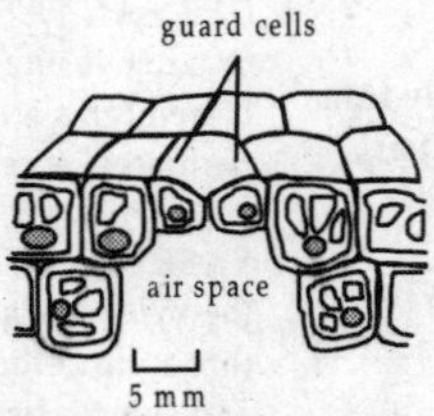

Stomata are openings in the epidermis, mainly on the lower surface of the leaf, that regulate gas exchange in the plant. They are formed by two specialized epidermal cells called *guard cells*, which regulate the diameter of the pore. Stomata are distributed in a distinct species-specific pattern within each epidermis.

Hairs (or trichomes) are appendages derived from epidermal cells. They exist in a variety of forms and are commonly found in all plant parts. Hairs function in protection, absorption, and secretion; for example,

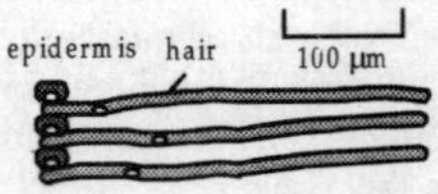

young, single-celled hairs develop in the epidermis of the cotton seed. When these grow, the walls will be secondarily thickened with cellulose to form cotton fibers.

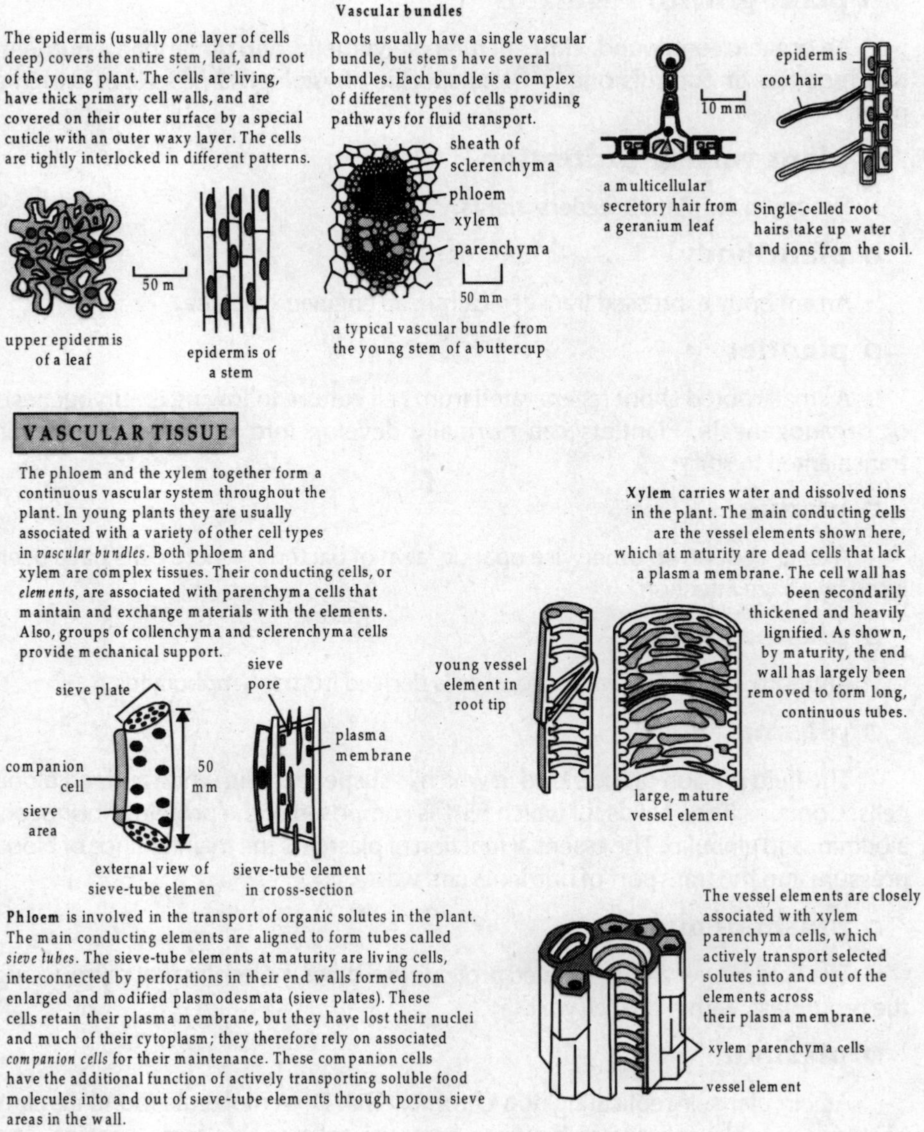

◻ plant genetic resources

The reproductive or vegetative propagating material of:

1. cultivated varieties (cultivars) in current use and newly developed varieties;
2. obsolete cultivars;
3. primitive cultivars (landraces);
4. wild and weed species, near relatives of cultivated varieties; and
5. special genetic stocks (including elite and current breeder's lines and mutants).

▫ plant growth regulator

An organic compound, either natural or synthetic, and other than a nutrient, that modifies or controls one or more specific physiological processes within a plant.

▫ plant variety protection

Synonym for plant breeders' rights.

▫ plantibody

An antibody expressed transgenically in an engineered plant.

▫ plantlet

A small rooted shoot regenerated from cell culture following embryogenesis or organogenesis. Plantlets can normally develop into normal plants when transplanted to soil.

▫ plaque

A clear spot on an otherwise opaque lawn of bacteria, where cells have been lysed by phage infection.

▫ plasma cells

Antibody-producing white blood cells derived from B lymphocytes.

▫ plasma

The fluid portion of the blood in which is suspended the white and red blood cells. Contains 8-9% solids, of which 85% is composed of the proteins fibrinogen, albumin, and globulin. The essential function of plasma is the maintenance of blood pressure and the transport of nutrients and waste.

▫ plasmalemma

The lipid bilayer and associated proteins and other molecules that surrounding the protoplast, within the cell wall.

▫ plasmid

An circular self-replicating non-chromosomal DNA molecule found in many bacteria, capable of transfer between bacterial cells of the same species, and occasionally of different species. Antibiotic resistance genes are frequently located on plasmids. Plasmids are particularly important as vectors for genetic engineering.

▫ plasmodesma

Fine protoplasmic thread that connects adjacent plant cells by passing through the plant cell wall. Exploited by viruses as a conduit for cell to cell movement.

▫ plasmolysis

Shrinkage of protoplasm caused by removal of water from a cell through osmosis when surrounded by a hypertonic solution.

❐ plastid

A general term for a number of plant cell organelles which carry non-nuclear DNA. Includes the pigment-carrying bodies:

1. chloroplasts in leaves,
2. chromoplasts in flowers, and
3. the starch-synthesizing amyloplasts in seeds.

❐ plastoquinone

One of a group of compounds involved in the transport of electrons as part of the process of photosynthesis.

❐ plate

1. Verb: to distribute a thin film of micro-organisms or plant cells onto a solid medium.
2. Noun: refers to the two segments of a Petri dish or similar.

❐ plating efficiency

The percentage of inoculated cells which give rise to cell colonies when seeded into culture vessels.

❐ pleiotropy

The simultaneous effect of a given gene on more than one apparently unrelated trait.

❐ ploidy

The number of complete sets of chromosomes per cell, e.g. one set: haploid, two sets: diploid, etc.

❐ plumule

The first bud of an embryo, or that portion of the young shoot above the cotyledons.

❐ pna

Abbreviation for peptide nucleic acid.

❐ point mutation

A change in DNA sequence at a specific locus. The smallest change involves the substitution, deletion or insertion of a single nucleotide. *See*: single nucleotide polymorphism.

❐ polar bodies

In female animals, the products of a meiotic division that do not develop into a functional ovum. The first polar body comprises one of the two products of the first meiotic divisions, which may fail to divide at the second division. The second polar body comprises one of the products of the second division.

❑ polar mutation

A mutation that influences the functioning of genes that are downstream from the site of mutagenesis but are in the same transcription unit.

❑ polar nuclei

Two centrally located nuclei in the embryo sac that unite with a second sperm cell in a triple fusion. In some plant species (particularly the monocotyledons), the product of this fusion develops into the endosperm.

❑ polar transport

A directed movement within plants of compounds (usually endogenous plant growth regulators) mostly in one direction; polar transport overcomes the tendency for diffusion in all directions.

❑ polarity

The observed differentiation of an organism, tissue or cell into parts having opposed or contrasted properties or form.

❑ pole cells

A group of cells in the posterior of *Drosophila* embryos that are precursors to the adult germ line.

❑ pollen culture

The *in vitro* culture and germination of pollen grains to generate haploid plants. *See*: anther culture; microspore culture.

❑ pollen grain

The mature microspore, produced in the pollen sac of angiosperms or the microsporangium of gymnosperms. Unicellular, with variable shape and size, and an elaborately structured wall.

❑ pollen

Mature microspores of seed plants.

❑ pollination

Transfer of pollen from anther to stigma in the process of fertilization in angiosperms; transfer of pollen from male to female cone in the process of fertilization in gymnosperms.

❑ poly-(A) polymerase

Enzyme that catalyses the addition of adenine residues to the 3' end of mRNA molecules, forming the characteristic poly-(A) tail.

❑ polyacrylamide gel electrophoresis

Ubiquitous method for separating nucleic acids and proteins on the basis of their molecular size. The method relies on the migration through an inert matrix (polyacrylamide gel) of electrically charged molecules as a result of the imposition of an electric field.

❑ polyacrylamide gel

Inert electrophoresis matrix, formed by the polymerization of acrylamide monomer in the presence of the cross-linker *N,N'*-methylene-*bis*-acrylamide. Gels are usually supported between two glass plates, which need to be removed for post-electrophoresis manipulations. Sometimes referred to incorrectly as acrylamide gels.

❑ polyadenylation

Post-transcriptional addition of multiple adenine residues to the 3' end of eukaryotic mRNA. Also called poly-(A) tailing. The adenine-rich 3' terminal segment is called a poly (A) tail.

❑ polycistronic

A single mRNA that contains the information necessary for the production of more than one polypeptide. Particularly characteristic of prokaryotic mRNAs.

❑ polyclonal antibody

A serum sample that contains a mixture of distinct immunoglobulin molecules, each recognizing a different antigenic determinant of a given antigen.

❑ polydactyly

An abnormality in which a person is born with more than the normal number of fingers or toes.

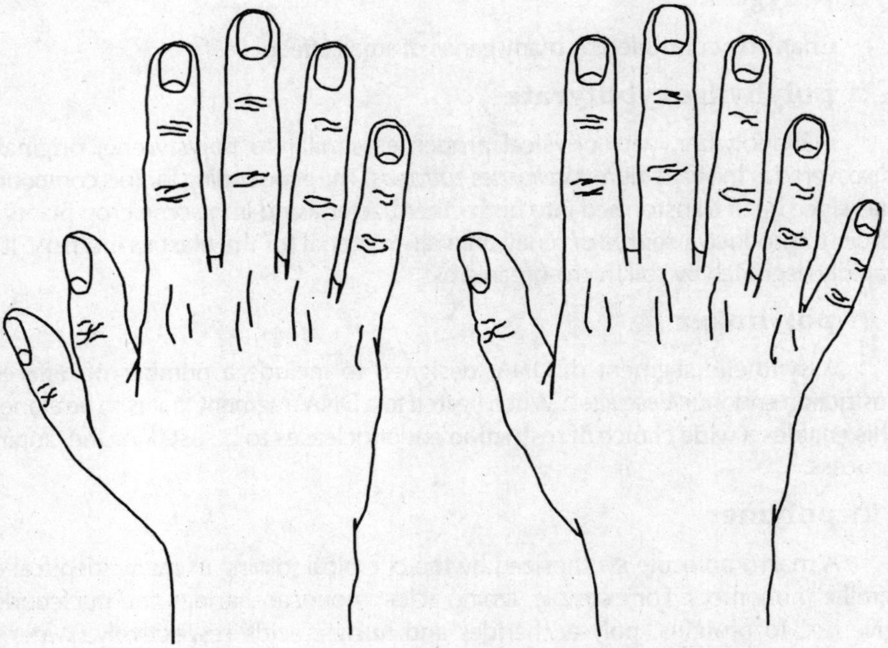

❐ polyembryony

The production of more than one embryo from a single egg cell (in animals) or from a range of embryogenic cell types (in plants). These embryos are genetically identical to one another.

❐ polyethylene glycol

A polymer having the general formula $HOCH_2(CH_2OCH_2)_nCH_2OH$ and available in a range of molecular weights. Thus PEG 1000 is a polyethylene glycol of average molecular weight 1000. PEG 4000 and 6000 are commonly used to promote cell or protoplast fusion, and to facilitate DNA uptake in the transformation of organisms such as yeast. Also used to concentrate solutions by withdrawing water from them via osmosis.

❐ polygalacturonase

An enzyme which catalyses the breakdown of pectin. A tomato engineered to contain an antisense-PG gene succeeded in delaying the onset of softening, by inhibiting the expression of PG. This allowed the fruit to be picked at a riper stage than is conventionally possible, and represented the first commercialized genetically engineered crop plant.

❐ polygene

One of a number of genes, each of small effect, which together act to determine the phenotype of a quantitative trait. The result is continuous variation in the trait and a seemingly non-Mendelian mode of inheritance.

❐ polygenic

Character controlled by many genes of small effect.

❐ polyhydroxybutyrate

A biopolymer, with physical properties similar to polystyrene, originally discovered in the bacterium *Alcaligenes eutropus*. The gene coding for this compound has since been transformed into both other bacteria and into some crop plants in order to produce a source of renewable raw material for the plastics industry. It is rapidly degraded by soil micro-organisms.

❐ polylinker

A synthetic segment of DNA, designed to include a number of different restriction endonuclease sites. When ligated to a DNA fragment that is to be cloned, this enables a wide choice of restriction endonucleases to be used for the cloning process.

❐ polymer

A macromolecule synthesized by the chemical joining of many identical or similar monomers. For example, amino acids, monosaccharides and nucleotides give rise to proteins, polysaccharides and nucleic acids respectively. Water is eliminated between the monomers as they link to form chains. The individual monomer units condensed within a chain are often referred to as residues, a term which is also employed for the bases incorporated in polynucleotides.

❐ polymerase

An enzyme that catalyses the formation of polymers from monomers. A DNA polymerase synthesizes DNA from deoxynucleoside triphosphates using a complementary DNA strand and a primer. An RNA polymerase synthesizes RNA from ribonucleoside triphosphates and a complementary DNA strand.

❐ polymerase chain reaction

A widespread molecular biology procedure that allows the production of multiple copies (amplification) of a specific DNA sequence, provided that the base pair sequence of each end of the target is known. It involves multiple cycles of DNA denaturation, primer annealing, and strand extension, and requires a thermostable DNA polymerase, deoxyribonucleotides, and specific oligonucleotides (primers).

The elegant technique of PCR, by which fragments of DNA can be made to replicate very rapidly, is illustrated.

POLYMERASE CHAIN REACTION

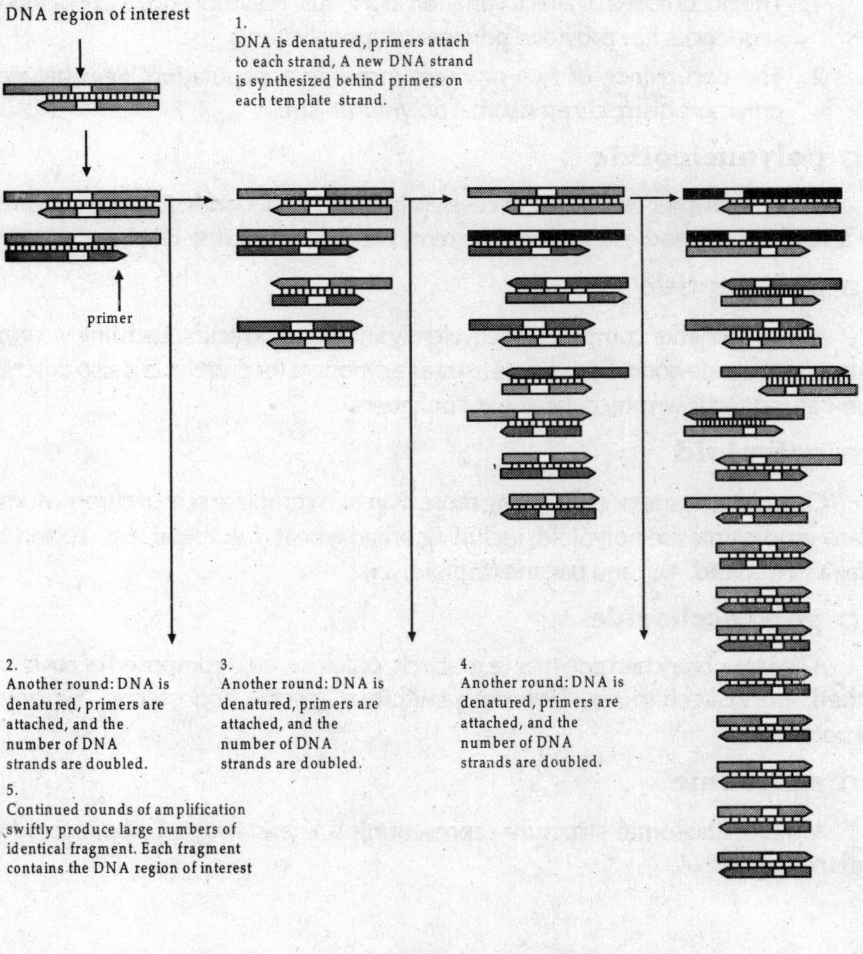

Figure Legend:

Polymerase chain reaction (PCR), is a common method of creating copies of specific fragments of DNA. PCR rapidly amplifies a single DNA molecule into many billions of molecules.

In one application of the technology, small samples of DNA, such as those found in a strand of hair at a crime scene, can produce sufficient copies to carry out forensic tests.

▫ polymerization

The chemical fusion of a number of identical or similar monomers to form a polymer. Common biological polymers are starch (polymerized monosaccharides), DNA (deoxyribonucleotides) and proteins (amino acids).

▫ polymery

The phenomenon whereby a number of genes at different loci (which may be polygenes) can act together to produce a single effect.

▫ polymorphism

1. The occurrence of allelic variation at a locus. Polymorphism in nucleotide sequences has provided powerful diagnostic tools.
2. The occurrence of two or more forms in a population. *See:* balanced polymorphism, chromosomal polymorphism.

▫ polynucleotide

A linear polymer composed of covalently linked nucleotides. Each link is formed by a single phosphodiester bond. The term is used to describe DNA and RNA.

▫ polypeptide

A linear polymer composed of covalently linked amino acids. Each link is formed by a single peptide bond. Sometimes used as a synonym for protein, but also describes non-natural and low-molecular-weight polymers.

▫ polyploid

Organism, tissue or cells having more than two complete sets of chromosomes. Many crop plants are polyploid, including bread wheat (hexaploid, 6x), cotton and alfalfa (tetraploid, 4x), and banana (triploid, 3x).

▫ polysaccharide

A linear or branched polymer (e.g. starch, cellulose, etc.) composed of covalently linked monosaccharides, including cellulose, pectin and starch. *Synonym*: carbohydrate.

▫ polysome

A multi-ribosomal structure representing a linear array of ribosomes held together by mRNA.

❐ polyspermy

The entry of several sperm nuclei into the egg during fertilization, although only one actually fuses with the egg nucleus.

❐ polytene chromosome

Giant chromosomes produced by interphase replication without division, and consisting of many identical chromatids arranged side by side.

❐ polyunsaturates

Oils in which some of the carbon-carbon bonds are not fully hydrogenated - i.e. of the form -CH=CH-, rather than $-CH_2-CH_2-$.

❐ polyvalent vaccine

A recombinant organism into which antigenic determinants have been cloned from a number of pathogens, for use as a vaccine.

❐ polyvinylpyrrolidone

An occasional constituent of plant tissue culture isolation media. PVP is of variable molecular weight and of general formula $(C_6H_9NO)_n$. Its antioxidant properties are used to prevent oxidative browning of excised plant tissues. Less frequently used as an osmoticum in culture media.

❐ population

A defined group of interbreeding organisms.

❐ population density

Number of cells or individuals per unit. The unit could be an area, or a volume of medium.

❐ population genetics

The branch of genetics that deals with frequencies of alleles and genotypes in breeding populations.

❐ porcine endogenous retrovirus

The provirus of a porcine retrovirus. The possibility that PERVs could be activated after xenotransplantation of pig organs into humans has raised concern that xenotransplantation may result in the transfer of novel infections to the human population.

❐ position effect

The influence of the location of a gene (particularly a transgene) on its expression and hence its effect on phenotype.

❐ positional candidate gene

A gene known to be located in the same region as a DNA marker that has been shown to be linked to a single-locus trait or to a QTL, and whose deduced function suggests that it could be the source of genetic variation in the trait in question.

❑ positional cloning

A strategy for gene cloning that relies on the identification of closely linked markers to the target trait, and then uses chromosome walking to identify, isolate and characterize the gene(s) responsible for the trait. The strategy is particularly appropriate when the biochemical basis of the target trait is unclear, thus precluding the use of a candidate gene approach.

❑ positive control system

A mechanism in which a regulatory protein(s) is required to turn on gene expression.

❑ positive selection

A method by which cells that carry a DNA insert integrated at a specific chromosomal location can be selected, since this integration confers a predictable phenotype.

❑ possible models for dna Replication

Three possible ways in which DNA can replicate are illustrated. The two original strands of DNA.

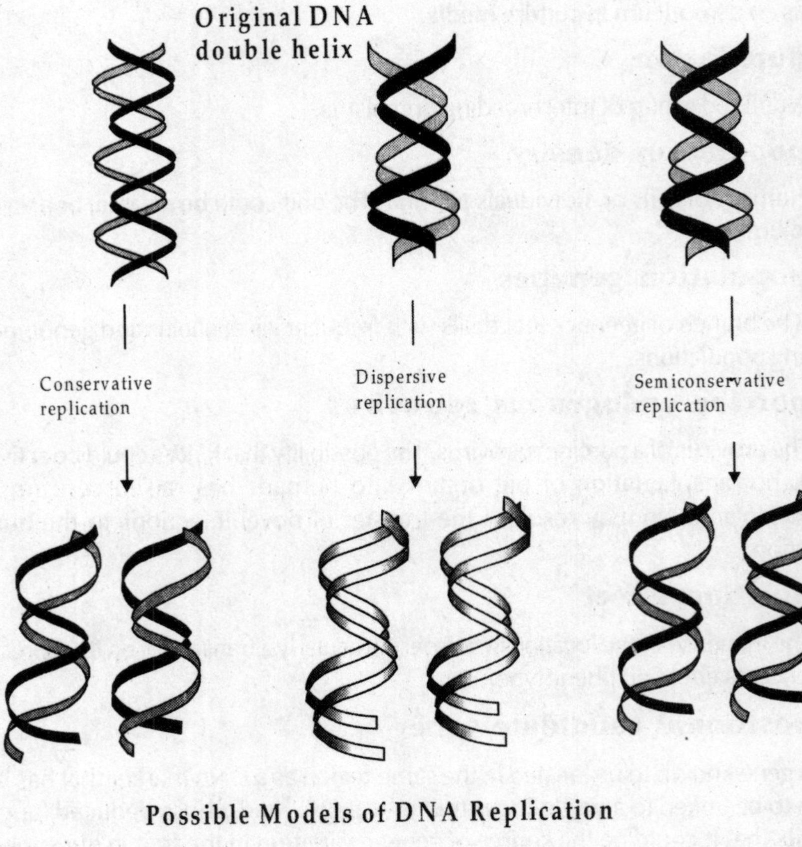

Possible Models of DNA Replication

To explain the phenomenon of heredity, biological information must be accurately copied and transmitted from each cell to all of its progeny. Three ways for DNA molecules to replicate may be considered, each obeying the rules of complementary base pairing.

- **Conservative replication** would leave intact the original DNA molecule and generate a completely new molecule.
- **Dispersive replication** would produce two DNA molecules with sections of both old and new DNA interspersed along each strand.
- **Semiconservative replication** would produce molecules with both old and new DNA, but each molecule would be composed of one old strand and one new one.

The replication is semiconservative. Each strand acts as a template for the synthesis of a new DNA molecule by the sequential addition of complementary base pairs, thereby generating a new DNA strand that is the complementary sequence to the parental DNA. Each daughter DNA molecule ends up with one of the original strands and one newly synthesized strand.

❐ post-replication repair

A recombination-dependent mechanism for repairing damaged DNA.

❐ post-translational modification

The addition of specific chemical residues to a protein after it has been translated. Common residues are phosphate groups (phosphorylation) and sugars (glycosylation).

❐ ppf

Abbreviation for photosynthetic photon flux.

❐ ppm

Abbreviation for parts per million.

❐ pr protein

Abbreviation for pathogenesis related protein.

❐ precautionary principle

The approach whereby any possible risk associated with the introduction of a new technology is avoided, until a full understanding of its impact on health, environment etc. is available. Particularly applied to the release of genetically modified organisms, since unlike many technologies, these cannot be recalled if problems arise.

❐ precocious germination

Early germination of a seed or embryo, prior to the full maturation of the embryo.

❐ pre-filter

A coarse filter used to screen out large particles from a fluid or gas, before it is passed through a much finer filter.

❑ pressure potential

The pressure generated within a cell, being the net difference between the cell's osmotic potential and the water potential of the external environment.

❑ pre-transplant

A stage in micropropagation - the rooting and hardening process prior to transfer to soil.

❑ preventive immunization

Infection with an antigen to elicit an antibody response that will protect the organism against future infections.

❑ Pribnow box

Consensus sequence near the mRNA start-point of prokaryotic genes.

❑ primary antibody

In an ELISA or other immunological assay, the antibody that binds to the target molecule.

❑ primary cell

A cell or cell line taken directly from a living organism, which is not immortalized.

❑ primary cell wall

The cell wall layer formed during cell expansion. Plant cells possessing only primary walls may divide or undergo differentiation.

❑ primary culture

A culture started from cells, tissues or organs taken directly from organisms. A primary culture may be regarded as such until it is sub-cultured for the first time. It then becomes a cell line.

❑ primary

First in order of time or development.

❑ primary growth

1. Apical meristem-derived growth; the tissues of a young plant.
2. Explant growth during the initial culture period.

❑ primary immune response

The immune response that occurs during the first encounter of a mammal with a given antigen.

❑ primary meristem

Meristem of the shoot or root tip giving rise to the primary plant body.

❑ primary structure

The linear sequence of residues making up a polymer such as a nucleic acid, polysaccharide or protein.

❐ primary tissue

A tissue that has differentiated from a primary meristem.

❐ primary transcript

The RNA molecule produced by transcription prior to any post-transcriptional modifications; also called a pre-mRNA in eukaryotes.

❐ primer

A short oligonucleotide annealed to a template of single-stranded DNA, providing a doubled stranded structure from which DNA polymerase will synthesize a new DNA strand to produce a duplex molecule.

❐ primer walking

A method for sequencing long (>1 kbp) cloned pieces of DNA. The initial sequencing reaction reveals the sequence of the first few hundred nucleotides of the cloned DNA. Using this, a new primer of about 20 nucleotides is synthesized, which is complementary to a sequence near the end of sequenced DNA, and used to sequence the next few hundred nucleotides of the cloned DNA. This procedure is repeated until the complete nucleotide sequence of the cloned DNA is determined.

❐ primordium

A group of cells which gives rise to an organ.

❐ primosome

A protein-replication complex that catalyses the initiation of synthesis of Okazaki fragments during discontinuous replication of DNA. It involves DNA primase and DNA helicase activities.

❐ probability

The frequency of occurrence of an event.

❐ proband

The individual in a family in whom an inherited trait is first identified.

❐ probe

A labelled DNA or RNA sequence used to detect the presence of a complementary sequence by hybridization with a nucleic acid sample.

❐ procambium

A primary meristem that gives rise to primary vascular tissues and, in most woody plants, to the vascular cambium.

❐ processed pseudo-gene

A copy of a functional gene which has no promoter, no introns and which, consequently, is not itself transcribed.

❑ production environment

All input-output relationships, over time, at a particular location. The relationships include biological, climatic, economic, social, cultural and political factors, which combine to determine the productive potential of a particular enterprise. Production environments are classified as *high-*, *medium-* and *low-input*

❑ production of acetyl-coa

The general pathways for the production of acetyl CoA from sugars and fats. The mitochondrion in eucaryotic cells is the place where acetyl CoA is produced from both types of major food molecules. It is therefore the place where most of the cell's oxidation reactions occur and where most of its ATP is made.

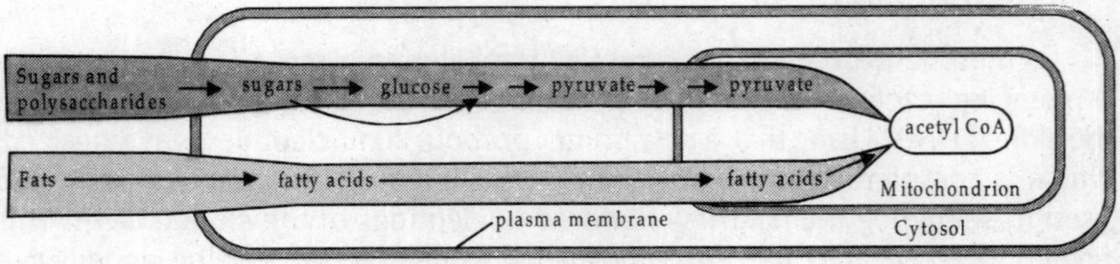

❑ production traits

Characteristics of animals, such as the quantity or quality of the milk, meat, fibre, eggs, work, etc., they (or their progeny) produce, which contribute directly to the value of the animals for the farmer, and that are identifiable or measurable at the individual level.

❑ productivity

The amount of economically significant product generated within a given period of time from a specified quantity of resource.

❑ pro-embryo

A group of cells arising from the division of the fertilized egg cell or somatic embryo before those cells which are to become the embryo are recognizable.

❑ progeny

Synonym of offspring.

❑ progeny testing

With respect to discrete loci, the inference of the allelic state of an individual from the pattern of segregation among its offspring. For a quantitative trait, the use of progeny performance to estimate the breeding value of an individual.

❑ progesterone

A hormone produced primarily by the corpus luteum, but also by the placenta. Its function in mammals is to prepare the inner lining of the uterus for the implantation of a fertilized egg. Also made by non-placental animals, including fish.

❐ prokaryote

A member of the large group of organisms, including bacteria and blue-green algae, in which the chromosome is not enclosed within a nucleus, but instead exists as a linear or circular strand. Prokaryotes do not undergo meiosis and do not have functional organelles such as mitochondria and chloroplasts.

❐ prolactin

A hormone, produced by the anterior pituitary gland, that stimulates and controls lactation in mammals.

❐ proliferation

Increase by frequent and repeated reproduction; growth by cell division.

❐ pro-meristem

The embryonic meristem that is the source of organ initials or foundation cells.

❐ promoter

1. A short DNA sequence, usually upstream of (5' to) the relevant coding sequence, to which RNA polymerase binds before initiating transcription. This binding aligns the RNA polymerase so that transcription will initiate at a specific site. The nucleotide sequence of the promoter determines the nature of the enzyme that attaches to it and the rate of RNA synthesis.
2. A chemical substance that enhances the transformation of benign cells into cancerous cells.

❐ pro-nuclear micro-injection

An early, low success-rate method to achieve transgenesis in animals, involving the micro-injection of many gene copies into one of the two pro-nuclei of a fertilized egg. Now being replaced by micro-injection into a culture of cloned embryos produced by nuclear transfer, which can be tested for expression of the transgene before transfer to recipient females.

❐ pro-nucleus

Either one of the two haploid gamete nuclei, just prior to their fusion in the fertilized egg.

❐ proofreading

The scanning of newly-synthesized DNA for structural defects, such as mismatched base pairs. A functional activity of most DNA polymerases.

❐ propagation of an action potential

(A) shows the voltages that would be recorded from a set of intracellular electrodes placed at intervals along the axon, whose width is greatly exaggerated in this schematic figure. Note that the action potential does not weaken as it travels. (B) shows the changes in the Na+ channels and the current flows, that give rise to the traveling disturbance of the membrane potential. The region of the axon with a

depolarized membrane. An action potential can only travel away from the site of depolarization because Na+-channel inactivation prevents the depolarization from spreading backward.

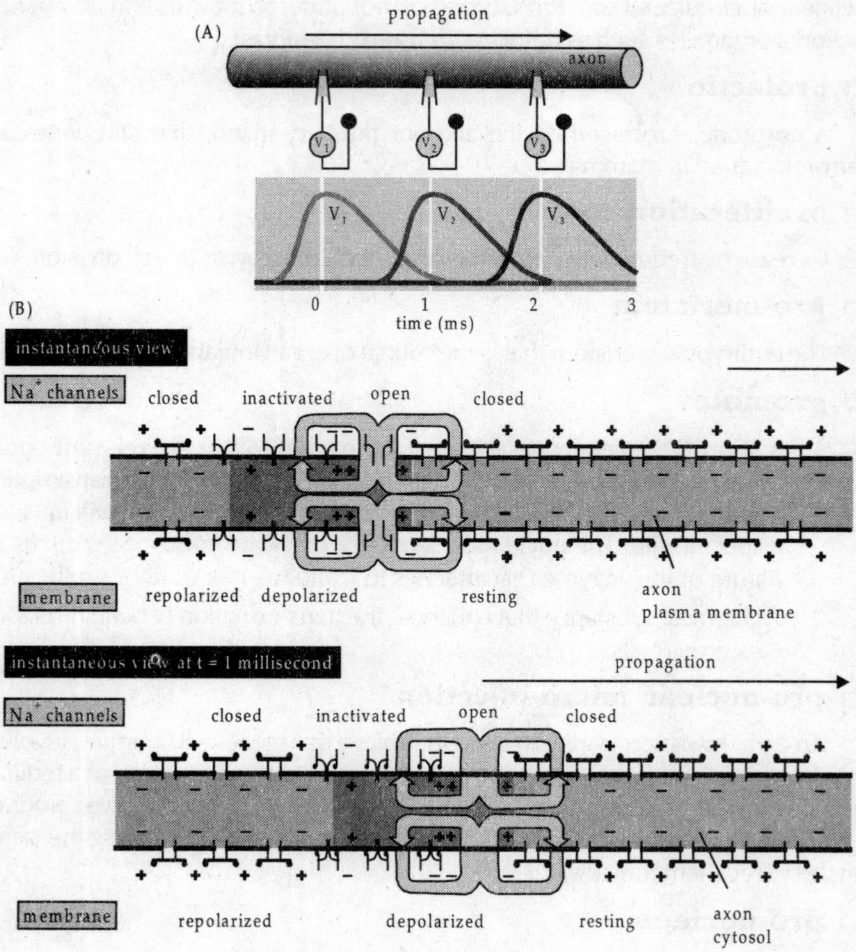

❑ propagation

The duplication of a whole plant from a range of vegetative materials; adapted for *in vitro* culture as micropropagation.

❑ propagule

Any structure capable of giving rise to a new plant by asexual or sexual reproduction, including bulbils, leafbuds, etc.

❑ pro-phage

The genome of a bacteriophage integrated into the chromosome of a lysogenic bacterial cell, and replicated along with its host chromosome.

▫ prophase

The first stage of nuclear division. The stage during which chromosome pairing occurs in the first division of meiosis (*see*: leptonema, zygonema, pachynema, diplonema, diakinesis). In mitosis and the second division of meiosis, the chromosomes shorten and thicken as a result of coiling.

▫ protamine

A class of small basic proteins that replace the histones in the chromosomes of some sperm cells.

▫ protease

An enzyme that catalyses the hydrolysis of proteins, cleaving the peptide bonds that link amino acids in protein molecules. *Synonym*: peptidase.

▫ protein

A large molecule composed of one or more chains of amino acids in a specific order; the order is determined by the base sequence of nucleotides in the gene that codes for the protein. Proteins are required for the structure, function, and regulation of the body's cells, tissues, and organs; and each protein has unique functions. Examples are hormones, enzymes, and antibodies.

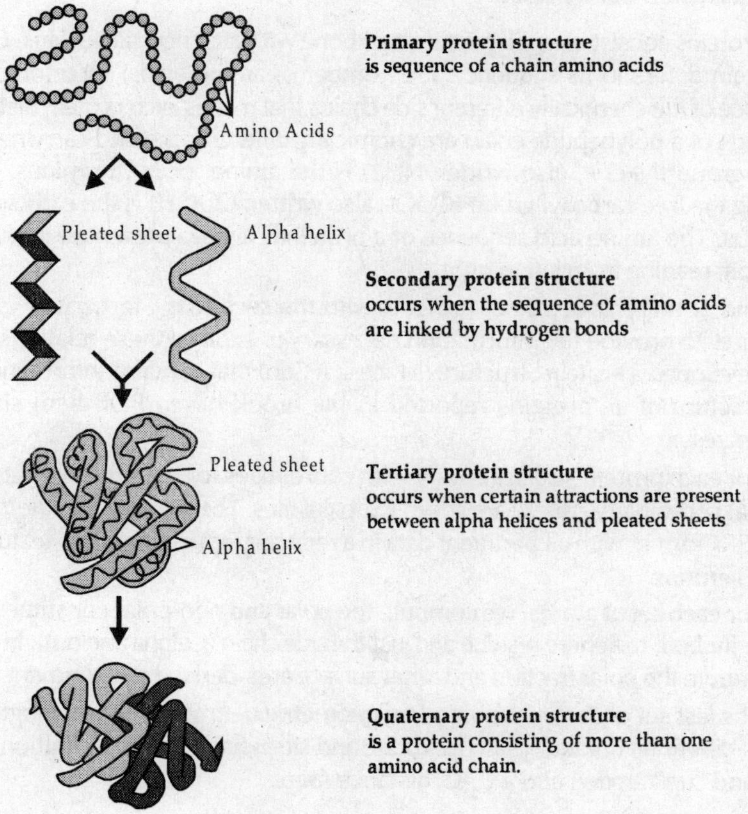

▫ protein crystallization

The production of a pure preparation of a protein. In this form, the three-dimensional structure of the molecule can be determined.

▫ protein engineering

Generating proteins with modified structures that confer novel properties such as higher catalytic specificity or thermal stability.

▫ protein kinase

An enzyme that catalyses the addition of a phosphate group(s) to a protein molecule at the sites of serine, threonine or tyrosine residues.

▫ protein metabolic step

One step in the chain of reactions that take place in an organism and dictate the composition of that organism.

▫ protein sequencing

The process of determining the amino acid sequence of a protein. Usually achieved following initially partial hydrolysis of the protein into smaller peptides by enzymatic digestion.

▫ protein structure

Proteins consist of a polypeptide backbone with attached side chains. Each type of protein differs in its sequence and number of amino acids; therefore, it is the sequence of the chemically different side chains that makes each protein distinct. The two ends of a polypeptide chain are chemically different: the end carrying the free amino group (NH3+, also written NH2) is the amino, or N-, terminus, and that carrying the free carboxyl group (COO-, also written COOH) is the carboxyl, or C-, terminus. The amino acid sequence of a protein is always presented in the N to C direction, reading from left to right.

The primary structure of proteins with the secondary, tertiary or quanteary structure. To provide the information necessary to explore these relationships, we have developed a Protein Structure database (PSdb) that contains information about the structure of all proteins reported in the brook haven PDB with side chain coordinates.

For each protein structure entry with coordinates for the side chain atoms (not C(alpha) coordinates), there are seven PSdb flat files. The primary file (the "seq" file) is in NBRF format with all pertinent data in a reduced representations "feature table" ("F;...") entries.

For each set of atoms, we compute the polar and non-polar constituents of the surface for both the entire residue and just the side chain (Calpha and out). In addition, we compute the polar fraction and other surface area-derived environment terms.

The last set of data is the standard geometric definitions: bond lengths, bond angles, dihedrals (including all backbone and side chain), plane definitions (center point and "up" arrow) and a $C_{\acute{a}}$-$C_{\acute{a}}$ distance map.

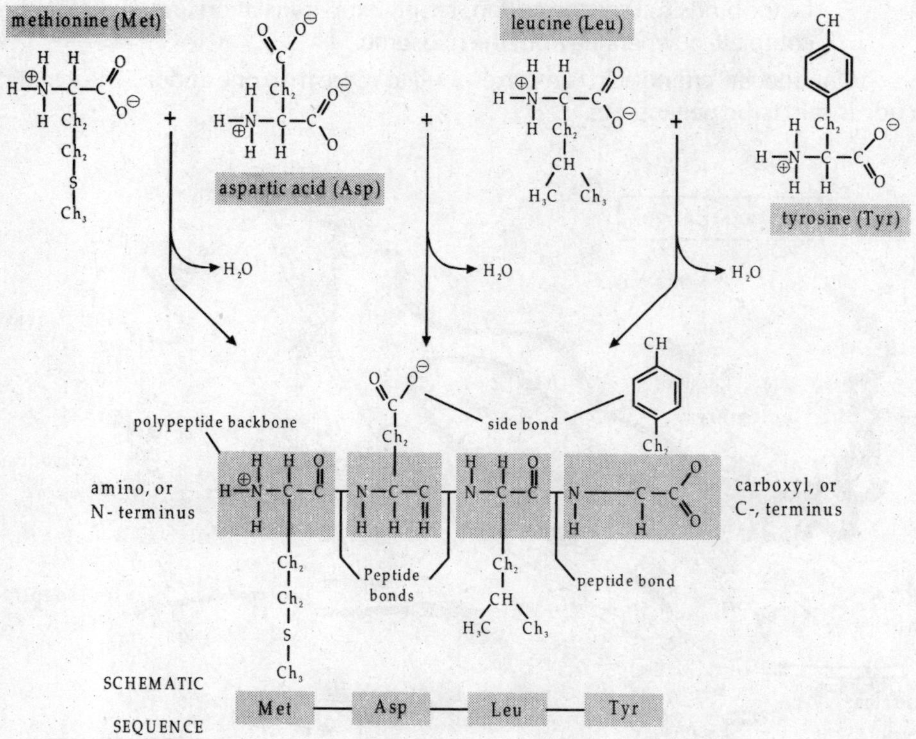

☐ protein synthesis

Process whereby DNA encodes for the production of amino acids and proteins. This process can be divided into two parts:

1. **Transcription:** Before the synthesis of a protein begins, the corresponding RNA molecule is produced by RNA transcription. One strand of the DNA double helix is used as a template by the RNA polymerase to synthesize a messenger RNA (mRNA). This mRNA migrates from the nucleus to the cytoplasm. During this step, mRNA goes through different types of maturation including one called splicing when the non-coding sequences are eliminated. The coding mRNA sequence can be described as a unit of three nucleotides called a codon.
2. **Translation:** The ribosome binds to the mRNA at the **start codon** (AUG) that is recognized only by the initiator tRNA. The ribosome proceeds to the elongation phase of protein synthesis. During this stage, complexes, composed of an amino acid linked to tRNA, sequentially bind to the appropriate codon in mRNA by forming complementary base pairs with the tRNA **anticodon**. The ribosome moves from codon to codon along the

mRNA. Amino acids are added one by one, translated into polypeptidic sequences dictated by DNA and represented by mRNA. At the end, a release factor binds to the stop codon, terminating translation and releasing the complete polypeptide from the ribosome.

One specific amino acid can correspond to more than one codon. The genetic code is said to be degenerate.

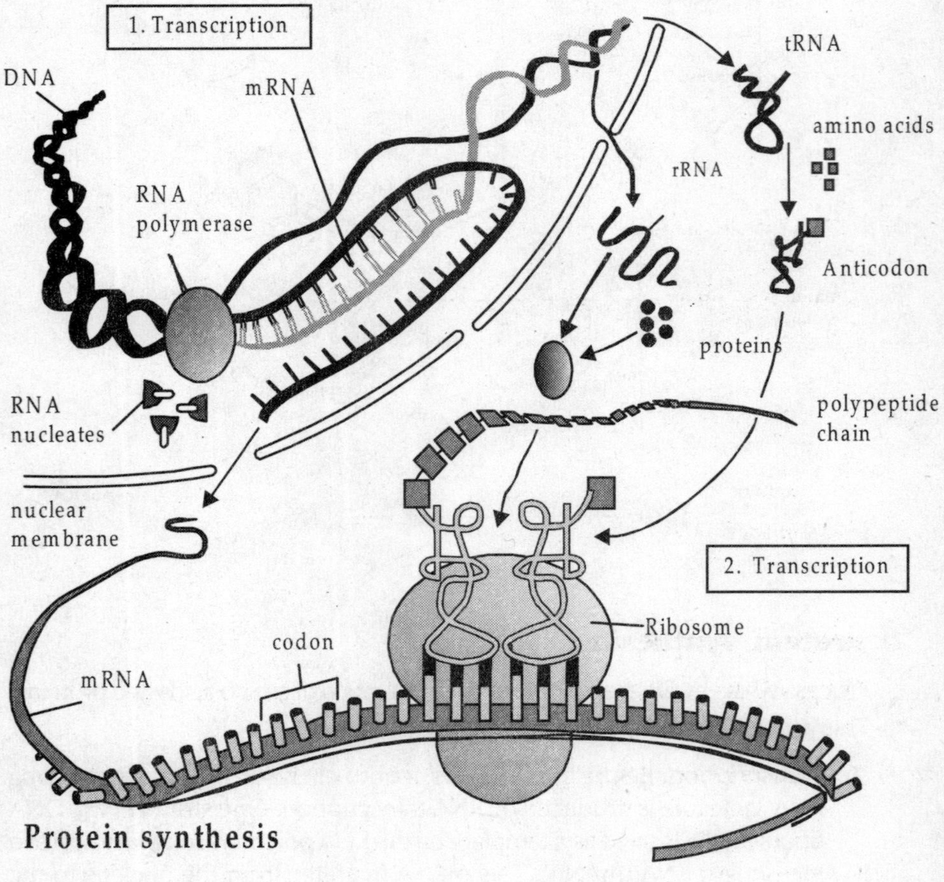

Protein synthesis

☐ proteinaceous infectious particle

Believed to be the agent responsible for the class of diseases called spongiform encephalopathies, including scrapie in sheep, bovine spongiform encephalopathy (BSE; mad cow disease) in cattle and CJD in humans. It is an abnormal form of a brain protein, and has no detectable nucleic acid content.

☐ proteolysis

Enzymatic degradation of a protein.

❐ proteolytic

Having the ability to degrade protein molecules.

❐ proteome

The complete complement of proteins made by a given species in all its tissues and growth stages.

❐ proteomics

An approach that seeks to identify and characterize complete sets of protein, and protein-protein interactions in a given species.

❐ protoclone

Regenerated plant derived from protoplast culture or a single colony derived from protoplasts in culture.

❐ protocol

The step-by-step experiments proposed to describe or solve a scientific problem, or the defined steps of a specific procedure.

❐ protocorm

A tuberous structure, formed following the germination of orchid seeds, from which develops a complete plant. It develops from an unorganized embryo in the seed, comprising only a few hundred cells. In culture, vegetative explants of several orchid species form round, smooth protocorms which can be multiplied indefinitely or induced to regenerate into a whole plant.

❐ protoderm

A primary meristem tissue that gives rise to epidermis.

❐ protogyny

The condition in which the female reproductive organs (carpels) of a flower mature before the male ones (stamens), thereby ensuring that self-fertilization does not occur.

❐ proto-oncogene

A normal gene that can be mutated to an oncogene. *Synonym*: cellular oncogene.

❐ protoplasm

The essential, complex living substance of cells, upon which all vital functions of nutrition, secretion, growth and reproduction depend.

❐ protoplast

A bacterial or plant cell for which the cell wall has been removed either chemically or enzymatically, leaving its cytoplasm enveloped by a peripheral membrane. Protoplasts are spherical and smaller than the elongate, angular shaped and often vacuolated cells from which they have been released.

▫ protoplast culture

The culturing *in vitro* of plant protoplasts. Where protoplasts can be regenerated into whole plants, they represent an attractive target for genetic manipulation.

▫ protoplast fusion

The induced or spontaneous coalescence of two or more protoplasts of the same or different species origin. Where fused protoplasts can be regenerated into whole plants, the opportunity exists for the creation of novel genomic combinations.

▫ prototroph

A nutritionally independent cell.

▫ pro-toxin

A latent, non-active precursor form of a toxin.

▫ protozoan

A microscopic, single-cell organism.

▫ provenance

The geographical and/or genetic origin of an individual.

▫ provirus

A double stranded DNA copy of the single RNA strand of a retrovirus, which has been integrated into a host genome.

▫ pseudo-affinity chromatography

A chromatographic technique in which a ligand is immobilized selectively to retain enzymes or other proteins.

▫ pseudo-autosomal region

A section at one end of the X and Y chromosomes for which there is sufficient homology that there is synapsis between them during meiosis.

▫ pseudocarp

A fruit that incorporates, in addition to the ovary wall, other parts of the flower, such as the receptacle (e.g. strawberry).

▫ pseudogene

An incomplete or mutated copy of a gene which is not transcribed because it lacks a continuous open reading frame. Those that lack introns are called processed pseudogenes and are most likely cDNA copies synthesized from mRNA by reverse transcriptase

▫ *Pseudomonas* spp.

A widely distributed Gram-negative bacterial genus. Many of the soil forms produce a pigment that fluoresces under ultraviolet light, hence the descriptive term fluorescent *Pseudomonas*.

p-site

Abbreviation for peptidyl-tRNA binding site.

psychrophile

A micro-organism that can grow at temperatures below 30 °C and as low as 0 °C. *See*: mesophile, thermophile.

puc

A widely used plasmid, containing as a marker a galactosidase gene.

pulsed-field gel electrophoresis

A procedure used to separate very large (50 kbp to several Mbp) DNA molecules by alteRNAting the direction of electric current in a pulsed manner across a gel.

punctuated equilibrium

The occurrence of speciation events in bursts, separated by long intervals of species stability.

pure line

A strain in which all members are genetically nearly identical and are indistinguishable by phenotype. Usually created by repeated generations of self-fertilization or close inbreeding.

purine

A double-ring, nitrogen-containing base present in nucleic acids. Adenine (A) and guanine (G) are the two purines normally present in DNA and RNA molecules.

pvp

1. Abbreviation for polyvinylpyrrolidone.
2. Abbreviation for plant variety protection.

pvr

Abbreviation for plant variety rights.

pwp

Abbreviation for permanent wilting point.

pyrethrins

Active constituents of pyrethrum (*Tanacetum cinerariifolium*) flowers, used as insecticides.

pyrimidine

A single-ring, nitrogen-containing base present in nucleic acids. Cytosine (C) and thymine (T) are present in DNA, whereas uracil (U) replaces T in RNA. Thymine is a synonym for 5-methyluracil.

☐ pyrogen

Bacterial substance that causes fever in mammals.

☐ pyrophosphate

A phosphate ion dimer; may be released on hydrolysis of ATP.

Q

❑ q

Denotes the longer of the two chromosome arms, e.g. human 10q is the long arm of human chromosome 10.

❑ q-beta replicase

A viral RNA polymerase secreted by a bacteriophage that infects *E. coli*. It has the property of being able to copy RNA sequences at a rapid rate.

❑ qsar

Abbreviation for quantitative structure-activity relationship.

❑ qtl

Abbreviation for quantitative trait locus.

❑ quadrivalent

A chromosome configuration visible in late prophase and metaphase of the first meiotic division, where four chromosomes are linked by chiasmata. Can occur in autotetraploids when four homologous chromosomes pair, or in diploids as a result of heterozygosity for a reciprocal translocation between two non-homologous chromosomes.

❑ quadruplex

The inheritance of alleles in autotetraploids. A genotype *AAAa* will produce gametes *AA*, *Aa* in the ratio 3:1.

❑ qualitative trait

A trait that shows discontinuous variation - i.e. individuals can be assigned to one of a small number of discrete classes.

❑ quantitative genetics

The area of genetics concerned with the inheritance of quantitative traits that show continuous variation, as opposed to qualitative traits. Since many of the critical targets in both plant and animal breeding are of this type, most practical improvement programs involve the application of quantitative genetics.

□ quantitative inheritance

Inheritance of measurable traits that depend on the cumulative action of many genes and/or involve a significant proportion of non-genetic determination.

□ quantitative structure-activity relationship

A computer modelling technique that enables the prediction of the likely activity of a molecule before it is synthesized. QSAR analysis relies on recognizing associations of molecular structures and activity from historical data.

□ quantitative trait

A measurable trait that shows continuous variation (e.g. height, weight, colour intensity, etc.) - i.e. the population cannot be classified into a few discrete classes.

□ quantitative trait locus

A locus where allelic variation is associated with variation in a quantitative trait. The presence of a QTL is inferred from genetic mapping, where the total variation is partitioned into components linked to a number of discrete chromosome regions.

□ quantum speciation

The rapid formation of new species, primarily by genetic drift.

□ quarantine

Isolation for a period after arrival in a new location, to allow any pre-existing disease symptoms to appear. Used in the context of regulations restricting the sale or shipment of living organisms, usually to prevent disease or pest invasion of an area.

□ quaternary structure

A level of protein structure where several individual molecules assemble together and form a functional cluster. A classic example is haemoglobin, a complex of four myoglobin-like units.

□ quiescent

A temporary suspension or reduction in the rate of activity or growth, while retaining the potential to resume prior activity. Applies particularly to cell division.

R

□ *r* genes

A class of plant genes conferring resistance to a specific strain (or group of strains) of a particular pathogen. Their primary function is to sense the presence of the pathogen and to trigger the defence pathways in the plant. *R* genes have been cloned from a number of plant species.

□ r_1

The first-generation offspring of a recombinant (genetically modified) organism. Not standard terminology.

□ race

A distinguishable group of organisms of a particular species. Criteria for distinctness can be one or a combination of geographic, ecological, physiological, morphological, genetic and karyotypic factors.

□ raceme

An inflorescence in which the main axis is elongated but the flowers are borne on pedicels that are about equal in length.

□ rachilla

Shortened axis of a spikelet.

□ rachis

Main axis of a spike; axis of fern leaf (frond) from which pinnae arise; in compound leaves, the extension of the petiole corresponding to the midrib of an entire leaf.

□ radiation hybrid cell panel

A somatic cell hybrid panel in which the chromosomes from the species of interest have been fragmented by irradiation prior to cell fusion. The resultant small fragments of chromosomes greatly increase the power of physical mapping in the species of interest.

□ radicle

The portion of the plant embryo which develops into the primary root.

❐ radioimmunoassay

An assay based on the use of a radioactively labelled antibody, where the amount of radiation detected indicates the amount of target substance present in the sample.

❐ radioisotope

An unstable isotope that emits ionizing radiation.

❐ ramet

An individual member of a clone, descended from the ortet.

❐ random amplified polymorphic dna

A PCR-based genotyping technique in which genomic template is amplified with single, short (usually 10-mer) randomly chosen primers. Typical patterns consist of a small number of amplified products of up to 2 kbp in length, which are separated by electrophoresis.

❐ random mutagenesis

A non-directed change of one or more nucleotide pairs in a DNA molecule.

❐ random primer method

A method for labelling DNA probes, mainly for Southern hybridization experiments. A mixture of short oligonucleotides is hybridized to a single-stranded DNA probe. In the presence of DNA polymerase and deoxyribonucleotides - one of which is labelled - DNA synthesis generates labelled copies of probe DNA.

❐ rapd

Abbreviation for random amplified polymorphic DNA.

❐ rate-limiting enzyme

The enzyme whose activity controls the output of final product from a multi-enzyme metabolic pathway.

❐ rational drug design

A systematic method of creating compounds by analysing their structure, function and stereochemical interactions.

❐ reading frame

The reading frame defines which sets of three nucleotides are read as triplets, and hence as codons, in DNA transcription. The start point is usually determined by the initiation codon, AUG. Thus the sequence AUGGCAAAA would be read as AUG/GCA/AAA not as A/UGC/CAA/AA. *See*: open reading frame.

❐ read-through

Transcription or translation that proceeds beyond the normal stopping point because of the absence of the usual transcription or translation termination signal of a gene.

❐ reca

A protein, found in most bacteria, that is essential for DNA repair and DNA recombination.

❐ recalcitrant

Of seeds, unable to survive drying and subsequent storage at low temperature.

❐ receptacle

Enlarged end of the pedicel or peduncle, to which other flower parts are attached.

❐ receptor

A trans-membrane protein located in the plasma membrane that can bind with a ligand on the extracellular surface, as a result of which it induces a change in activity on the cytoplasmic surface. More generally, a site in a molecule that allows the binding of a ligand.

❐ receptor-binding screening

A biotechnology-based method for drug discovery, which relies on the fact that many drugs act by binding to specific receptors on or in cells. Since receptors *in vivo* bind to hormones or to other cells, and thereby control the cell's behaviour, a receptor bound with a drug will likely affect the normal activity of the cell.

❐ recessive and dominant inheritance

Mechanisms of recessive and dominant inheritance.

Recessive Inheritance: Both parents carry a normal gene (**N**), and a faulty, recessive, gene (**n**). The parents, although carriers, are unaffected by the faulty gene. Their offspring are affected, not affected, or carriers. This type of inheritance was first shown by Mendel.

Dominant Inheritance: One parent has a single, faulty dominant gene (**D**), which overpowers its normal counterpart (**d**), affecting that parent. When the affected parent mates with an unaffected and non-carrier mate (**dd**), the offspring are either affected or not affected, but they are not carriers.

Recessive inheritance

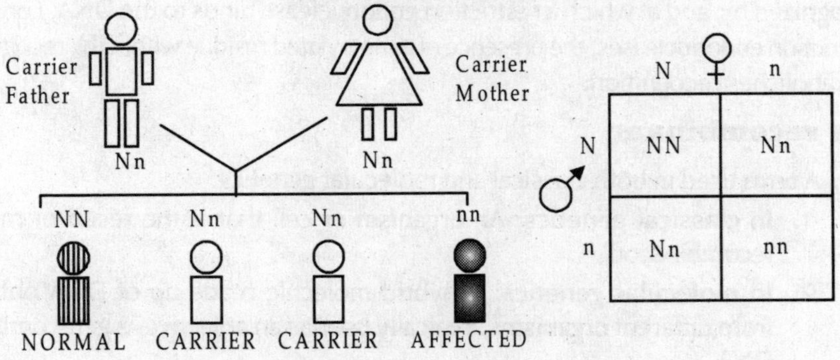

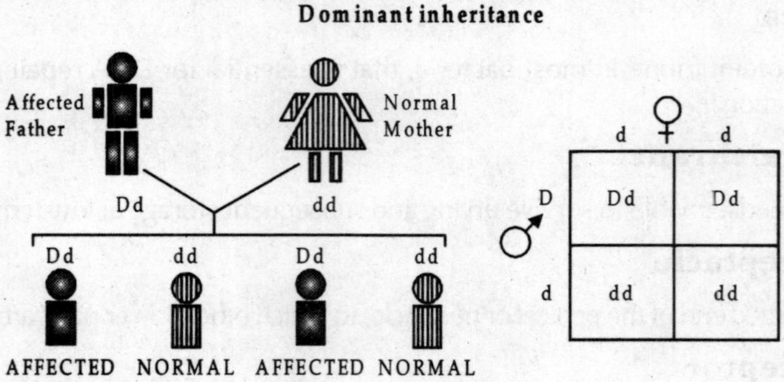

❐ recessive

Describing an allele whose effect with respect to a particular trait is not evident in heterozygotes.

❐ recessive allele

Allelic state of a gene, where homozygosity is required for the expression of the relevant phenotype.

❐ recessive oncogene

A single copy of this gene is sufficient to suppress cell proliferation; the loss of both copies of the gene contributes to cancer formation. *Synonym*: anti-oncogene recessive-acting oncogene.

❐ reciprocating shaker

A platform shaker used for agitating culture flasks, with a back and forth action at variable speeds.

❐ recognition sequence

Synonym of recognition site.

❐ recognition site

A nucleotide sequence, typically 4-8bp long and often palindromic, that is recognized by, and at which a restriction endonuclease binds to the DNA. For some restriction endonucleases, the presence of a methylated residue within the recognition site abolishes recognition.

❐ recombinant

A term used in both classical and molecular genetics.

1. **In classical genetics:** An organism or cell that is the result of meiotic recombination.
2. **In molecular genetics:** A hybrid molecule made up of DNA obtained from different organisms. Typically used as an adjective, e.g. recombinant DNA.

❐ recombinant DNA technology

A set of techniques for manipulating DNA, including: the identification and cloning of genes; the study of the expression of cloned genes; and the production of large quantities of gene product.

❐ recombinant DNA

The result of combining DNA fragments from different sources.

❐ recombinant human

A prefix denoting molecules made through the use of recombinant DNA technology.

❐ recombinant protein

A protein encoded by a cloned gene. *Synonym*: heterologous protein.

❐ recombinant RNA

RNA molecules joined *in vitro* by T4 RNA ligase.

❐ recombinant toxin

A single multifunctional toxic protein encoded by a recombinant gene.

❐ recombinant vaccine

A vaccine produced from a cloned gene.

❐ recombinase

A class of enzymes that are able to alter the arrangement of DNA sequences in a site-specific way.

❐ recombination fraction

The proportion of recombinant (with respect to two loci) gametes arising from meiosis. Linkage maps are based on estimates of recombination fraction between all pair-wise combinations of loci. *See*: map distance. *Synonyms*: recombination frequency, crossing-over unit.

❐ recombination frequency

Synonym: recombination fraction

❐ recombination

The production of a DNA molecule with segments derived from more than one parent DNA molecule. In eukaryotes, this is achieved by the reciprocal exchange of DNA between non-sister chromatids within an homologous pair of chromosomes during prophase of the first meiotic division.

❐ recombinational hot spot

A chromosomal region where recombination appears to occur more frequently than expected.

❑ reconstructed cell

A viable transformed cell resulting from genetic engineering.

❑ reduction division

The first division of meiosis in which the chromosome number is reduced from the somatic to the gametic number.

❑ refugium

An area set aside to provide protection/escape from ecological consequences occurring elsewhere.

❑ regeneration

The growth of new tissues or organs to replace those injured or lost. In plant tissue culture, regeneration refers to the development of organs or plantlets from an explant. *See*: conversion; micropropagation; organogenesis.

❑ regulator

Substance regulating growth and development of cells, organs, etc.

❑ regulatory gene

A gene with the primary function of controlling the rate of synthesis of the products of one or several other genes or pathways.

❑ regulatory protein coordinating gene expression

The action of the glucocorticoid receptor is illustrated. On the left is shown a series of genes, each of which has various gene activator proteins bound to its regulatory region. However, these bound proteins are not sufficient on their own to activate transcription efficiently. On the right is shown the effects of adding an additional gene regulatory protein.

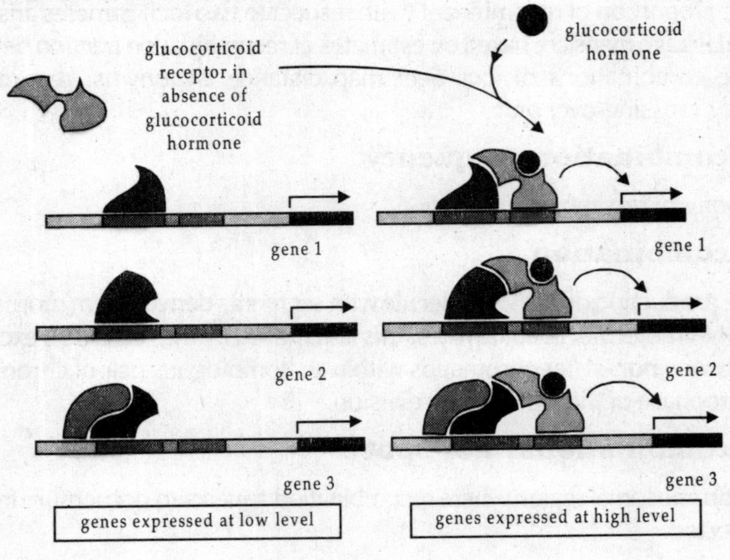

❑ regulatory sequence

A DNA sequence involved in regulating the expression of a gene, e.g. a promoter or operator region (in the DNA molecule).

❑ rejuvenation

1. Reversion from adult to juvenile stage.
2. The process of regular reproduction of seed stocks or collections in gene banks, in order to ensure continued viability.

❑ relative sizes of cells and their components

The sizes of cells and of their component parts, and the units in which they are measured.

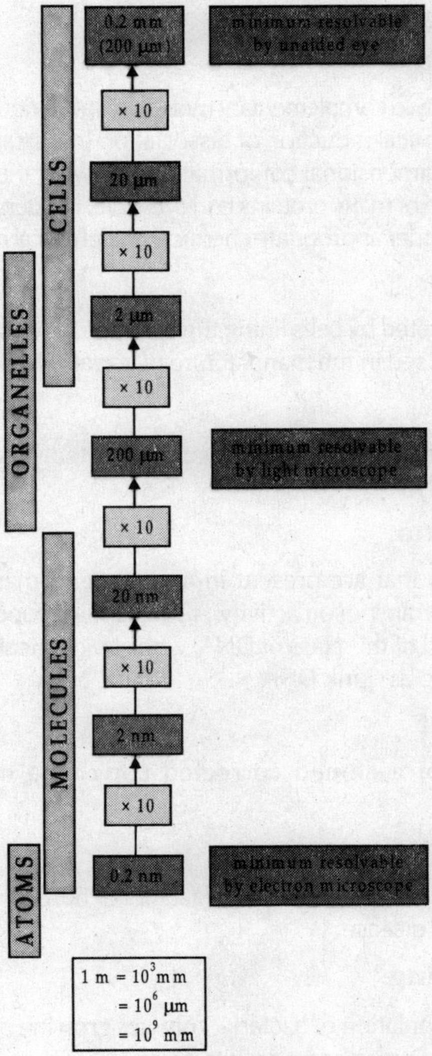

❐ relaxed plasmid

A plasmid that replicates independently of the bacterial chromosome and is present in 10-500 copies per cell.

❐ release factor

1. A soluble protein that recognizes termination codons in mRNAs and terminates translation in response to these codons.
2. A hormone, produced by the hypothalamus, which stimulates the release of a hormone from the anterior pituitary gland into the bloodstream.

❐ remediation

The cleanup or containment of a hazardous waste disposal site to the satisfaction of the applicable regulatory agency. This can sometimes be accomplished with naturally occurring or engineered micro-organisms or plants.

❐ renaturation

The reforming of two complementary molecules into a double-stranded structure, following heat or chemical induction of dissociation (denaturation). Of protein, the resumption of three-dimensional conformation, allowing the molecule to function normally. Denaturation of many proteins is irreversible, but denatured DNA molecules will renature readily under appropriate chemical and physical conditions.

❐ rennin

An enzyme, secreted by cells lining the stomach in mammals, responsible for the clotting of milk. Used in the manufacture of certain dairy products.

❐ repeat unit

A sequence of nucleotides that occurs repeatedly, often in a head-to-tail arrangement (tandemly).

❐ repetitive dna

DNA sequences that are present in a genome in many copies, some of it originating from retrotransposon activity. A substantial proportion of all eukaryotic genomes is composed of this class of DNA, whose biological function is uncertain. Sometimes referred to as 'junk DNA'.

❐ replacement

The addition of a cloned corrected copy of a defective gene. *See*: homogenotization.

❐ replacement therapy

The administration of metabolites, co-factors or hormones that are deficient as the result of a genetic disease.

❐ replica plating

Duplicating a population of bacterial colonies growing on agar medium in one Petri plate to agar medium in another Petri plate.

❑ replicase

A viral enzyme necessary for the replication of the virus in the host cell.

❑ replication fork

Y shaped structure associated with DNA replication. It represents the point at which the strands of double-stranded DNA are separated so that replication can proceed.

❑ replication

The *in vivo* synthesis of double-stranded DNA by copying from a single-stranded template.

❑ replicative form

The molecular configuration of viral nucleic acid that is the template for replication in the host cell.

❑ replicon

The portion of a DNA molecule which can be replicated from a single origin of replication. Plasmids and the chromosomes of bacteria, phages and other viruses usually have a single origin of replication so that their entire genome constitutes a single replicon. Eukaryotic chromosomes have multiple origins of replication, so comprise several replicons. Also used to describe a DNA molecule capable of independent replication.

❑ replisome

The complete replication apparatus, present at a replication fork, that carries out the replication of DNA.

❑ reporter gene

A gene that encodes a product that can be readily assayed. Used as a marker to confirm the incorporation of a transgene into a cell, organ or tissue, and as a means of testing the efficiency of specific promoters.

❑ repressible enzyme

An enzyme whose activity can be diminished by the presence of a regulatory molecule.

❑ repressible gene

A gene whose expression can be diminished or extinguished by the presence of a regulatory molecule.

❑ repression

Inhibition of transcription by preventing RNA polymerase from binding to the transcription initiation site.

❑ repressor

A protein which binds to a specific DNA sequence upstream from the transcription initiation site of a gene and prevents RNA polymerase from commencing mRNA synthesis.

❑ repressor protein switching genes on and off

If the level of tryptophan inside the cell is low, the tryptophan repressor protein does not bind tryptophan and thus cannot bind to its control region.

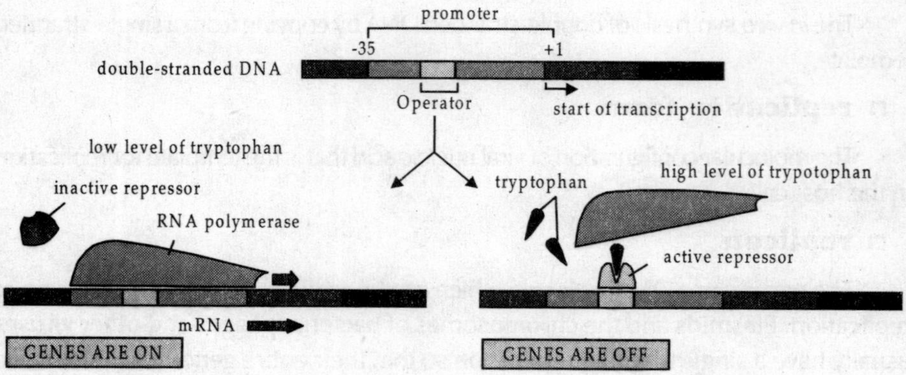

❑ reproduction

1. Sexual reproduction: the regular alternation of meiosis and fertilization which provides for the production of offspring. The main biological significance of sexual reproduction lies in the phenomenon of recombination.
2. Asexual or agamic reproduction: the development of a new individual from a single cell or group of cells in the absence of meiosis.

❑ repulsion

A double heterozygote in which the dominant (or wild-type) allele at one locus and the recessive (or mutant) allele at a second linked locus occur on the same chromosome (genetic constitution *Ab/aB*). *Synonym*: *trans* configuration. *Opposite*: coupling, *cis* configuration.

❑ residue

Materials remaining after degradation and/or attempted removal, e.g. pesticide residues in food.

❑ resistance factor

A plasmid that confers antibiotic resistance to a bacterium.

❑ resistance

The ability to withstand abiotic (high temperature, drought etc.) or biotic (disease) stress, or a toxic substance. Often in the context of genetic determination of resistance.

▫ rest period

A physiological condition of viable seeds, buds or bulbs that prevents growth even in the presence of otherwise favourable environmental conditions.

▫ restitution nucleus

A single nucleus arising from a failure of nuclear division, either during meiosis, in which a gamete is formed with the unreduced chromosome number; or at mitosis to give a cell with a doubled chromosome number.

▫ restriction endonuclease

A class of enzymes that cut DNA after recognizing a specific sequence. The three types of restriction endonuclease are: I. Where the cut occurs within a random sequence at sites >1kbp from the recognition sequence, and has both restriction and methylation activities. II: Cuts within, or near a short, usually palindromic recognition sequence. A separate enzyme methylates the same recognition sequence. III: Cuts 24-26bp downstream from a short, asymmetrical recognition sequence, requires ATP and has both restriction and methylation activities. Type II enzymes are the class used for most molecular biology applications.

▫ restriction enzyme

Synonym of restriction endonuclease.

▫ restriction enzyme: action of ecori

Restriction enzymes, also called restriction nucleases *(EcoRI in this example)*, surrounds the DNA molecule at the point it seeks*(sequence GAATTC)*. It cuts one strand of the DNA double helix at one point and the second strand at a different, complementary point (between the G and the A base).

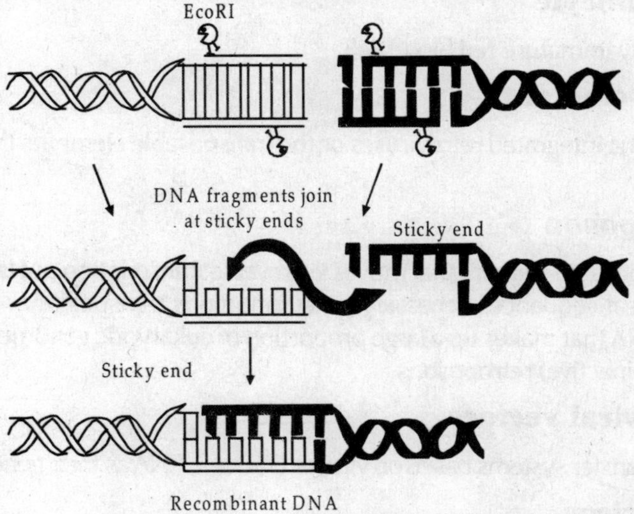

Restriction Enzyme
Action of EcoRI

The separated pieces have single stranded "sticky-ends," which allow the complementary pieces to combine. The newly joined pieces are stabilized by DNA ligase.

EcoRI, one of many restriction enzymes, is obtained from the bacteria Escherichia coli.

❐ restriction exonuclease

A class of enzymes that degrade DNA or RNA, starting from either the 5'- or the 3'-end.

❐ restriction fragment

A shortened DNA molecule generated by the cleavage of a larger molecule by one or more restriction endonucleases.

❐ restriction fragment length polymorphism

A class of genetic marker based on the detection of variation in the length of restriction fragments generated when DNA is treated with restriction endonucleases. Differences in fragment lengths arise due to genetic variation with respect to the presence or absence of specific recognition site(s). RFLPs were initially detected by Southern hybridization but are now detected by electrophoresis of digested PCR product.

❐ restriction map

The linear arrangement of restriction endonuclease recognition sites along a DNA molecule.

❐ restriction site

Synonym of recognition site.

❐ reticulocyte

A slightly immature red blood cell.

❐ retro-element

Any of the integrated retroviruses or the transposable elements that resemble them.

❐ retroposon

A transposable element that moves via reverse transcription but lacks the long terminal repeat sequences necessary for autonomous transposition. Much of the repetitive DNA that makes up a large proportion of eukaryotic genomes consists of silenced (i.e. inactive) retroposons.

❐ retroviral vectors

Gene transfer systems based on viruses that have RNA as their genetic material.

❐ retrovirus

A class of eukaryotic RNA viruses that, by using reverse transcription, can form double-stranded DNA copies of their genomes, which can integrate into the chromosomes of an infected cell. Pathogenic retroviruses include HIV and the causative agents of many vertebrate animal cancers.

❑ retrovirus replication

Process by which retroviruses use a host cell to replicate.

Retroviruses contain viral RNA and several copies of reverse transcriptase (DNA polymerase). After infecting a cell, the reverse transcriptase is used to make the initial copies of viral DNA from viral RNA. Once a DNA strand has been synthesized, a complementary viral DNA strand is made. These double strand copies of viral DNA are inserted into the host-cell chromosome and host-cell RNA polymerase is used to make virus-related RNA. These RNA strands serve as templates for making new copies of the viral chromosomal RNA and serve also as mRNA. mRNA is translated into viral proteins that are used to make the virus envelope. New viral particles are assembled, bud from the plasma membrane, and are released. An example of this process is illustrated in the replication of the retrovirus, HIV (human immunodeficiency virus).

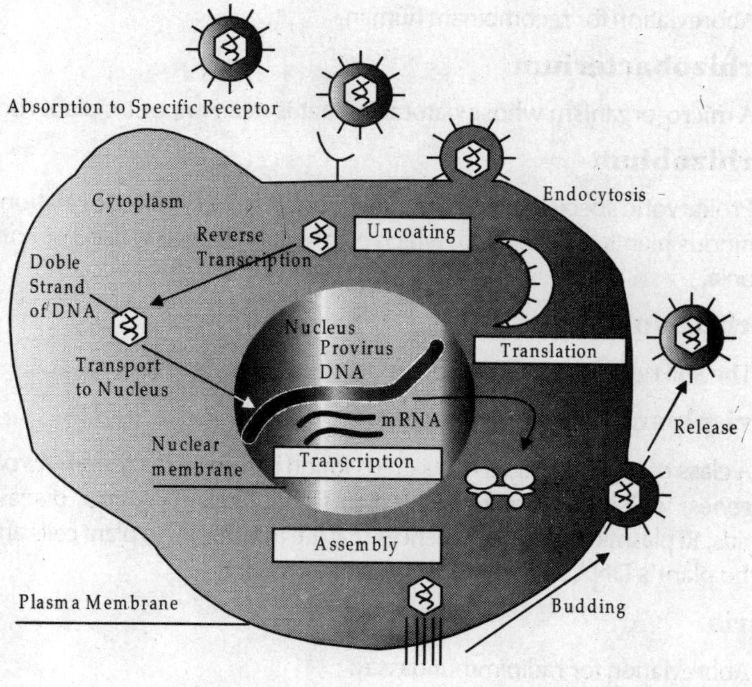

❑ reversal transfer

Transfer of a culture from a callus-supporting medium to a shoot-inducing medium.

❑ reverse transcriptase

An enzyme that uses an RNA molecule as a template for the synthesis of a complementary DNA strand.

❑ reverse transcription

The synthesis of DNA from a template of RNA, accomplished by reverse transcriptase.

❑ reversion

Restitution of a mutant gene to the wild-type condition, or at least to a form that gives the wild-type phenotype; more generally, the appearance of a trait expressed by a remote ancestor.

❑ rfrf

Abbreviation for replicative form.

❑ rflp

Abbreviation for restriction fragment length polymorphism.

❑ rh

Abbreviation for recombinant human.

❑ rhizobacterium

A micro-organism whose natural habitat is near, on, or in, plant roots.

❑ rhizobium

Prokaryotic species which are able to establish a symbiotic relationship with leguminous plants, as a result of which elemental nitrogen is fixed or converted to ammonia.

❑ rhizosphere

The soil region in the immediate vicinity of growing plant roots.

❑ ri plasmid

A class of large conjugative plasmids found in the soil bacterium *Agrobacterium rhizogenes,* which can infect certain plants and cause hairy root disease. Like Ti plasmids, Ri plasmids include sequences that are transferred to plant cells and inserted into the plant's DNA as part of the infection process.

❑ ria

Abbreviation for radioimmunoassay.

❑ ribonuclease

Any enzyme that catalyses the hydrolysis of RNA.

❑ ribonucleic acid

An organic acid polymer composed of adenosine, guanosine, cytidine and uridine ribonucleotides. The genetic material of some viruses, but more generally is the molecule, derived from DNA by transcription, that either carries information (messenger RNA), provides sub-cellular structure (ribosomal RNA), transports amino acids (transfer RNA), or facilitates the biochemical modification of itself or other RNA molecules.

❐ ribose

A monosaccharide found in all ribonucleosides, ribonucleotides and RNA. Its close analogue, 2-deoxyribose, is similarly found in all deoxyribonucleosides, deoxyribonucleotides and DNA.

❐ ribosomal binding site

A sequence of nucleotides near the 5' end of a bacterial mRNA molecule that facilitates the binding of the mRNA to the small ribosomal sub-unit. Also called the Shine-Delgarno sequence.

❐ ribosomal dna

The coding locus for ribosomal RNA. This is generally a large and complex locus, typically composed of a large number of repeat units, separated from one another by the intergenic spacer. A repeat unit comprises a gene copy for each individual ribosomal RNA component, separated from one another by the internal transcribed spacer.

❐ ribosomal rna

The RNA molecules that are essential structural and functional components of ribosomes, where protein synthesis occurs. Different classes of rRNA molecule are identified by their sedimentation (S) values. *E. coli* ribosomes contain one 16S rRNA molecule (1541 nucleotides long) in one (small) ribosomal sub-unit, and a 23S rRNA (2904 nucleotides) and a 5S rRNA (120 nucleotides) in the other (large) sub-unit. These three rRNA molecules are synthesized as part of a large precursor molecule which also contains the sequences of a number of tRNAs. Special processing enzymes cleave this large precursor to generate the functional molecules. Constitutes about 80% of total cellular RNA.

❐ ribosome

The sub-cellular structure that contains both RNA and protein molecules and is the site for the translation of mRNA into protein. Ribosomes comprise large and small sub-units.

❐ ribosome-inactivating protein

A class of plant proteins that inhibit normal ribosome function, and are thus highly toxic. Type 1 RIPs consist of single polypeptide chain proteins; type 2 (e.g. ricin) consist of two proteins linked by a disulphide bridge, one the toxin and the other a lectin that attaches to recognition sites on a target cell.

❐ ribosomes

A cytoplasmic organelle composed of ribonucleic acid and protein that functions in the synthesis of protein. Ribosomes interact with messenger RNA and transfer RNA to join together amino acid units into a polypeptide chain according to the sequence determined by the genetic code.

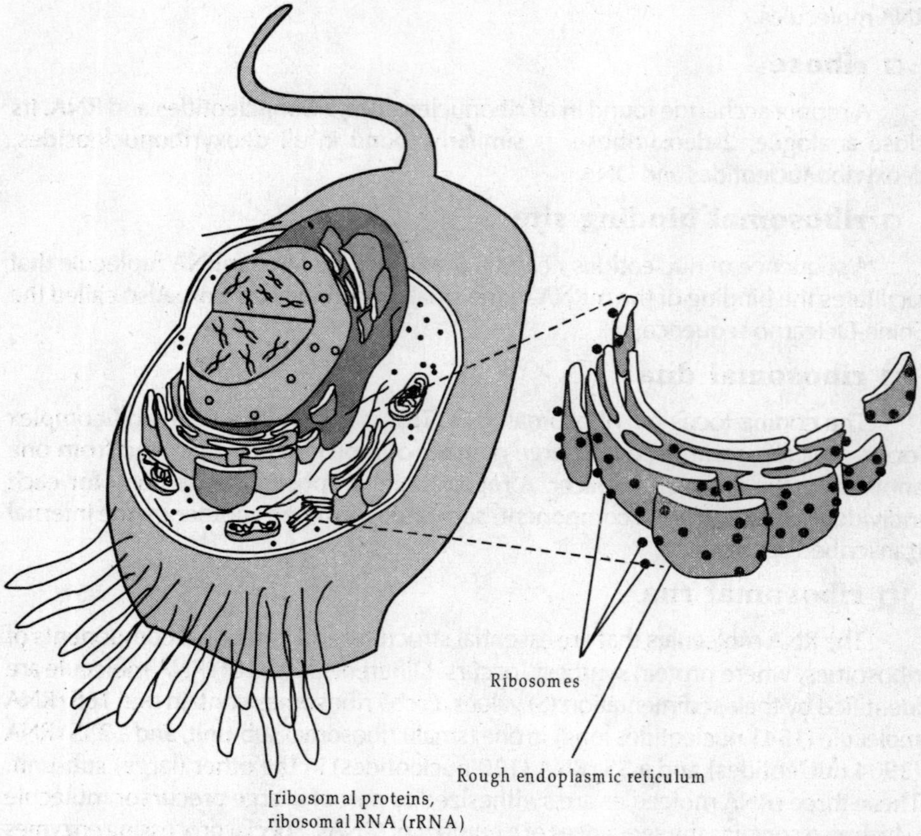

❒ ribozyme

An RNA molecule that can catalyse chemical cleavage of itself or of other RNAs.

❒ ribulose

A keto-pentose sugar ($C_5H_{11}O_5$) involved in the carbon dioxide fixation pathway of photosynthesis.

❒ ribulose biphosphate

A five-carbon sugar combined with carbon dioxide to form a six-carbon intermediate in the first stage of the dark reaction of photosynthesis.

❒ rinderpest

Cattle plague; a viral infection of cattle, sheep and goats.

❒ rip

Abbreviation for ribosome-inactivating protein.

❐ risk analysis

A process consisting of three components: risk assessment, risk management and risk communication performed to understand the nature of unwanted, negative consequences to human and animal health, or the environment.

❐ risk assessment

A scientifically based process consisting of the following steps: i) hazard identification; ii) hazard characterization; iii) exposure assessment; and iv) risk characterization.

❐ risk communication

The interactive exchange of information and opinions throughout the risk analysis process concerning hazards and risks, risk-related factors and risk perceptions, among risk assessors, risk managers, consumers, industry, the academic community and other interested parties, including the explanation of risk assessment findings and the basis of risk management decisions.

❐ risk management

The process, distinct from risk assessment, of weighing policy alternatives, in consultation with all interested parties, considering risk assessment and other factors relevant for the health protection of consumers and for the promotion of fair trade practices, and, if needed, selecting appropriate prevention and control options.

❐ r-loops

Single-stranded DNA regions in RNA-DNA hybrids formed *in vitro* under conditions where RNA-DNA duplexes are more stable than DNA-DNA duplexes.

❐ rna - (ribonucleic acid)

Ribonucleic acid, the most common form of nucleic acid used for storage of chemical energy, processing genetic information from genes (DNA) via messenger RNA (mRNA) into proteins. RNA is one of the most ancient form of molecular structures with enzymatic activity. As a matter of fact, protein biosynthesis is entirely controlled by RNA molecules including mRNA (genetic information), transfer RNA (tRNA) for translating the DNA code into amino acid code, and ribosomal RNA (rRNA) that provide the enzymatic linkage (chemical bond formation) of amino acids into proteins. The use of RNA probably precedes the use of proteins and most modern genomes (except for some viruses) are made of DNA instead of RNA but can only be read by proteins. DNA is the more chemically stable of the two forms of nucleic acids.

Ribonucleic acid Genetic material found in the nucleus and cytoplasm of a cell. RNA acts as a "middle-man", converting genetic information from DNA to proteins. Three types of RNA are; mRNA (messenger RNA, which contains the specific sequence of nucleotides necessary to dictate amino acid sequence in proteins), tRNA (transfer RNA, which serves as the "adaptor" to position the appropriate amino acid next to a growing polypeptide chain during protein synthesis), and rRNA (ribosomal RNA, which is the RNA component of ribosomes).

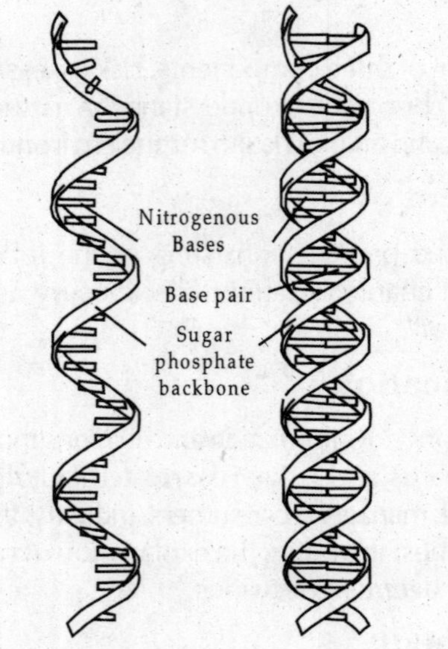

RAN
Ribonucleic acid

DNA
Deoxyribonucleic acid

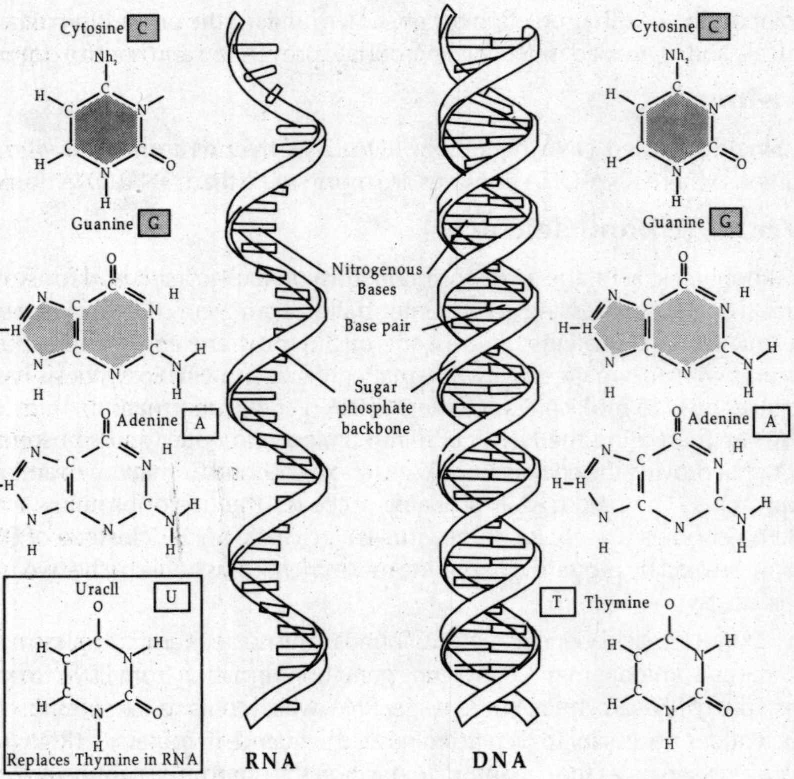

RNA
Ribonucleic acid

DNA
Deoxyribonucleic acid

Nitrogenous Bases

Nitrogenous Bases

❐ rna

Abbreviation for ribonucleic acid.

❐ rna editing

Post-transcriptional processes that alter the information encoded in RNAs.

❐ rna polymerase

A polymerase enzyme that catalyses the synthesis of RNA from a DNA template.

❐ rna synthesis and processing

Process by which non-coding sequences of base pairs (introns) are subtracted from the coding sequences (exons) of a gene in order to transcribe DNA into messenger RNA (mRNA.)

In chromosomes, DNA acts as a template for the synthesis of RNA in a process called transcription. In most mammalian cells, only 1% of the DNA sequence is copied into a functional RNA (mRNA). Only one part of the DNA is transcribed to produce nuclear RNA, and only a minor portion of the nuclear RNA survives the RNA processing steps.

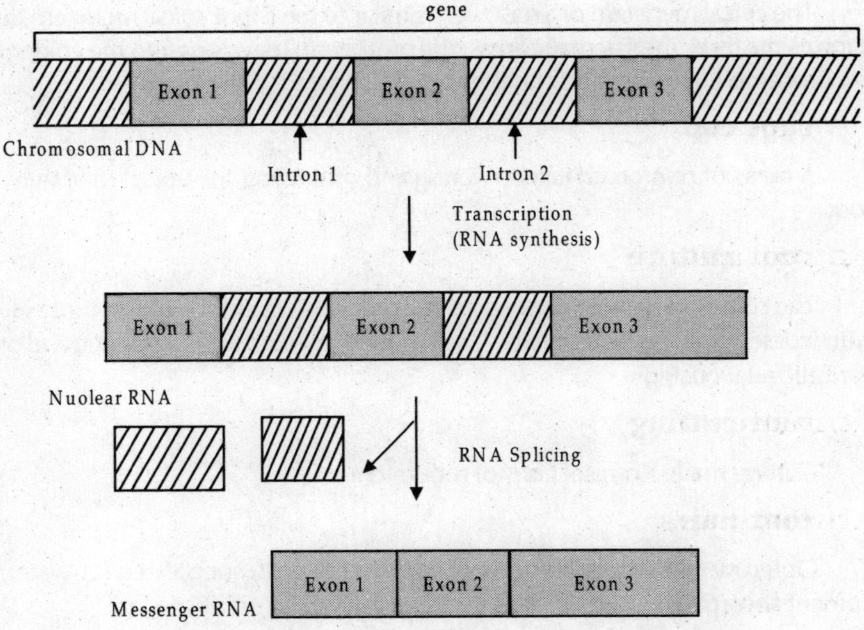

RNA synthesis and processing

One of the most important stages in RNA processing is RNA splicing. In many genes, the DNA sequence coding for proteins, or "exons", may be interrupted by stretches of non-coding DNA, called "introns". In the cell nucleus, the DNA that includes all the exons and introns of the gene is first transcribed into a complementary RNA copy called "nuclear RNA," or nRNA. In a second step, introns are removed from nRNA by a process called RNA splicing. The edited sequence is called "messenger RNA," or mRNA.

The mRNA leaves the nucleus and travels to the cytoplasm, where it encounters cellular bodies called ribosomes. The mRNA, which carries the gene's instructions, dictates the production of proteins by the ribosomes.

❐ rnaase

Abbreviation for ribonuclease.

❐ rnase

Abbreviation for ribonuclease.

❐ rol genes

A family of genes, present on the Ri plasmid of *Agrobacterium rhizogenes*, that when transferred to a plant upon infection by the bacterium, induce the formation of roots. Used as a means of root induction on different species and cultivars of micropropagated fruit trees.

❐ root apex

The apical meristem of a root; very similar to the shoot apical meristem in that it forms the three meristematic areas: the protoderm (develops into the epidermis); the procambium (the stele); and the growth meristem (the cortex).

❐ root cap

A mass of reinforced cells covering and protecting the apical meristem of a root.

❐ root culture

The culture of isolated apical or lateral root tips to produce *in vitro* root systems with indeterminate growth habits. Used to study mycorrhizal, symbiotic and plant-parasitic relationships.

❐ root cutting

Cutting made from sections of roots alone.

❐ root hairs

Outgrowths from epidermal cell walls of the root, specialized for water and nutrient absorption.

❐ root nodule

A small round mass of cells attached to the roots of leguminous plants, containing symbiotic nitrogen-fixing bacteroids, particularly *Rhizobium* spp.

❐ root

The descending axis of a plant, normally below ground, which serves to anchor the plant and to absorb and conduct water and mineral nutrients.

❒ root tuber

Thickened root that stores carbohydrates.

❒ root zone

The volume of soil or growing medium containing the roots of a plant. In soil science, the depth of the soil profile in which roots are normally found.

❒ rootstock

The trunk or root material to which buds or scions are inserted in grafting.

❒ rotary shaker

Rotating apparatus with a platform on which liquid media or cultures can be continuously shaken.

❒ Roundup-readyä

Describing transgenic crop varieties that carry the bacterial gene which detoxifies the herbicide glyphosate, thereby making them resistant to its application.

❒ rrna

Abbreviation for ribosomal RNA.

❒ rubp

Abbreviation for ribulose biphosphate.

❒ ruminant

Animal having a rumen - a large digestive sac in which fibrous plant material is fermented by commensal microbes, prior to its digestion in a "true" stomach (the *abomasum*). Common farm ruminants are cattle and sheep.

❒ runner

A lateral stem that grows horizontally along the ground surface and gives rise to new plants either from axillary or terminal buds. *Synonym*: stolon.

❒ rust

A generic descriptor for various serious fungal plant pathogens, which infect the leaves and stems of crops. The appearance of spores is reminiscent of metallic rust, although the colour varies, according to species, from yellow to reddish-brown.

S

❑ s phase

The phase in the cell cycle during which DNA synthesis occurs.

❑ s_1 mapping

A method to characterise post-transcriptional modifications in RNA (removal of introns etc.) by hybridizing RNA with single-stranded DNA and treating with S_1 nuclease.

❑ S_1 nuclease

An enzyme obtained from the filamentous fungus *Aspergillus oryzae* which specifically degrades RNA or single-stranded DNA into its constituent mononucleotides, and cleaves nicked double-stranded DNA at the nick.

❑ saccharifaction

Following liquefaction, the hydrolysis of polysaccharides by glucoamylase to maltose and glucose.

❑ saline resistance

Synonym for salt tolerance.

❑ salmonella

A genus of rod-shaped, Gram-negative bacteria that are a common cause of food poisoning.

❑ salt tolerance

The ability of a plant in soil or in culture to withstand a concentration of common salt (sodium chloride) which is damaging or lethal to most other plants. Breeding and selection for increased tolerance and resistance in crop plants is of great current interest. *Synonym*: saline resistance. An organism with extreme salt tolerance is a halophyte.

❑ sap

Fluid content of the xylem and phloem cells of plants. Fluid content of the vacuole generally referred to as cell sap.

❒ saprophyte

An organism (generally a fungus) that depends on dead plant or animal tissue for its source of nutrition and metabolic energy.

❒ satellite dna

Highly repetitive DNA in plant and animal genomes, consisting of millions of copies of sequences typically in the range 5-500 bp long. Thousands of copies occur tandemly (head-to-tail) at each of many sites. It can be isolated from the rest of the genomic DNA by density gradient centrifugation.

❒ satellite rna

A small, self-splicing RNA molecule that accompanies several plant viruses, including tobacco ringspot virus.

❒ sc

Abbreviation for synaptonemal complex.

❒ sca

Abbreviation for specific combining ability.

❒ scaffold

The central proteinaceous core structure of condensed eukaryotic chromosomes. The scaffold is composed of non-histone chromosomal proteins.

❒ scale up

Conversion of a process, such as fermentation of a micro-organism, from a small laboratory scale to a larger industrial scale.

❒ scanning electron microscope

An electron-beam-based microscope used to examine, in a three dimensional screen image, the surface structure of prepared specimens.

❒ scar

Abbreviation for sequence characterized amplified region.

❒ scarification

The chemical or physical treatment given to certain seeds having hard, impermeable seed coats in order to puncture or weaken the seed coat sufficiently to permit water uptake and germination.

❒ sce

Abbreviation for sister chromatid exchange.

❒ scion

A twig or bud used for grafting onto another plant or rootstock.

❒ scion-stock interaction

The effect of a rootstock on a scion (and *vice versa*) in which a particular scion grafted onto a specific s performs differently than it would either on its own roots or on a different rootstock.

❒ sclerenchyma

A strengthening tissue in plants, composed of cells with heavily lignified cell walls.

❒ scp

Abbreviation for single-cell protein.

❒ scrapie

A spongiform encephalopathy disease of sheep.

❒ screen

Preliminary characterization of a sample collection on the basis of a set of simple established criteria (biochemical, anatomical, physiological, etc.). Often applied to the process of selection for specific purposes, such as for disease resistance or for improved agronomic performance in crop plants.

❒ sds

Abbreviation for sodium dodecyl sulphate.

❒ sds-page

Abbreviation for sodium dodecyl sulphate polyacrylamide gel electrophoresis.

❒ secondary antibody

In an ELISA or other immunological assay system, the antibody designed to bind to the primary antibody, and to which a label is generally attached.

❒ secondary cell wall

The innermost layer of cell wall, giving rigidity to the cells. Characterized by its highly organized microfibrillar structure, and only formed in certain cells after cell elongation has ceased.

❒ secondary growth

Type of growth characterized by an increase in the thickness of stem and root, and resulting from the formation of secondary vascular tissues by the vascular cambium.

❒ secondary immune response

The rapid immune response that occurs during the second (and subsequent) encounters of the immune system of a mammal with a specific antigen. *See*: primary immune response.

❐ secondary messenger

A chemical compound within a cell that is responsible for initiating the response to a signal from a chemical messenger (such as a hormone) that cannot enter the target cell itself.

❐ secondary metabolism

The production by living organisms of substances not essential for primary metabolic functions or physiology. Their role is associated with interaction with the environment, for example for defence, as elicitors or as attractants. Some of these have useful pharmacological or nutritional properties, while others are toxic.

❐ secondary metabolite

Product of secondary metabolism.

❐ secondary phloem

Phloem tissue formed by the vascular cambium during secondary growth in a vascular plant.

❐ secondary root

A branch or lateral root.

❐ secondary structure

Localized three dimensional conformations adopted by macromolecules, in particular nucleic acids and polypeptides. These arise as a result of the action of non-covalent forces generated by interactions between residues which are brought into close contact with one another. Examples are alpha-helix regions and beta-pleated sheets in proteins, and hairpin loops in nucleic acids. *See*: primary structure, tertiary structure, quaternary structure.

❐ secondary thickening

Deposition of secondary cell wall materials which result in an increase in thickness in stems and roots.

❐ secondary vascular tissue

Vascular tissue (xylem and phloem) formed by the vascular cambium during secondary growth in a vascular plant.

❐ secretion

The transport of a molecule from the inside of a cell through the cell membrane.

❐ seed

Botanically, the matured ovule without accessory parts. Colloquially, anything which may be sown; i.e. seed potatoes (which are vegetative tubers); seed of wheat (karyopses) etc.

❐ seed storage proteins

Proteins accumulated in large amounts in protein bodies within seeds. They act as a source of amino acids during germination. Of interest in biotechnology: 1.

1. As a major source of human and animal nutritional protein.
2. As a model expression system. Since they are produced in large amounts relative to other proteins, and are stored in stable, compact bodies in the plant seed, it may be possible to engineer transgenes which are expressed in the same way as seed storage proteins - i.e. in large amounts and in a convenient form.

❐ segment-polarity gene

A gene that functions to define the anterior and posterior components of body segments in *Drosophila*.

❐ segregant

An individual derived from a cross between two unlike parents.

❐ segregation

For genes, the separation of allele pairs from one another and their resulting assortment into different cells at meiosis. For chromosomes, the separation and re-assortment of the two homologues in anaphase of the first meiotic division. For individuals, the occurrence of different genotypes and/or phenotypes among offspring, resulting from chromosome or allele separation in their heterozygous parents.

❐ selectable

Having a gene product that, when present, enables the identification and preferential propagation of a particular genotype.

❐ selectable marker

A gene whose expression allows the identification of a specific trait or gene in an organism.

❐ selection

1. Differential survival and reproduction of phenotypes.
2. A system for either isolating or identifying specific genotypes in a mixed population.

❐ selection coefficient

A measure of the intensity of selection at a locus, commonly abbreviated as *s*. It represents the proportionate reduction in the gametic contribution of a particular genotype, compared with the (generally most favoured) standard genotype.

❐ selection culture

A selection based on difference(s) in environmental conditions or in culture medium composition, such that preferred variant cells or cell lines (presumptive or putative mutants) are favoured over other variants or the wild type.

❐ selection differential

The difference between the mean of the individuals selected to be parents and the mean of the overall population; it represents the average superiority of the selected parents; commonly abbreviated as *S*.

❐ selection pressure

The intensity of selection acting on a population of organisms or on cells in culture. Its effectiveness is measured in terms of differential survival and reproduction, and consequently in changes in allele frequency in a population.

❐ selection response

The difference between the mean of the individuals selected to be parents and the mean of their offspring. Predicted response is calculated as the product of narrow-sense heritability and selection differential.

❐ self-incompatibility

In plants, the inability of the pollen to fertilize ovules (female gametes) of the same plant.

❐ self-replicating elements

Extrachromosomal DNA elements that have origins of replication for the initiation of their own DNA synthesis.

❐ self-sterility

Synonym of self-incompatibility.

❐ sem

Abbreviation for scanning electron microscope.

❐ semen sexing

Synonym of sperm sexing.

❐ semi-conservative replication

During DNA duplication, each strand of a parent DNA molecule acts as a template for the synthesis of a new complementary strand. Thus, one half of a pre-existing DNA molecule is conserved during each round of replication.

❐ semi-continuous culture

Cells in an actively dividing state which are maintained in culture by periodically draining off the medium and replenishing it with fresh medium.

❐ semi-permeable membrane

A natural or synthetic material which selectively allows the passage of certain ions or molecules.

❐ semi-sterility

The condition of partial fertility. Often associated with chromosomal aberrations or the result of mutagenesis.

❐ senescence

A late stage in the development of multicellular organisms, during which irreversible loss of function and degradation of biological components occur. The physiological ageing process in which cells and tissues deteriorate and finally die.

▫ sense rna

The RNA transcript of the coding strand DNA (often represented as the (+)-strand).

▫ sensitivity

In diagnostic tests, the smallest amount of the target molecule that the assay can detect.

▫ sepsis

Destruction of tissue by pathogenic micro-organisms or their toxins, especially through infection of a wound.

▫ septum

A dividing wall or partition, which splits a structure into separate cells or compartments.

▫ sequence characterized amplified region

A molecular marker obtained by the conversion to a sequence-tagged site of a single random amplified polymorphic DNA product.

▫ sequence divergence

The percent difference in the nucleotide sequence between related nucleic acid sequences, or in the amino acid sequence in a comparison between related proteins.

▫ sequence hypothesis

The concept that genetic information exists as a linear DNA code, and that DNA and gene product sequence are collinear.

▫ sequence

The linear order of nucleotides along a DNA or RNA molecule, and the process of obtaining this. Genome sequencing aims to generate the linear order of all nucleotides present in the nuclear DNA of an organism.

▫ sequences of base pairs mapping

Genetic maps are representations of disease traits, physiological traits, or random genes assigned to particular chromosomes and mapped relative to one anothre. This model of a genetic map shows the location of 8 markers (called here A-H) along the chromosome. The triangle maps a small part of the chromosome in greater detail, as may be needed for an intensively studied partion. Distances in genetic maps are measured in centimorgans (cM, about 1 million base pairs).

Physical maps are not representations but overlapping collections of DNA fragments. DNA is snipped into fragments by the action of restriction enzymes, then cloned and stored in a variety of forms such as plasmids in bacteria or YACs in yeast. These tiny fragments (measured in kilobases, kb) may then be analyzed by various means to discover the base-by-base sequençe of DNA..

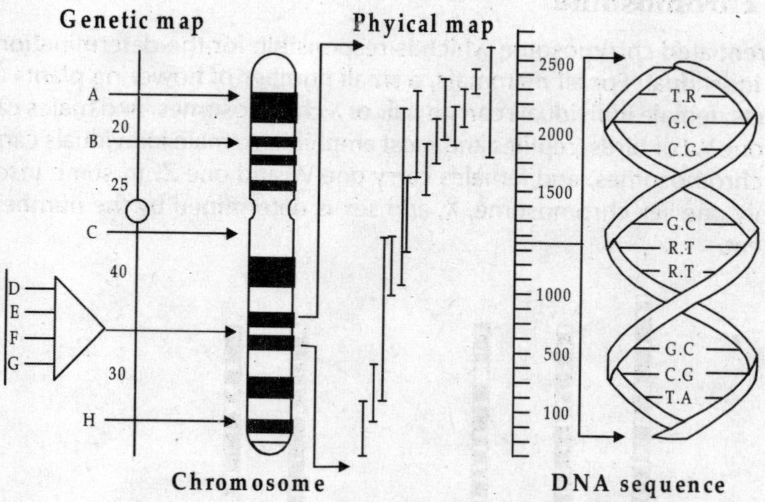

Sequences of base pairs mapping

❑ sequence-tagged site

Short unique DNA sequence (200-500 bp long) that can be amplified by PCR and is thus tagged to the site on the chromosome from which it was amplified.

❑ serial division

Splitting of excised shoot-tip material growing *in vitro*, in order to induce the development of greater numbers of plantlets.

❑ serial float culture

A technique whereby immature anthers are floated on a liquid medium, and continue their development through to the release of pollen.

❑ serology

The study of serum reactions between an antigen and its antibody. Mainly used to identify and distinguish between antigens, such as those specific to particular micro-organisms or viruses.

❑ serum albumin

A globular protein obtained from blood and body fluids. Bovine and human serum albumins are abbreviated BSA and HSA respectively.

❑ serum

Blood plasma that has had its clotting factor removed.

❑ sewage treatment

A widespread biotechnological processes in developed economies. Methods vary widely, but all are designed for the biological break-down of human and animal waste in order to allow safe discharge into the environment.

❑ sex chromosome

Differentiated chromosome which is responsible for the determination of sex of the individual. For all mammals, a small number of flowering plants and many insects, female individuals carry a pair of X chromosomes, and males carry one X and one Y. For birds, reptiles and most amphibians, male individuals carry a pair of W chromosomes, and females carry one W and one Z. In some insects there is only one sex chromosome, X, and sex is determined by the number of these present.

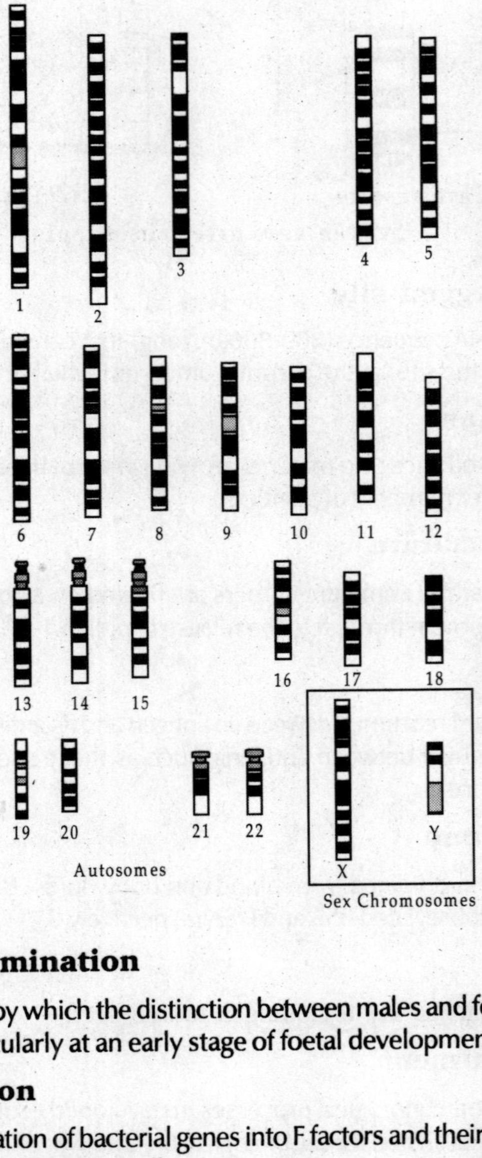

❑ sex determination

Any method by which the distinction between males and females is established in a species, particularly at an early stage of foetal development.

❑ sex duction

The incorporation of bacterial genes into F factors and their subsequent transfer, by conjugation, to a recipient cell.

▫ sex factor

A bacterial episome (e.g. the F plasmid in *E. coli*) that enables the cell to be a donor of genetic material. The sex factor may be propagated in the cytoplasm, or it may be integrated into the bacterial chromosome.

▫ sex hormones

Steroid hormones that control sexual development in animals.

▫ sex linkage

Referring to genes present on one of the sex chromosomes, thus genetically linked to the sex of the individual.

▫ sex linked inheritance: drosophilia

In sex-linked inheritance, alleles on sex chromosomes are inherited in predictable patterns.

For example, in *Drosophila* the locus for eye color is located on the X chromosome.

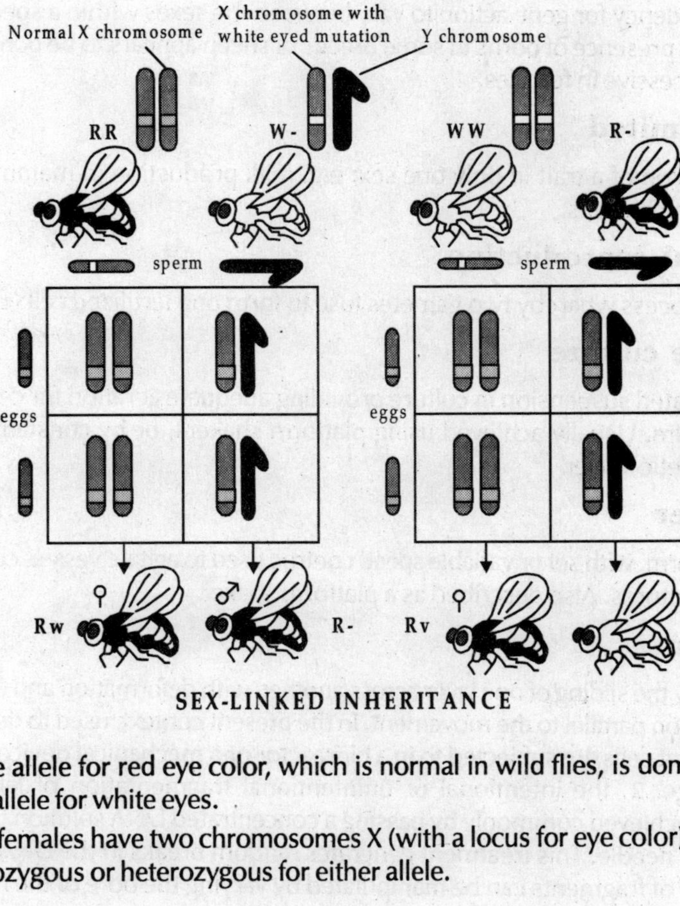

SEX-LINKED INHERITANCE

The allele for red eye color, which is normal in wild flies, is dominant to the mutant allele for white eyes.

As females have two chromosomes X (with a locus for eye color), they might be homozygous or heterozygous for either allele.

Males, who carry only one X chromosome, are always hemizygous. They carry only the one X chromosome inherited from their mother, and it determines their eye color.

In the left hand example, homozygous red eyed females (RR) mate with hemizygous white eyed males (w-). In the offspring, all the daughters are red eyed heterozygotes (Rw) and all sons are red eyed hemizygotes (R-).

In the right hand, homozygous white eyed females (ww) mate with hemizygous red eyed males (R-).

In the offspring, all the daughters are red eyed heterozygotes (Rw) and all sons are white eyed hemizygotes (w-).

❐ sex mosaic

Synonym of gynandromorph.

❐ sexed embryos

Embryos separated according to sex.

❐ sex-influenced dominance

The tendency for gene action to vary between the sexes within a species. For example, the presence of horns in some breeds of sheep appears to be dominant in males but recessive in females.

❐ sex-limited

Expression of a trait in only one sex; e.g. milk production in mammals; egg production in chickens.

❐ sexual reproduction

The process whereby two gametes fuse to form one fertilized cell (zygote).

❐ shake culture

An agitated suspension in culture providing adequate aeration for cells in the liquid medium. Usually achieved using platform shakers, or by constant stirring with a magnetic stirrer.

❐ shaker

A platform, with set or variable speed control, used to agitate vessels containing liquid cell cultures. Also described as a platform shaker.

❐ shear

Literally the sliding of one layer across another, with deformation and fracturing in the direction parallel to the movement. In the present context, used to describe 1. the forces that cells are subjected to in a bioreactor or a mechanical device used for cell breakage. 2. the intentional or unintentional fragmentation of large DNA molecules, achieved commonly by passing a concentrated DNA solution through a hypodermic needle. This treatment generates random breaks in the DNA, and the average size of fragments can be manipulated by varying the bore of the needle.

❒ shine-dalgarno sequence

A conserved sequence of prokaryotic mRNAs that is complementary to a sequence near the 5' terminus of the 16S ribosomal RNA and is involved in the initiation of translation. *See*: ribosomal binding site.

❒ shoot differentiation

The development of growing points, leaf primordia and finally shoots from a shoot tip, axial bud, or callus surface.

❒ shoot tip

The terminal bud (0.1 - 1.0 mm) of a plant, which consists of the apical meristem (0.05 - 0.1 mm) and the immediately surrounding leaf primordia and developing leaves, and adjacent stem tissue.

❒ shoot-tip graft

A shoot tip or meristem tip grafted onto a prepared seedling or micropropagated rootstock in culture. Meristem tip grafting is mainly used for *in vitro* virus elimination from *Citrus* spp. and other plants.

❒ short interspersed nuclear element

Families of short (150-300 bp), moderately repetitive DNA elements of eukaryotic genomes. They appear to be DNA copies of certain tRNA molecules, created presumably by the unintended action of reverse transcriptase during retroviral infection.

❒ short-day plant

A plant which will not flower until triggered to do so by exposure to one or a number of dark periods equal to or longer than its critical period. Other plant species are long-day and some are daylength neutral. Genetic variation for daylength sensitivity is present in many crop species.

❒ shotgun genome sequencing

A strategy for sequencing a whole genome, in which the genomic DNA is initially fragmented into pieces small enough to be sequenced. Specialized computer software is then used to piece together the individual sequences to create long contiguous tracts of sequenced DNA.

❒ shuttle vector

A plasmid capable of replicating in two different host organisms because it carries two different origins of replication and can therefore be used to transfer genes from one to the other.

❒ sib-mating

The deliberate crossing of siblings. Generally done where self-incompatibility prevents the production of self-fertilized progeny.

□ siderophore

A low molecular weight entity that binds very tightly to iron. Siderophores are synthesized by a variety of soil micro-organisms to ensure that the organism is able to obtain sufficient amounts of iron from the environment.

□ sieve cell

A long and slender sieve element in vascular plants, characterized by relatively unspecialized sieve areas and tapering end walls that lack sieve plates.

□ sieve element

The phloem cell concerned with longitudinal conduction of food materials.

□ sieve plate

Perforated wall area in a sieve tube element, through which strands connecting sieve tube protoplasts can pass.

□ sieve tube

A tube within the phloem tissue of a plant, composed of joined sieve elements.

□ sigma factor

The sub-unit of prokaryotic RNA polymerases responsible for the initiation of transcription at specific initiation sequences.

□ signal sequence

A stretch of 15-30 amino acid residues at the N terminus of a protein, which is thought to enable the protein to be secreted (pass through a cell membrane). The signal sequence is removed as the protein is secreted. *Synonyms*: signal peptide, leader peptide.

□ signal transduction

The biochemical events that conduct the signal of a hormone or growth factor from the cell exterior, through the cell membrane, and into the cytoplasm. This involves a number of molecules, including receptors, ligands and messengers.

□ signal-to-noise ratio

A specifically produced response (signal) compared to the response level (noise) when no specific stimulus (activity) is present.

□ silencing

Loss of gene expression either through an alteration in the DNA sequence of a structural gene, or its regulatory region; or because of interactions between its transcript and other mRNAs present in the cell (see: antisense RNA).

□ sine

Abbreviation for short interspersed nuclear element.

❑ **single copy**

A gene or DNA sequence which occurs only once per (haploid) genome. Many structural genes are single copy.

❑ **single node culture**

Culture of separate lateral buds, each carrying a piece of stem tissue.

❑ **single nucleotide polymorphism**

A genetic marker resulting from variation in sequence at a particular position within a DNA sequence. SNPs are commonly the result of transition changes (A for G, T for C), but also transversions (G or A for T or C) and single base deletions. Such variation is extensive throughout all genomes, and offers the particular advantage of being detectable without the need for gel electrophoresis.

❑ **single primer amplification reaction**

A PCR-based genotyping technique in which genomic template is amplified with a single primer.

❑ **single-cell protein**

Protein produced by micro-organisms, particularly yeast. Used as either a feed or a food additive.

❑ **single-strand conformational polymorphism**

A technique for detection of mutations in a defined DNA sequence. Single-stranded polynucleotides are electrophoretically separated on non-denaturing gels. Intrachain base pairing results in a limited number of conformers stabilized by intrachain loops, and mutated DNA shows on electrophoresis an altered assortment of such conformers.

❑ **single-strand dna binding protein**

A protein that coats single-stranded DNA, preventing renaturation and so maintaining the DNA in an extended state.

❑ **single-stranded dna**

DNA molecules separated from their complementary strand, either by its absence or following denaturation.

❑ **single-stranded nucleic acid**

Nucleic acid molecules consisting of only one polynucleotide chain. The genomes of many viruses are single-stranded DNA molecules, as are most biologically effective RNAs. Many RNA molecules do include double-stranded regions formed by the intra-strand base-pairing of self-complementary sequences, and these determine the 3-dimensional shape (conformation) that they adopt *in vivo*.

❑ **sire**

Male animal chosen for breeding.

❑ sister chromatid exchange

Reciprocal interchanges of the two chromatid arms within a single chromosome.

❑ site-specific

A term used to describe any process or enzyme which acts at a defined sequence within a DNA or RNA molecule.

❑ site-specific mutagenesis

The induction of mutations, by molecular biology techniques, in one or more specific nucleotides within a defined coding sequence in order to create altered forms of the gene product. Used to define the active sites of proteins and for protein engineering.

❑ six-base cutter

Type II restriction endonucleases whose recognition site and cleavage site consist of a characteristic sequence of six nucleotide pairs.

❑ sky (spectral karyotype)

A visualization of all of the chromosomes in the genome all together with each chromosome labeled with a diffėrent sheds. The SKY technique is useful for identifying chromosome abnormalities.

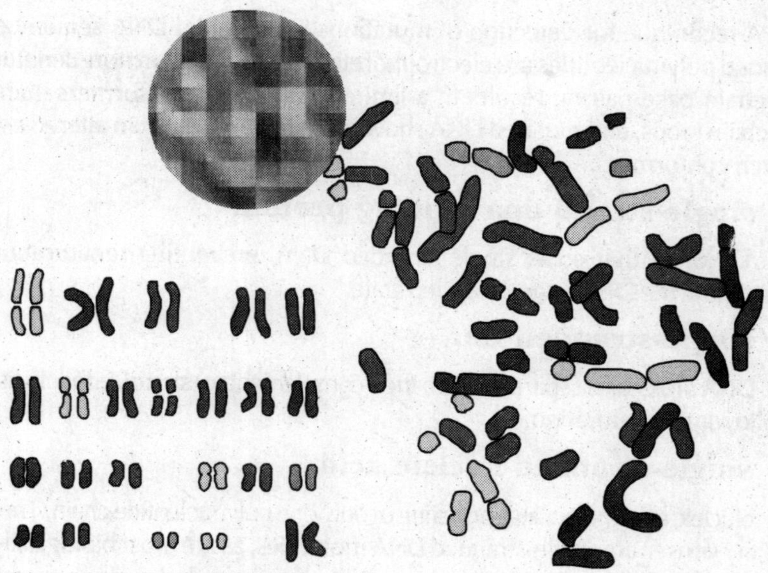

❑ small nuclear ribonucleoprotein

A complex of small nuclear RNA and nuclear protein, heavily involved in the post-transcriptional processing of mRNA, especially the removal of introns. snRNPs are a major component of spliceosomes.

❐ small nuclear rna

RNA transcripts of 100-300 bp that associate with proteins to form small nuclear ribonucleoprotein particles. Most snRNAs are components of the spliceosomes.

❐ snp

Abbreviation for single nucleotide polymorphism.

❐ snrna

Abbreviation for small nuclear RNA.

❐ snrnp

Abbreviation for small nuclear ribonucleoprotein.

❐ sodium dodecyl sulphate

A detergent used to solubilize protein and DNA from biological materials. Specific use in sodium dodecyl sulphate polyacrylamide gel electrophoresis.

❐ sodium dodecyl sulphate polyacrylamide gel electrophoresis

A widely employed electrophoretic method for the separation of proteins from biological samples. The sodium dodecyl sulphate gives a uniform charge density to the surface of proteins or nucleic acids, so that their rate of migration through the gel is determined largely by their molecular weight.

❐ soil amelioration

The improvement of poor soils. Includes the fungal and bacterial break down of plant organic matter, to form humus; the release of minerals - such as phosphates - to the soil, making them available to plants; the fixation of nitrogen. Can sometimes include an element of bioremediation.

❐ soil-less culture

Growing plants in nutrient solution without soil.

❐ solid medium

Nutrient medium solidified by the addition of a gelling agent, commonly agar.

❐ somaclonalariation

Epigenetic or genetic changes induced during the callus phase of plant cells cultured *in vitro*. Sometimes visible as changed phenotype in plants regenerated from culture.

❐ somatic cell

Cells not involved in sexual reproduction, i.e. not germ cells.

❐ somatic cell embryogenesis

The process of differentiation of somatic embryos either from explant cells (direct embryogenesis), or from callus generated from explants (indirect embryogenesis). *Synonym*: asexual embryogenesis.

❐ somatic cell gene therapy

The delivery of a transgene(s) to a somatic tissue in order to correct a physiological defect.

❐ somatic cell hybrid panel

A panel of cells created by cell fusion, typically involving a reference species (e.g. hamster) and the species of interest (e.g. sheep) with each member of the panel containing a different mixture of chromosomes from the two species. By relating the presence or absence of cloned fragments (via *in situ* hybridization) or PCR products to the presence or absence of particular chromosomes from the species of interest, such panels can be used for physical mapping.

❐ somatic cell variant

A somatic cell with unique characters not present in the other cells, and which could be selected for by an appropriate screen.

❐ somatic embryo

An organized embryo-like structure. Although morphologically similar to a zygotic embryo it is initiated from somatic plant cells. Under *in vitro* conditions, somatic embryos go through developmental processes similar to embryos of zygotic origin. Each somatic embryo is potentially capable of developing into a normal plantlet.

❐ somatic hybridization

Naturally occurring or induced fusion of somatic protoplasts or cells of two genetically different parents. The difference may be as wide as interspecific. Wide synthetic hybrids formed in this way (i.e. not via gametic fusion) are known as cybrids. Not all cybrids contain the full genetic information (nuclear and non-nuclear) of both parents.

❐ somatic hypermutation

The high frequency of mutation that occurs in the gene segments encoding the variable regions of immunoglobulins during the differentiation of B lymphocytes into antibody producing plasma cells.

❐ somatic reduction

Halving of the chromosomal number of somatic cells; a possible method of producing "haploids" from somatic cells and calli by artificial means.

❐ somatic

Referring to cell types, structures and processes other than those associated with the germ line.

❑ sonication

Disruption of cells or DNA molecules by high frequency sound waves.

❑ sos response

The synthesis of a whole set of DNA repair, recombination and replication proteins in bacteria suffering severe DNA damage (e.g. following exposure to UV light).

❑ source dna

The DNA from an organism that contains a target gene, and used as the starting material in a cloning experiment.

❑ source organism

A bacterium, plant or animal from which DNA is purified and used in a cloning experiment.

❑ southern blot

A nitrocellulose or nylon membrane to which DNA fragments previously separated by gel electrophoresis, have been transferred by capillary action.

❑ southern blotting: gel transfer

Detection of specific DNA fragments by gel-transfer hybridization (Southern blotting). (A) The mixture of double-stranded DNA fragments generated by restriction nuclease treatment of DNA is separated according to length by electrophoresis. (B) A sheet of either nitrocellulose paper or nylon paper is laid over the gel; and the separated DNA fragments are transferred to the sheet by blotting. The gel is supported on a layer of sponge in a bath of alkali solution, and the buffer is sucked through the gel and the nitrocellulose paper by paper towels stacked on top of the nitrocellulose. As the buffer is sucked through, it denatures the DNA and transfers the single-stranded fragments from the gel to the surface of the nitrocellulose sheet, where they adhere firmly. This transfer is necessary to keep the DNA firmly in place while the hybridization procedure (D) is carrried out. (C) The nitrocellulose sheet is carefully peeled off the gel. (D) The sheet containing the bound single-stranded DNA fragments is placed in a sealed plastic bag together with buffer containing a radioactively labeled DNA probe specific for the required DNA sequence. The sheet is exposed for a prolonged period to the probe under conditions favoring hybridization. (E) The sheet is removed from the bag and washed thoroughly, so that only probe molecules that have hybridized to the DNA on the paper remain attached. After autoradiography, the DNA that has hybridized to the labeled probe will show up as bands on the autoradiograph. An adaptation of this technique to detect specific sequences in RNA is called Northern blotting. In this case mRNA molecules are electrophoresed through the gel and the probe is usually a single-stranded DNA molecule.

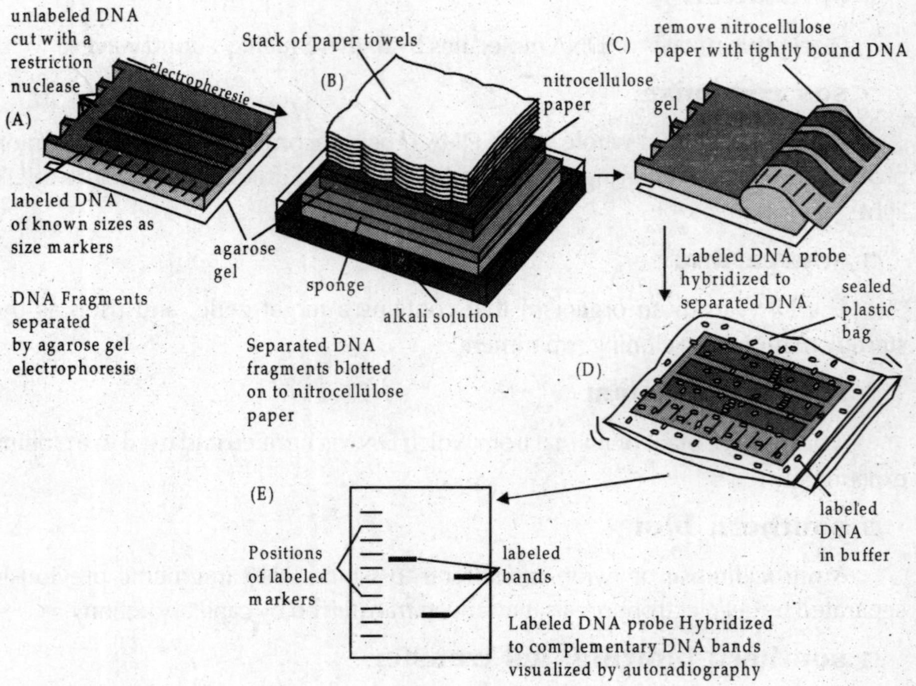

❑ southern hybridization

A procedure in which a cloned, labelled segment of DNA is hybridized to DNA restriction fragments on a Southern blot.

❑ spacer sequence

A DNA sequence separating neighbouring genes; spacer sequences are not usually transcribed.

❑ spar

Abbreviation for single primer amplification reaction.

❑ sparger

A device that introduces, into a bioreactor, air in the form of fine bubbles.

❑ spatial autocorrelation statistics

A set of statistical parameters aimed to depict the spatial (geographical) pattern of genetic diversity in a population.

❑ speciation

The evolutionary differentiation of a pre-existing species into one or more distinct species.

❐ species

A class of individuals capable of interbreeding, but which is reproductively isolated from other such groups having many characteristics in common. A somewhat arbitrary and sometimes blurred classification; but still quite useful in many situations.

❐ specific combining ability

A component of genetic variance calculable where a number of genotypes are intercrossed in all possible combinations. The SCA measures the deviation of the performance of a particular cross from the average general combining ability of its two parents.

❐ specificity

For diagnostic tests, the ability of a probe to react precisely and uniquely with its target molecule.

❐ spent medium

After sub-culture, medium which is discarded because it has been depleted of nutrients, dehydrated, or accumulated toxic metabolic products.

❐ sperm competition

Competition between different spermatozoa to fertilize the egg cell of a single female.

❐ sperm sexing

The separation of mammalian sperm into those bearing an X chromosome and those bearing a Y chromosome, in order to be able to produce, via artificial insemination or *in vitro* fertilization, animals of a specified sex. Methods for achieving this include the inactivation of X-bearing or Y-bearing sperm by antibodies recognizing sex-specific sperm surface peptides, and fluorescence-activated cell sorting.

❐ spermatid Immature spermatozoon

One of the four cells formed at the end of the second meiotic division in spermatogenesis.

❐ spermatocyte

The premeiotic parental cell of the spermatids; the *primary* spermatocyte before the initiation of the first meiotic division; the *secondary* spermatocyte after completion of the first meiotic division, but before the initiation of the second division.

❐ spermatogenesis

The series of cell divisions in the testis as a result of which the formation and the maturation of the male gametes (i.e. sperm) are achieved.

❐ spermatogonium

Primordial male germ cell. These can either divide by mitosis to produce daughter cells, or enter a growth phase and differentiate into a primary spermatocyte.

❐ spermatozoon

The mature, mobile gametic cell of male animals, produced in the testis.

❐ spharoblast

Nodule of wood which can give rise to adventitious shoots with juvenile characteristics.

❐ spheroplast

A microbial or plant cell from which most of the cell wall has been removed, usually by enzymatic treatment. Strictly, in a spheroplast, some of the cell wall remains, while in a protoplast the cell wall has been completely removed. In practice, the two words are often used interchangeably.

❐ spike

1. An inflorescence in which the main axis is elongated and the flowers are sessile.
2. The deliberate addition of a known quantity of a known substance to an analytical sample, used to validate the analytical technique.

❐ spikelet

The unit of inflorescence in grasses, made up of a small group of florets.

❐ spindle

An intracellular fibrous structure, involved in the control of chromosome movement in mitosis and meiosis.

❐ spliceosome

A complex of small nuclear ribonucleoproteins and other proteins that assemble on an immature mRNA and catalyse the excision of an intron. *See*: splicing.

❐ splicing

1. During the maturation of eukaryotic mRNA, the process that removes intron sequences and covalently joins exon sequences. *Synonym*: editing.
2. In recombinant DNA technology, the term refers to the ligation of two fragments of DNA together.

❐ splicing junction

The DNA sequence immediately surrounding the boundary between an exon and an intron. There is a degree of sequence conservation in these regions, allowing the identification of introns in newly sequenced genes.

❐ split gene

In eukaryotes, the encoding DNA of many structural genes is made up of exons and introns. This commonly found pattern of interruption in the coding sequence is referred to as a 'split gene'.

❒ spontaneous mutation

A mutation occurring in the absence of any known mutagen.

❒ sporangium

A reproductive structure in plants that produces spores. A megasporangium produces megaspores, which give rise to the female gametophyte; in seed plants it is represented by the ovule. A microsporangium produces microspores, which give rise to the male gametophyte; it is represented in seed plants by the pollen sac.

❒ spore

1. A reproductive cell that develops into an individual without union with other cells; some spores such as meiospores are the product of the germ line, but others are asexual in nature.
2. A small, protected resting body, often synthesized by micro-organisms when nutrient levels are low.

❒ sporocyte

A diploid germ line cell that is the parent of the four haploid spores generated by meiosis.

❒ sporophyll

A leaf that bears sporangia.

❒ sporophyte

The diploid generation in the life cycle of a plant, and that produces haploid spores by meiosis.

❒ sport

An individual plant, or portion thereof, showing a recognizably different phenotype from the parent, presumably as a result of spontaneous mutation. Novel traits displayed by some sports can become of great agricultural worth, but generally they are disadvantageous.

❒ ssdna

Abbreviation for single-stranded DNA.

❒ ssr

Abbreviation for simple sequence repeat.

❒ stacked genes

Refers to the insertion of two or more genes into the genome of an organism. An example would be a plant carrying a *Bt* transgene giving insect resistance, and a *bar* transgene giving resistance to a specific herbicide.

❒ staggered cuts

Symmetrically cleaved phosphodiester bonds that lie on both strands of double-stranded DNA, but are not opposite one another.

❐ stamen

Floral structure made up of an anther and a filament. The stamen is the male organ of a flower.

❐ standard deviation

A statistical measure of variability in a population of individuals or in a set of data.

❐ standard error

A statistical measure that indicates the predictive accuracy over all individuals of a mean value derived from a sample population.

❐ starch

The major plant carbohydrate storage substance, particularly but not exclusively found in seeds, and used both as food and feed source and for various industrial processes. A large water-insoluble heterogenous group of polysaccharides, consisting of various proportions of the two glucose polymers, amylose and amylopectin. Starch is broken down into simple metabolisable sugars *in vivo* by the action of amylases.

❐ start codon

The codon which specifies the first amino acid of a polypeptide chain and at which the ribosome starts the process of translation. In bacteria, this is either AUG (translated as *n*-formyl methionine) or, rarely, GUG (valine). In eukaryotes, it is always AUG and is translated as methionine. The start codon sets the reading frame for translation.

❐ starter culture

Micro-organisms that are deliberately added to foods to alter flavour, colour, texture, smell, or taste.

❐ stationary culture

A culture maintained without agitation.

❐ stationary phase

The plateau of the growth curve, during which cell number remains relatively constant, following the logarithmic phase.

❐ steady state

In a continuous fermentation process, the condition under which the number of cells removed with the outflow is exactly balanced by the number of newly synthesized cells.

❐ stele

The central vascular cylinder, inside the cortex, of roots and stems of higher plants.

▢ stem cell

An undifferentiated somatic cell that is capable of either division to give rise to daughter stem cells, or differentiating into any specialized cell type given the appropriate signals. Cultured stem cells are critical to the concept of therapeutic cloning.

▢ stem

The main body of the above-ground portion of a tree, shrub, herb or other plant; the ascending axis, whether above or below ground, of a plant.

▢ steps leading from gene to protein

The final level of a protein in the cell depends on the efficiency of each step and on the rates of degradation of the RNA and protein molecules. (A) In eucaryotic cells, the initial RNA molecule produced by transcription (the primary transcript) contains both intron and exon sequences. Its two ends are modified, and the introns are removed by an enzymatically catalyzed RNA splicing reaction. The resulting mRNA is then transported from the nucleus to the cytoplasm, where it is translated into protein. Although these steps are depicted as occurring one at a time, in a sequence, in reality they often occur simultaneously. For example, the RNA cap is typically added and splicing typically begins before the primary transcript has been completed. (B) In procaryotes, the production of mRNA molecules is simpler. The 5´ end of an mRNA molecule is produced by the initiation of transcription by RNA polymerase, and the 3´ end is produced by the termination of transcription. Since procaryotic cells lack a nucleus, transcription and translation take place in a common compartment. In fact, translation of a bacterial mRNA often begins before its synthesis has been completed.

(A) EUCARYOTES

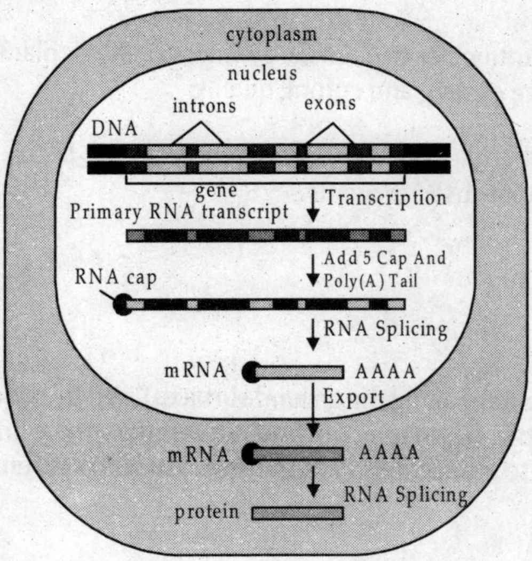

(B) PROCARYOTES

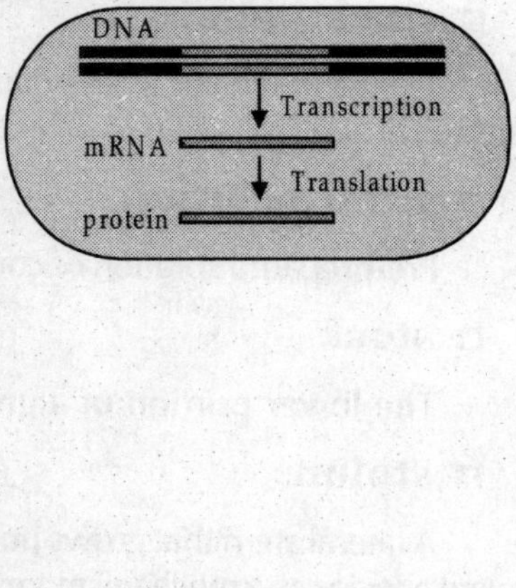

❐ sterile

1. Medium or object free of viable micro-organisms.
2. Incapable of producing viable gametes.

❐ sterile room

Dedicated space for the carrying out of activities that require sterile conditions. Can usually be achieved more economically with a laminar air-flow cabinet.

❐ sterility

Complete or partial failure of an individual to produce functional gametes or viable zygotes under a given set of environmental conditions.

❐ sterilize

1. The elimination of micro-organisms, using heat, irradiation, filtration or chemicals.
2. The operation of making an animal incapable of producing offspring.

❐ Steward bottle

Flask developed for the growth of cells and tissues in a liquid medium, in which they can be periodically submerged during rotation.

❐ stigma

Receptive portion of the style, to which pollen adheres.

❐ stirred-tank fermenter

A growth vessel in which cells or micro-organisms are mixed by mechanically-driven impellers.

❐ stock plant

The source plant from which cuttings or explants are obtained. Stock plants should be well maintained to optimize explant and cutting quality.

❐ stock solution

Pre-prepared solution of commonly used reagents.

❐ stock

The lower portion of a graft.

❐ stolon

A lateral stem that grows horizontally along the ground surface. Used by some plant species as a mechanism for dispersal, since stolon nodes can differentiate into normal stems and roots, giving rise to a daughter plant removed from the parent.

❐ stoma

1. Any of various small openings or pores in an animal body, especially an opening resembling a mouth in various invertebrates.

2. A pore in the epidermis of the leaf or stem of a plant, which allows the exchange of gases, including water vapour, to and from the intercellular spaces. Sometimes used loosely to refer to the pore along with its associated pair of guard cells.

❒ stomatal complex

Includes the stoma, together with its guard cells and, when present, any related subsidiary cells.

❒ stomatal index

A measurement of the surface density of stomata. This parameter has been found useful in comparing leaves of different sizes. Relative humidity and light intensity during leaf development affect the value of stomatal index.

❒ stop codon

A set of three nucleotides for which there is no corresponding tRNA molecule to insert an amino acid into the polypeptide chain. Protein synthesis is hence terminated and the completed polypeptide released from the ribosome. Three stop codons are known: UAA (ochre), UAG (amber) and UGA (opal).

❒ str

Abbreviation for sequence tandem repeat.

❒ strain

A group of individuals derived by descent from a single individual within a species.

❒ stratification

Subjection of moist seeds to a period of low temperature (+2 to +4 °C) to break dormancy.

❒ streptavidin

A microbial protein with a high affinity for the B complex vitamin biotin. The specific interaction of these two molecules has been exploited in labelling technology and in applications where a specific molecule needs to be captured or purified.

❒ stress

Non-optimal conditions for growth. Stresses may be imposed by biotic (pathogens, pests) or abiotic (environment, such as heat, drought etc.) factors.

❒ stringency

Reaction conditions (notably temperature, salt concentration and pH) that affect the annealing process of single-stranded DNA or RNA to make double-stranded DNA or RNA, or DNA/RNA hybrids. At high stringency, duplexes form only between strands with perfect complementarity; lower stringency allows the annealing of strands with some degree of mismatch.

▢ stringent plasmid

A plasmid that can only replicate at the same time as does the main bacterial chromosome, and is present as a single or, at most, several copies per cell.

▢ stroma

The supporting connective tissue of an organ or plastid.

▢ structural gene

A gene that encodes a polypeptide, with either enzymatic or structural functions, and that is required for the normal metabolism and growth of a cell or organism.

▢ structure of dna

The structure of DNA is illustrated by a right handed double helix, with about 10 nucleotide pairs per helical turn. Each spiral strand, composed of a sugar phosphate backbone and attached bases, is connected to a complementary strand by hydrogen bonding (non- covalent) between paired bases, adenine (A) with thymine (T) and guanine (G) with cytosine (C).

Adenine and thymine are connected by two hydrogen bonds (non-covalent) while guanine and cytosine are connected by three.

This structure was first described by James Watson and Francis Crick in 1953.

THE STRUCTURE OF DNA

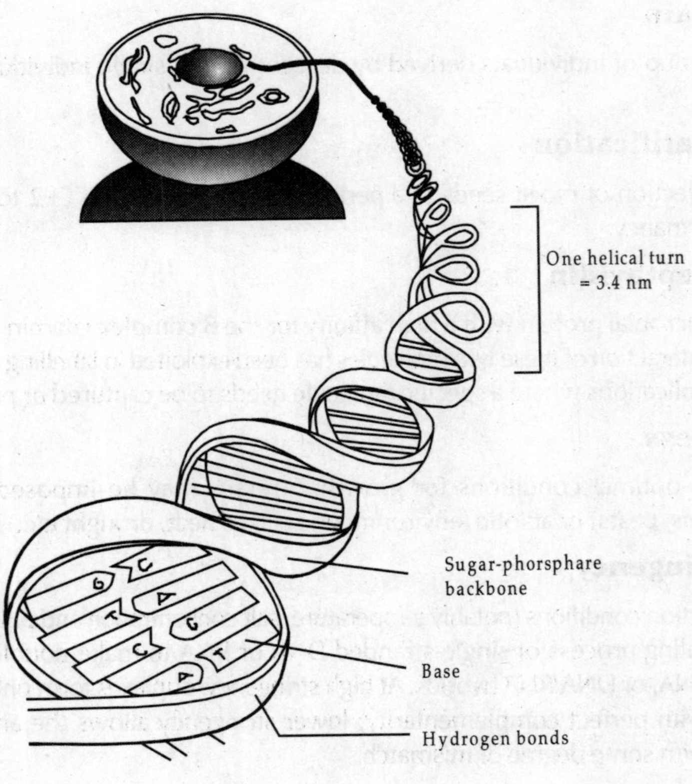

❐ structure-functionalism

The scientific tradition that stresses the relationship between a physical structure and its function, e.g. the related disciplines of anatomy and physiology.

❐ sts

Abbreviation for sequence-tagged site.

❐ style

Slender column of tissue that arises from the top of the ovary and terminates in the stigma, and through which the pollen tube must grow to achieve fertilization.

❐ sub-clone

A procedure in which a large cloned DNA molecule is divided into smaller fragments, each one of which is then separately cloned.

❐ sub-culture

Division and transfer of a portion of a culture to fresh medium. Sometimes used to denote the adding of fresh liquid to a suspension culture.

❐ sub-culture interval

The time between consecutive sub-cultures of cells.

❐ sub-culture number

The number of times cells, etc., have been sub-cultured.

❐ subgenomic promoter

A promoter added to a virus for a specific heterologous gene, resulting in the formation of mRNA for that gene alone.

❐ subspecies

Population(s) of organisms sharing certain characteristics that are not present in other populations of the same species.

❐ sub-strain

Derived from a strain by the isolation of an individual or group of individuals having properties or markers not shared by the strain as a whole.

❐ substrate

1. A compound that is altered by an enzyme.
2. Food source for growing cells or micro-organisms.
3. Material on which a sedentary organism lives and grows.

❐ sub-unit vaccine

One or more immunogenic proteins, either purified from the pathogen itself or produced from a cloned pathogen gene. A vaccine composed of a purified antigenic determinant that is separated from the virulent organism.

□ sucker

A shoot that arises from an underground root or stem. Of particular significance to grafted plants, since the sucker will be genotypically rootstock, rather than scion.

□ suckering

Type of vegetative propagation where lateral buds grow out to produce an individual that is a clone of the parent.

□ sucrose density gradient centrifugation

A procedure used to fractionate nucleic acids on the basis of their size.

□ superbug

Jargon for a particular engineered strain of *Pseudomonas*, in which various hydrocarbon-degrading genes, derived from different plasmids, were combined into one genotype. This provided the basis for the precedent-setting legal decision that declared that genetically engineered organisms were patentable.

□ supercoil

The conformation of a double-stranded DNA molecule placed under torsional stress as a result of interactions with proteins. The stress is accommodated by a twist imposed on the duplex. A left-handed supercoil favours unwinding of the double helix; a right-handed supercoil favours tighter winding.

□ supercoiled plasmid

The predominant *in vivo* form of most plasmids, in which the DNA is coiled around histone-like proteins. When supporting proteins are stripped away during DNA extraction from the bacterial cell, the plasmid molecule also tends to supercoil around itself *in vitro*.

□ supergene

A group of tightly linked genes that are co-inherited, and may be functionally related.

□ supernatant

The liquid phase remaining after insoluble materials are pelleted by centrifugation or precipitation.

□ suppressor mutation

A mutation that reverses the effect of an earlier mutation, e.g. a mutation in a gene for a tRNA that permits it to read and override an amber mutation.

□ suppressor-sensitive mutant

An organism that can grow in the presence, but not in the absence of a second genetic factor (the suppressor).

❑ survey of nucleotides part 1

The basic building block of DNA. A nucleotide consists of one of four nitrogenous bases - adenine, guanine, cytosine, or thymine - attached to a phosphate-sugar group. In DNA the sugar group is deoxyribose, while in RNA (a DNA-related molecule that helps to translate genetic information into proteins), the sugar group is ribose, and the base uracil substitutes for thymine. Each group of three nucleotides in a gene is known as a codon. A nucleic acid is a long chain of nucleotides joined together, and therefore is sometimes referred to as a "polynucleotide."

Deoxyribonucleic Acid (DNA)

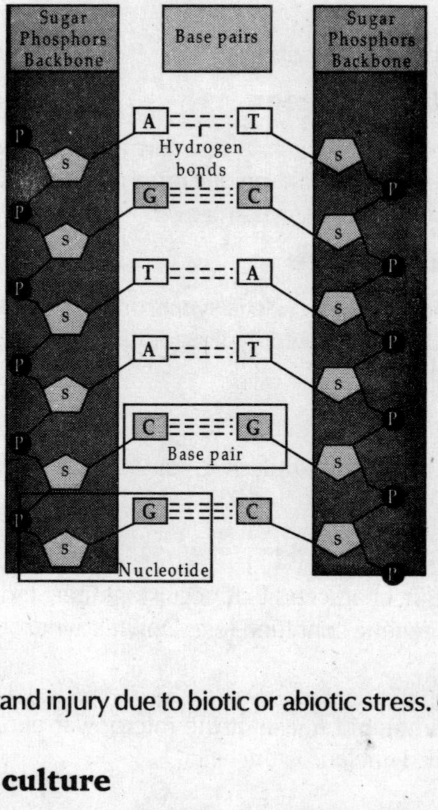

❑ susceptible

Inability to withstand injury due to biotic or abiotic stress. *Opposite*: resistance, tolerance.

❑ suspension culture

A type of culture in which cells and/or clumps of cells grow and multiply while suspended in a liquid medium.

❑ symbiont

An organism living in symbiosis with another, dissimilar organism.

❑ symbiosis

The close association of two different kinds of living organisms where there is benefit to both or where both receive an advantage from the association. A prominent example is the colonization of *Rhizobium* spp. inside the roots of leguminous plants.

❑ sympatric speciation

The evolution of new species by populations that inhabit the same or overlapping geographic regions.

❑ sympodial

A type of plant development in which the terminal bud of the stem stops growing due either to its abortion, or to its differentiation into a floral meristem. Frequently, the uppermost lateral bud then takes over the further axial growth of the stem.

❑ synapsis

Synonym of chromosome pairing.

❑ synaptonemal complex

A ribbon-like proteinaceous structure formed between paired homologous chromosomes at the end of the first meiotic prophase. The SC binds the chromatids along their length, and facilitates crossing over.

❑ synchronous culture

A culture in which the cell cycle is synchronized for the majority of the cells present. Synchrony can be induced by the addition of drugs which arrest the cell cycle at specific stages.

❑ syncytium

A group of cells in which cytoplasmic continuity is maintained; the effect is of a multinucleate cell.

❑ syndrome

A group of specific characters that occur together, and are characteristic of a particular disease or genetic condition (e.g. Down's syndrome).

❑ synergid

One of the two haploid nuclei at the micropylar end of the embryo sac of higher plants. The third nucleus is the egg.

❑ synergism

An interaction between two organisms (e.g. *Rhizobium* and legumes) in which the growth of one is helped by the other.

❑ syngamy

Synonym of fertilization.

❑ synkaryon

The initial hybrid nucleus of the zygote, formed by the fusion of the gametic nuclei upon fertilization. A hybrid nucleus formed by the fusion of two different somatic cells during somatic cell hybridization is called a heterokaryon.

□ synteny

The occurrence of two or more loci on the same chromosome, without regard to their genetic linkage. Increasingly used to describe the conservation of gene order between related species.

□ synthesis of biological polymers

The synthesis of polysaccharides, proteins, and nucleic acids. Synthesis of each kind of biological polymer involves the loss of water in a condensation reaction. Not shown is the consumption of high-energy nucleoside triphosphates that is required to activate each monomer prior to its addition. In contrast, the reverse reaction< (hydrolysis). water of addition simple the by polymers.

T

▫ t

Abbreviation for thymine.

▫ t cell

Lymphocytes which pass through the thymus gland during maturation. Different kinds of T cells play important roles in the immune response.

▫ t cell receptor

An antigen-binding protein, located on the surface of mammalian killer T cells, which mediates the cellular immune response. T cell antigen encoding genes are assembled from gene segments by somatic recombination processes that occur during lymphocyte differentiation.

▫ t_0, t_1 and t_2

Successive generations of plants following a transformation event. The parent transformed plant is T_0, its immediate progeny is T_1, and the progeny of the T_1 are T_2 plants etc. Of particular interest is the stability of transgene expression from T_0 to T_2, and beyond.

▫ t4 dna ligase

An enzyme, present in bacteria infected with bacteriophage T4, which catalyses the joining (ligation) of, and repairs nicks in, duplex DNA molecules. Ligation activity requires that one DNA molecule has a 5'-phosphate group, and that the other has a free 3'-hydroxyl group.

▫ tailing

The *in vitro* addition, to the 3'-hydroxyl ends of a double-stranded DNA molecule, of multiple copies of a single nucleotide by the enzyme terminal transferase.

▫ tandem repeat

Two (or more) contiguous identical DNA sequences. The orientation can be either head-to-tail, or head-to-head.

▫ tank bioreactor

A fermentation vessel designed to grow large scale quantities of a micro-organism (bacteria, yeast or fungi). Most tank bioreactors are designed to be stirred mechanically, since this allows effective distribution throughout the culture of gas and nutrients. Alternative bioreactors use fibre or membrane surfaces to immobilize the cultured cells.

❐ tap root

Root system in which the primary root has a much larger diameter than any lateral roots (e.g. carrot).

❐ *Taq* polymerase

A heat-stable DNA polymerase isolated from the thermophilic bacterium *Thermus aquaticus,* widely used in PCR.

❐ target

In diagnostic tests, the molecule or nucleic acid sequence assayed in a sample. In mutagenesis, the gene sequence that needs to be altered to generate the desired change in phenotype.

❐ target site duplication

A short sequence of DNA duplicated when a transposable element inserts at a new locus; usually found at each end of the insertion.

❐ targeted drug delivery

A method of delivering the activated form of a drug molecule to the site in the body where it is needed, rather than allowing it reach the target by uncontrolled diffusion.

❐ targeting vector

A cloning vector carrying a DNA sequence capable of participating in a recombinational event at a specified chromosomal location in the host cell.

❐ tata box

A widely conserved adenine- and thymine-rich DNA sequence found 25-30 bp upstream of the transcription initiation point of many eukaryotic genes. The TATA box is implicated in the promotion of gene transcription as it acts as a binding site for RNA polymerase. Analogous to the Pribnow box in prokaryotic promoters.

❐ tautomeric shift

The transfer of a hydrogen atom from one position in an organic molecule to another position. Tautomers can have widely different biological activities, as the shift can induce a significant change in the conformation of the molecule.

❐ tautomerism

A type of isomerism in which the two isomers arising from a tautomeric shift are in equilibrium.

□ t-cell-mediated (cellular) immune response

The synthesis of antigen-specific T cell receptors and the development of killer T cells in response to an encounter of immune system cells with an unrecognized immunogenic molecule.

□ t-dna

The DNA segment of the Ti plasmid, present in pathogenic *Agrobacterium tumefaciens*, that is transferred to plant cells and inserted into the plant's DNA as part of the infection process. Wild type T-DNA encodes enzymes that induce the plant to synthesize specific opines that are required for bacterial growth. In engineered T-DNAs, these genes are replaced by a transgene(s).

□ telomerase

An enzyme that maintains the structure of the telomere by adding the required repetitive sequences to the ends of eukaryotic chromosomes.

□ telomere

The structure found at the end of eukaryotic chromosomes containing specialized repetitive (and widely conserved across species) DNA sequences, which are necessary to assure the completion of a cycle of DNA replication.

□ telophase

The last stage in each mitotic or meiotic division, in which the chromosomes coalesce at each pole of the dividing cell.

□ temperate phage

A phage (virus) that invades but does not normally destroy (lyse) the host bacterial cell. Under specific circumstances, the lytic cycle is induced, resulting in the release of infective phage particles.

□ temperature-sensitive mutant

An organism that can grow at one temperature but not at another.

□ temperature-sensitive protein

A protein that is functional at one temperature but loses function at another (usually higher) temperature.

□ template

An RNA or single-stranded DNA molecule, used by polymerases to generate a complementary nucleotide strand.

□ terminal bud

A branch tip, an undeveloped shoot containing rudimentary floral buds or leaves, enclosed within protective bud scales.

□ terminal transferase

An enzyme that catalyses the addition of nucleotides to the 3' end of a DNA molecule.

❒ terminalization

Repelling movement of the centromeres of bivalents in the diplotene stage of the meiotic prophase, that appears to move visible chiasmata toward the ends of the bivalents.

❒ termination signal

In transcription, a nucleotide sequence that specifies RNA chain termination.

❒ terminator

1. A DNA sequence just downstream of the coding segment of a gene, which is recognized by RNA polymerase as a signal to stop synthesizing mRNA.
2. A term used in GMO technology for a transgenic method which genetically sterilizes the progeny of the planted seed, thereby preventing the use of farm-saved seed.

❒ terminator gene

A specific variety-level genetic use restriction technology. A patented technique.

❒ terminator region

A DNA sequence that signals the end of transcription.

❒ tertiary structure

The three-dimensional conformation taken up by complete macromolecules as a result of intramolecular interactions, such as hydrogen-bonding. *See*: primary structure, secondary structure, quaternary structure.

❒ test cross

A cross between a genetically unknown individual and a recessive tester to determine whether the individual in question is heterozygous or homozygous for a certain allele. It can also used as a method to test for linkage, i.e. to estimate recombination fraction.

❒ testis

Male sex organ where spermatozoa mature and are stored.

❒ testosterone

Male hormone, synthesized in the testis of mammals; used to induce sex reversal in fish.

❒ tetracycline

An antibiotic that interferes with protein synthesis in prokaryotes. A gene encoding resistance to tetracycline has been widely used as a marker to distinguish between transformed and non-transformed cells in the production of transgenic plants.

❐ tetrad

The four haploid cells formed after the second meiotic division in plants (pollen tetrads) or fungi (ascospores).

❐ tetraploid

An organism, or a tissue whose cells contain four haploid sets of chromosomes.

❐ tetrasomic

Pertaining to a nucleus or an organism with four members of one of its chromosomes, whereas the remainder of its chromosome complement is diploid. Chromosome formula: $2n + 2$.

❐ tetratype

In fungi, a tetrad of spores that contains four different types; e.g. *AB, aB, Ab* and *ab*.

❐ tgge

Abbreviation for thermal gel gradient electrophoresis.

❐ thallus

Plant body without true roots, stems, or leaves.

❐ therapeutic agent

A compound used for the treatment of a disease or for improving the well-being of an organism. therapeutic cloning The potential use of stem cells to grow, *in vitro*, tissue or organs for use in transplantation. Because these cells would be obtained from, and would therefore be genetically identical to the patient's own cells, problems of transplant rejection would be overcome. The technique would also remove the difficulty of identifying an organ donor.

❐ thermal gel gradient electrophoresis

A method for separating DNA fragments according to their mobility under increasingly denaturing conditions imposed by heat.

❐ thermal shock

Exposure to reduced or increased temperature for a significant period.

❐ thermolabile

Not resistant to heat, often in the context of a molecule which is unstable upon heating.

❐ thermophile

An organism which is adapted to high temperatures, such as in hot springs and geysers, smoker vents on the sea floor, and domestic hot water pipes. A wide range of bacteria, fungi and simple plants and animals can grow at temperature up to 50 °C; thermophiles grow and reproduce at above 50 °C. They can be classified, according to their optimal growth temperature, into simple thermophiles (50-65 °C); thermophiles (65-85 °C), and extreme thermophiles (>85 °C).

❑ thermosensitivity

Loss of biological activity of a molecule at high temperature.

❑ thermostable

A molecule which retains its biological activity at some specified higher temperature.

❑ thermotherapy

Exposure to elevated temperatures, a technique mainly used for virus or mycoplasma elimination, taking advantage of the higher thermostability of the host over its pathogen.

❑ thinning

1. Removal of older stems to promote new growth.
2. Removal of excess fruits to improve the size and quality of the remaining fruits.
3. Removal of seedlings spaced too closely for optimum growth.

❑ thymidine kinase

An enzyme that allows a cell to utilize an alternate metabolic pathway for incorporating thymidine into DNA. Used as a selectable marker to identify transfected eukaryotic cells.

❑ thymidine

The deoxyribonucleoside resulting from the combination of the base thymine (T) and the sugar 2-deoxy-D-ribose.

❑ thymidine triphosphate

Abbreviation: TTP; dTTP is strictly correct but rarely used.

❑ thymidylic acid

Synonym for thymidine 5'-monophosphate, a deoxyribonucleotide containing the nucleoside thymidine.

❑ thymine

(Abbreviation: T). One the bases found in DNA.

❑ ti plasmid

Tumour-inducing plasmid. A large plasmid present in pathogenic *Agrobacterium tumefaciens*, responsible for the induction of tumours in plant with crown gall disease. Engineered forms of this plasmid are central to the production of transgenics in many crop species.

❐ tissue

A group of cells of similar structure which sometimes performs a special function.

❐ tissue culture

The *in vitro* culture of cells, tissues or organs in a nutrient medium under sterile conditions.

❐ titre

1. The concentration of infectious virus particles present in a suspension.
2. A measure of antibody concentration, given by the highest dilution of the sample that results either in a useable immunoassay, or in the formation of visible precipitate when challenged by the appropriate antigen.

❐ tk

Abbreviation for thymidine kinase.

❐ tmp

Abbreviation for the deoxyribonucleotide thymidine 5'-monophosphate. *See*: thymidylic acid.

❐ tolerance

Incomplete resistance to a given biotic or abiotic stress. Tolerant genotypes are less inhibited by the stress, but are not immune.

❐ tonoplast

The cytoplasmic membrane bordering the vacuole of plant cells. It plays a prominent role in regulating the osmotic pressure exerted by the cell sap.

❐ totipotency

The ability of a cell or tissue to be induced to regenerate into a complete organism.

❐ toxicity

The extent to which a toxic compound negatively affects a given trait.

❐ toxin

A compound produced by one organism, which is deleterious to the growth and/or survival of another organism of the same or different species.

❐ tracer

A substance (typically a radioactive isotope or a fluorescent dye) that can be detected by physical means, and which is used to analyse the progress of a chemical reaction or a biological process.

□ tracheid

An elongated, tapering xylem cell, with lignified pitted walls, adapted for solute conduction and physical support. Found in conifers, ferns and related plants.

□ trait

One of the many characteristics that define an organism. The phenotype is a description of one or more traits.

□ *trans* heterozygote

A double heterozygote that contains two mutations arranged in the *trans* configuration.

□ *trans*-acting

1. A term describing substances that are diffusable and that can affect spatially separated entities within cells.
2. A genetic element (e.g. a promoter sequence) that is effective only when present in the *trans* configuration.

□ *trans*-acting factor

Any of the multiple ancillary DNA-binding proteins that interact with the *cis*-regulatory DNA sequences to control gene expression.

□ transcapsidation

The partial or full coating of the nucleic acid of a virus particle with the coat protein of a different virus.

□ transcript

An RNA molecule that has been synthesized from a specific DNA template. In eukaryotes, the primary transcript produced by RNA polymerase is often processed or modified in order to form functional mRNA, rRNA or tRNA.

□ transcription factor

A protein that regulates the transcription of genes.

□ transcription

Synthesis of RNA from a DNA template via RNA polymerase.

□ transcription unit

A segment of DNA that contains signals for the initiation and termination of transcription, and is transcribed into one RNA molecule.

□ transcriptional anti-terminator

A protein that prevents RNA polymerase from terminating transcription at specific transcription termination sequences.

□ transcriptional roadblock

A DNA-binding protein which affects the rate at which RNA polymerases transcribe genes. The protein/DNA complex interferes with the passage of the elongation complex. In some cases these obstacles are readily bypassed, but in others a significant level of pausing or termination occurs, and this can then act as a control point for gene expression.

❑ transduction

1. Genetic: the transfer by means of a viral vector of a DNA sequence from one cell to another.
2. Signal: any process that helps to produce biological responses to events in the environment (e.g. transduction of hormone binding into cellular events by hormone receptors).

❑ transfection

The infection of a cell with isolated viral DNA (or RNA), resulting in the production of intact viral particles.

❑ transfer and cloning of the insulin gene

Example of the technique of DNA cloning into a plasmid: Insertion of the gene coding for insulin into a bacterial plasmid, which in turn carries the gene into a replicating bacterial cell that produces human insulin.

Plasmid: Plasmids are small circles of DNA found in bacteria cells, separate from the bacterial chromosome and smaller than it. They are able to pass readily from one cell to another, even when the cells are clearly from different species, far apart on the evolutionary scale. Consequently, plasmids can be used as **vectors,** permitting the reproduction of a foreign DNA by using the bacterial replicating system.

cDNA: Human genes composed of coding and non- coding sequences. The copy of the coding sequences is called cDNA. It can be obtained from the reverse transcription of messenger RNA. The transcription and translation of the insulin cDNA will allow the production of a functional insulin molecule.

Transfer of the Insulin gene into a plasmid vector.

- The plasmid is cut across both strands by a restriction enzyme, leaving loose, sticky ends to which DNA can be attached.
- Special linking sequences are added to the human cDNA so that it will fit precisely into the loose ends of the opened plasmid DNA ring.
- The plasmid containing the human gene, also called a recombinant plasmid, is now ready to be inserted into another organism, such as a bacterial cell.

Cloning the Insulin gene

The recombinant plasmids and the bacterial cells are mixed up. Plasmids enters the bacteria in a process called **transfection.** With the recombinant DNA molecule successfully inserted into the bacterial host, another property of plasmids can be exploited - their capacity to replicate. Once inside a bacterium, the plasmid containing the human cDNA can multiply to yield several dozen copies. When the bacteria divide, the plasmids are divided between the two daughter cells and the plasmids continue to reproduce. With cells dividing rapidly (every 20 minutes), a bacterium containing human cDNA (encoding for insulin, for example) will shortly produce many millions of similar cells (clones) containing the same human gene.

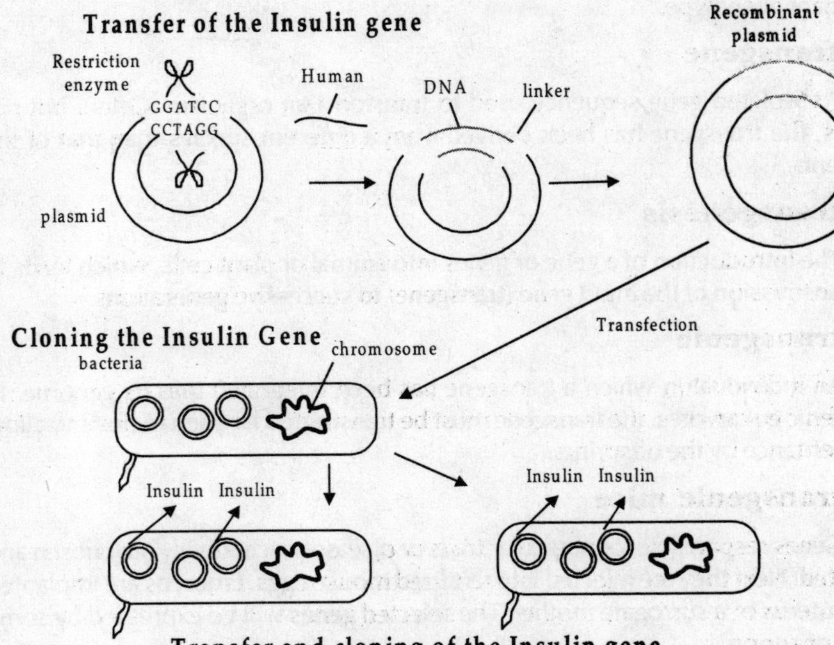

Transfer and cloning of the Insulin gene

▫ transferase

A class of enzymes that catalyses the transfer of a group of atoms from one molecule to another.

▫ transformant

A cell or organism that has been genetically altered through the integration of a transgene(s). Primary: the first generation following the transformation event. Secondary: progeny of the primary transformant.

▫ transformation

1. The uptake and integration of DNA in a cell, in which the introduced DNA is intended to change the phenotype of the recipient organism in a predictable manner.
2. The conversion, by various means, of cultured animal cells from controlled to uncontrolled cell growth, typically through infection with a tumour virus or transfection with an oncogene.

▫ transformation efficiency or frequency

The fraction of a cell population that takes up and integrates the introduced transgene; expressed as the number of transformed cells recovered divided by the total number of cells in a population.

▫ transforming oncogene

A gene that, upon transfection, converts a previously immortalized cell to the malignant phenotype.

▫ transgene

An isolated gene sequence used to transform an organism. Often, but not always, the transgene has been derived from a different species than that of the recipient.

▫ transgenesis

The introduction of a gene or genes into animal or plant cells, which leads to the transmission of the input gene (transgene) to successive generations.

▫ transgenic

An individual in which a transgene has been integrated into its genome. In transgenic eukaryotes, the transgene must be transmitted through meiosis to allow its inheritance by the offspring.

▫ transgenic mice

Genes responsible for particular traits or disease susceptibility are chosen and extracted. Next they are injected into fertilized mouse eggs. Embryos are implanted in the uterus of a surrogate mother. The selected genes will be expressed by some of the offspring.

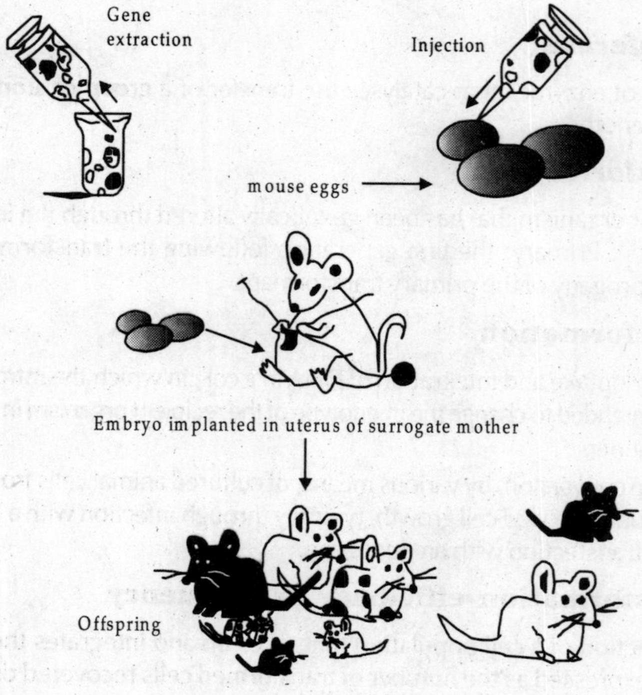

Transgenic Mice

Since the first gene transfers into mice were successfully executed in 1980, transgenic mice have allowed researchers to observe experimentally what happens to an entire organism during the progression of a disease. Transgenic mice have become models for studying human diseases and their treatments.

❒ transgressive variation

The appearance, in a segregating generation, of individuals showing expression of a trait outside the extremes defined by the parent of the cross that was used to generate the population.

❒ transient expression

Short-term activity of a transgene following its introduction into target tissue. Transient expression usually implies non-integration of the transgene into the host genome.

❒ transition stage

The period between juvenile and reproductive stages of growth.

❒ transition

The substitution in DNA or RNA of one purine by another purine, or of one pyrimidine by another pyrimidine.

❒ transition-state intermediate

In a chemical reaction, an unstable and high-energy configuration assumed by reactants on the way to making products. Enzymes are thought to bind and stabilize the transition state, thus lowering the energy of activation needed to drive the reaction to completion.

❒ translation

The process of polypeptide synthesis in which the amino acid sequence is determined by mRNA, mediated by tRNA molecules, and carried out on ribosomes.

❒ translocation

An abnormality in which a piece of one chromosome (which carries the genes) breaks off and binds to another chromosome. APL is characterized by specific translocations of genes from chromosome number 17 onto number 15.

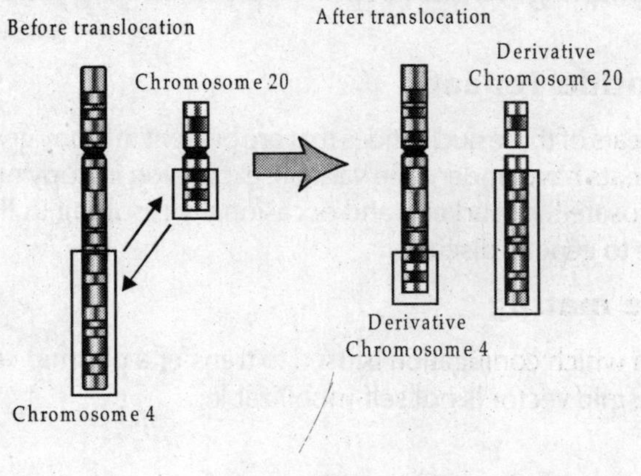

◻ transposable (genetic) element

A DNA element that can move from one location in the genome to another.
transposase An enzyme encoded by a transposon gene that catalyses the movement of a DNA sequence to a different site in a DNA molecule.

◻ transposition

The process whereby a transposon or insertion sequence inserts itself into a new site on the same or another DNA molecule. The exact mechanism is not fully understood and different transposons may transpose by different mechanisms. Transposition in bacteria does not require extensive DNA homology between the transposon and the target DNA.

◻ transposon

Synonym of transposable genetic element.

◻ transposon tagging

A method of gene isolation that exploits the disruption of normal gene expression that is the result of an insertion of a transposon within, or close to the target. Since the sequence of the transposon is known, this can be used as a DNA probe to define the DNA fragment containing the target gene. Large-scale experiments to generate populations of gene mutations are colloquially referred to as gene machines.

◻ transversion

The substitution in DNA or RNA of one purine by a pyrimidine or *vice versa*.

◻ tribrid protein

A fusion protein that has three segments, each encoded by parts of different genes.

◻ trichome

A short filament of cells, resulting in a hair-like structure.

◻ tri-hybrid

The hybrid offspring of a cross between parents carrying contrasting alleles at three loci.

◻ trinucleotide repeat

Tandem repeats of three nucleotides that are present in many genes. Commonly trinucleotide repeats have undergone variable expansion in copy number, forming the basis of microsatellite markers, and occasionally resulting in the formation of alleles giving rise to genetic disease.

◻ tripartite mating

A process in which conjugation is used to transfer a plasmid vector to a target cell when the plasmid vector is not self-mobilizable.

triplet

A sequential group of three nucleotides in DNA or RNA.

triploid

A cell, tissue or organism containing three times the haploid number of chromosomes.

trisomy

The presence in a diploid cell or organism of an extra chromosome of one homologue (chromosome formula: $2n + 1$).

triticale

The hybrid man-made species formed by the crossing of tetraploid or hexaploid wheat with diploid rye.

trna

Abbreviation for transfer RNA. Small RNA molecules that transfer amino acids to the ribosome during protein synthesis. Each tRNA binds just one species of amino acid and recognizes a specific codon in the mRNA, thus implementing the genetic code.

tropism

Plant response to an external stimulus, resulting in the bending/turning of stem or root growth. Typical tropisms are phototropism (light), geotropism (gravity) or hydrotropism (water).

true-to-type

Conforming to the phenotype of the breed/variety.

trypsin

A proteolytic enzyme used *in vivo* for the digestion of peptides. It acts by hydrolysing peptide bonds on the carboxyl side of the amino acids arginine and lysine.

trypsin inhibitor

Substances inactivating trypsin, typically found in seed tissue of certain plants, where they are thought to have evolved as anti-feedant agents against insect predators.

ttp

Abbreviation for thymidine 5'-triphosphate. TTP is required for DNA synthesis since it is a direct precursor molecule.

tubulin

The major protein component of the microtubules of eukaryotic cells.

tumble tube

A glass tube mainly used *in vitro* to agitate and consequently aerate suspension cultures. The tube, which is commonly attached to a slowly revolving platform, is closed at both ends, with a side-neck opening.

▫ tumor-suppressor gene

A gene that regulates cell growth. If such a gene becomes dysfunctional, and potentiating damage occurs to the cell, then uncontrolled growth and a cancer may result.

▫ tumour virus

A virus capable of transforming a cell to a malignant phenotype.

▫ tunica

The outer one- to four-cell layer region of the apical meristem, where cell division is anticlinal, i.e. perpendicular to the surface.

▫ turbidostat

An open continuous culture in which a pre-selected biomass density is uniformly maintained by automatic removal of excess cells. The fresh medium flows in response to an increase in the turbidity (usually corresponding to cell density) of the culture.

▫ turgid

Swollen, distended; referring to a cell that is extended as a result of adequate water uptake. Loss of turgidity in plant cells is a sign of water deficit.

▫ turgor pressure

The pressure within a cell resulting from the absorption of water into the vacuole and the imbibition of water by the protoplasm.

▫ turion

An underground bud or shoot from which an aerial stem arises.

▫ twin

One of two individuals originating from the same zygote.

▫ types of weak non-covalent bonds part 1

The principal types of weak non-covalent bonds that hold macromolecules together.

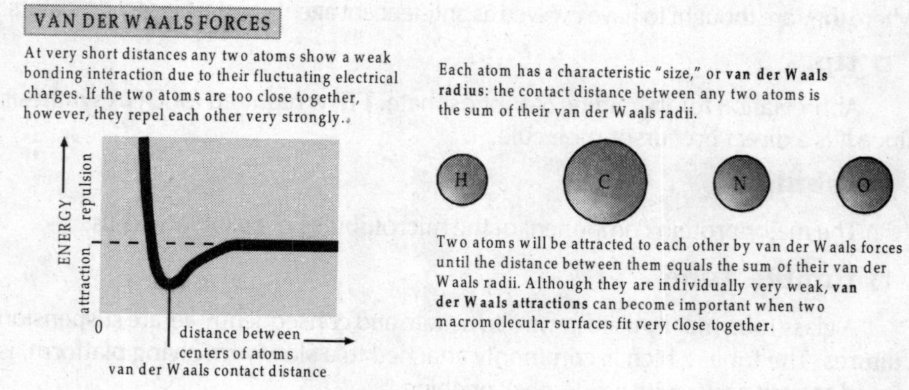

WEAK CHEMICAL BONDS

Organic molecules can interact with other molecules through short-range noncovalent forces.

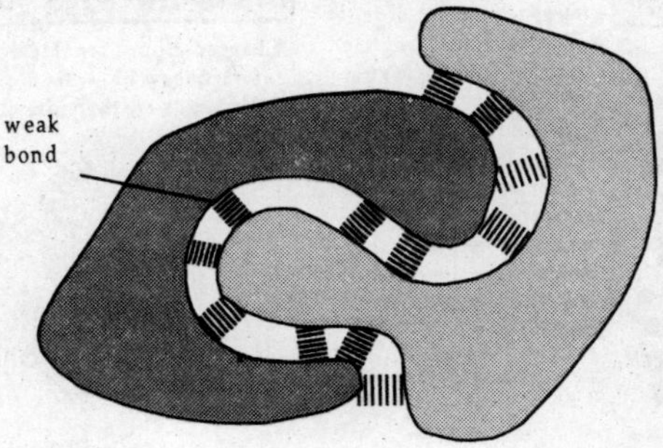

Weak chemical bonds have less than 1/20 the strength of a strong covalent bond. They are strong enough to provide tight binding only when many of them are formed simultaneously.

HYDROGEN BONDS

As already described for water (see Panel 2-2, pp. 50–51) **hydrogen bonds** form when a hydrogen atom is "sandwiched" between two electron-attracting atoms (usually oxygen or nitrogen).

Hydrogen bonds are strongest when the three atoms are in a straight line:

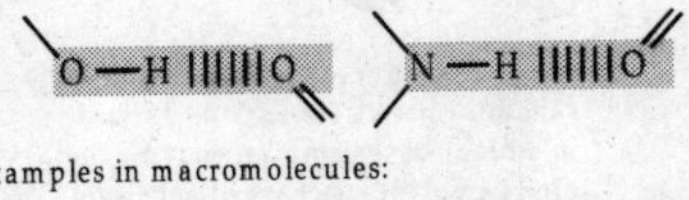

Examples in macromolecules:

Amino acids in polypeptide chains hydrogen-bonded together.

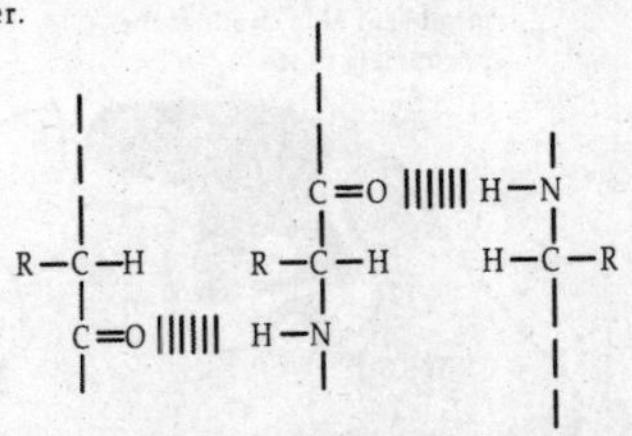

Two bases, G and C, hydrogen-bonded in DNA or RNA.

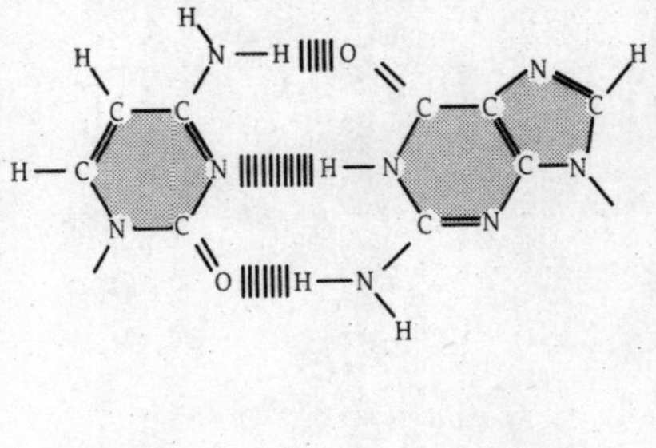

HYDROGEN BONDS IN WATER

Any molecules that can form hydrogen bonds to each other can alternatively form hydrogen bonds to water molecules. Because of this competition with water molecules, the hydrogen bonds formed between two molecules dissolved in water are relatively weak.

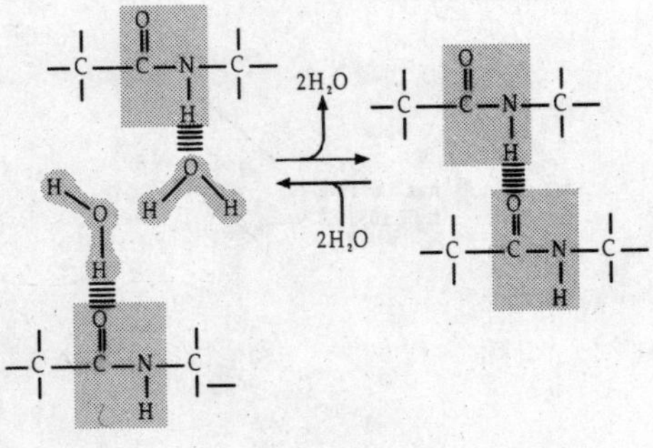

❑ types of weak non-covalent bonds part 2

The principal types of weak non-covalent bonds that hold macromolecules together.

HYDROPHOBIC FORCES

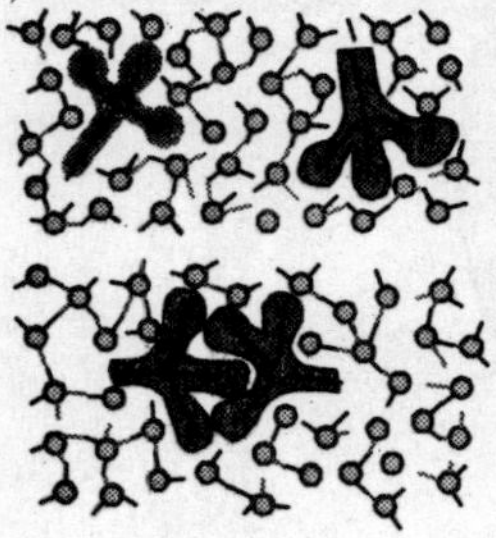

Water forces hydrophobic groups together in order to minimize their disruptive effects on the hydrogen-bonded water network. Hydrophobic groups held together in this way are sometimes said to be held together by "hydrophobic bonds," even though the attraction is actually caused by a repulsion from the water.

IONIC BONDS IN AQUEOUS SOLUTIONS

Charged groups are shielded by their interactions with water molecules. Ionic bonds are therefore quite weak in water.

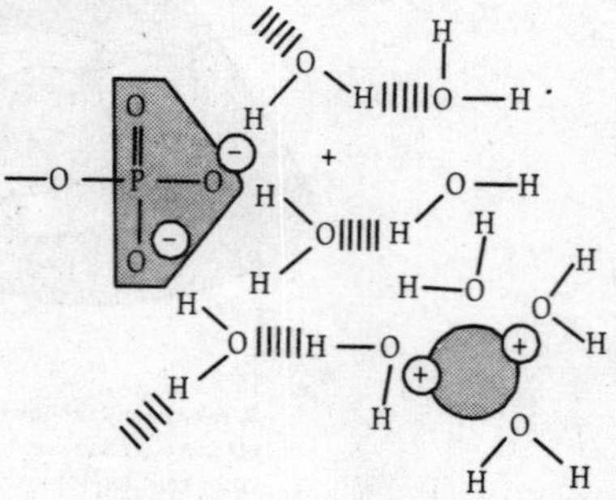

Similarly, other ions in solution can cluster around charged groups and further weaken ionic bonds.

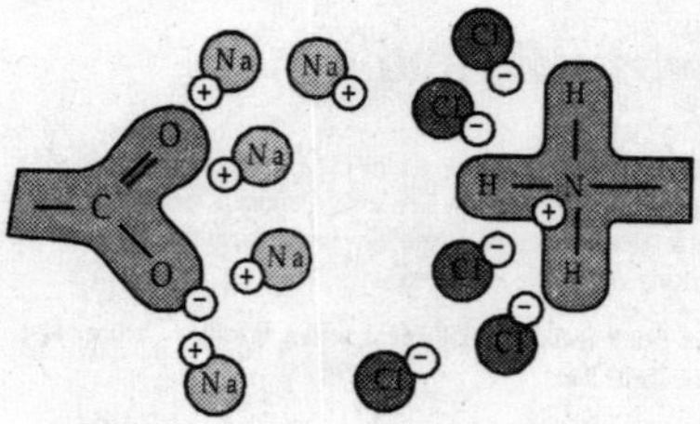

IONIC BONDS

Ionic interactions occur either between fully charged groups (ionic bond) or between partially charged groups.

The force of attraction between the two charges, d+ and d–, falls off rapidly as the distance between the charges increases.

In the absence of water, ionic forces are very strong. They are responsible for the strength of such minerals as marble and agate.

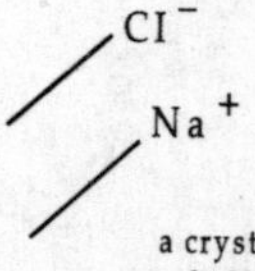

hand drawn fig 2.105/2.07

a crystal of salt, NaCl

Despite being weakened by water and salt, ionic bonds are very important in biological systems; an enzyme that binds a positively charged substrate will often have a negatively charged amino acid side chain at the appropriate place.

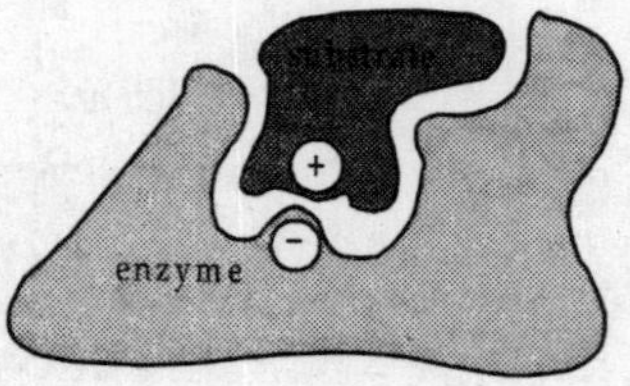

U

□ u

Abbreviation for uracil.

□ ubiquitin

A small protein, present in all eukaryotic cells, which plays an important role in tagging proteins destined for proteolytic cleavage (because they are damaged or no longer needed).

□ ump

Abbreviation for the (ribo)nucleotide uridine 5'-monophosphate. *See*: uridylic acid.

□ understock

Host plant for a grafted scion, a branch or shoot from another plant; an understock may be a fully grown tree or a stump with a living root system.

□ undifferentiated

Undifferentiated cells are those which have not been committed to become part of a specialized tissue.

□ unencapsidated

A virus not enclosed by a coat protein or capsid.

□ unequal crossing over

Abnormal meiotic event, in which one chromatid contains a duplication and the other a deletion. Often arises in a region containing repeated DNA sequences, which can pair out of register.

□ unicellular

Tissues, organs or organisms consisting of a single cell.

□ uniparental inheritance

The inheritance of genes exclusively from one parent, e.g. chloroplast DNA is inherited either maternally (many angiosperms) or paternally (most gymnosperms).

❑ unisexual

Higher organisms (animals or plants) possessing either male or female reproductive organs, but not both.

❑ univalent

An unpaired chromosome at the first division of meiosis.

❑ universal donor cell

Cells that, after introduction into a recipient, will not induce an immune response that leads to their rejection.

❑ universality

Referring to the genetic code, the triplet codons are translated to the same amino acid, with minor exceptions, in virtually all species.

❑ unorganized growth

In vitro formation of tissues with few differentiated cell types and no recognizable structure; typical structure of calli formed in tissue culture.

❑ upstream

1. The stretch of DNA lying in the 5' direction from the site being considered. Where the reference point is the initiation of transcription, the first transcribed base is designated +1 and upstream nucleotides are marked with minus signs, e.g. -1, -10;
2. In chemical engineering, those phases of a manufacturing process that precede the biotransformation step. Refers to the preparation of raw materials for a fermentation process. Also called upstream processing.

❑ uracil

(Abbreviation: U). One the bases found in RNA.

❑ uridine

The (ribo)nucleoside resulting from the combination of the base uracil (U) and the sugar D-ribose.

❑ uridine triphosphate (uridine 5'-triphosphate)

(Abbreviation: UTP). Required for RNA synthesis since it is a direct precursor molecule.

❑ utilization of farm animal genetic resources

The use and development of animal genetic resources for the production of food in a sustainable system of agriculture.

V

❒ vaccine

A preparation of dead or attenuated (weakened) pathogens, or of derived antigenic determinants, that can induce the formation of antibodies in a host, and thereby produce host immunity against the pathogen. *See*: sub-unit vaccine, viral vaccine, DNA vaccine, inoculum.

❒ vaccinia

The cowpox virus used to vaccinate against smallpox and, experimentally, as a carrier of genes for antigenic determinants cloned from other disease organisms.

❒ vacuole

A fluid-filled membrane-bound cavity inside many plant cells, in which various plant products and by-products are stored.

❒ variable domain

Regions of antibody molecules that have different amino acid sequences in different antibody molecules. These regions are responsible for the antigen-binding specificity of the antibody.

❒ variable expressivity

Variation in the phenotype caused by different alleles of the same gene and/or by the action of other genes and/or by the action of non-genetic factors.

❒ variable number tandem repeat

A DNA sequence, present as tandem repeats, for which the number of copies varies greatly between unrelated genotypes.

❒ variable surface glycoprotein

One of a battery of antigenic determinants expressed by a micro-organism to elude immune detection.

❒ variance

A statistical term representing a measure of the dispersion of data from the overall mean. Used to quantify the variability of a population.

❑ variant

An individual that is genetically distinct from others in the population.

❑ variation

Differences between individuals within a population or among populations.

❑ variegation

The occurrence, within a single tissue, organ or organism, of mosaicism. Usually referring to plants showing either both green and albino colouration in a leaf, or flecks of contrasting colour in a flower. The origin of variegation can be through viral infection, nutritional deficiency, or genetic instability caused by transposon activity.

❑ variety

1. A naturally occurring subdivision of a species, with distinct morphological characters.
2. A defined strain of a crop plant, selected on the basis of phenotypic (sometimes genotypic) homogeneity.

❑ vascular bundle

A strand of tissue containing primary xylem and primary phloem (and procambium if present) and frequently enclosed by a bundle sheath of parenchyma or fibres.

❑ vascular cambium

In biennials and perennials, cambium giving rise to secondary phloem and secondary xylem.

❑ vascular plant

Plant species possessing organized vascular tissues.

❑ vascular

Plant tissue specialized for the conduction of water or nutrients

❑ vascular system

1. A specialized network of vessels for the circulation of fluids throughout the body tissue of an animal.
2. The system of vascular tissue in plants.

❑ vector

1. An organism, usually an insect, that carries and transmits pathogens.
2. A small DNA molecule (plasmid, virus, bacteriophage, artificial or cut DNA molecule) that can be used to deliver DNA into a cell. Vectors must be capable of being replicated and contain cloning sites for the introduction of foreign DNA.

❐ velocity density gradient centrifugation

A procedure used to separate **macromolecules** based on their rate of movement through a density gradient.

❐ velogenetics

The combined use of marker-assisted selection and embryo technologies such as OPU, IVM and IVF, in order to increase the rate of genetic improvement in animal populations.

❐ vermiculite

Material made from expanded mica used as a rooting medium and as a soil additive.

❐ vernalization

Chilling juvenile plants for a minimum period in order to induce flowering. Some plants require veRNAlization to flower, but others have no such requirement.

❐ vessel

A series of xylem elements whose function is to conduct water and nutrients in plants.

❐ vessel element

A type of cell occurring within the xylem of flowering plants. Many are water-conducting vessels.

❐ viability test

Assay of the number or percent of living cells or plants in a population following a specific treatment. Often used to describe quality of seed following long-term storage.

❐ viability

The capability to live and develop normally.

❐ viable

Capable of normal completion of life cycle.

❐ vibrio

Comma-shaped bacterium.

❐ *vir* genes

A set of genes on a Ti plasmid that prepare the T-DNA segment for transfer into a plant cell.

❐ viral coat protein

A protein present in the layer surrounding the nucleic acid core of a virus.

❒ viral oncogene

A viral gene that promotes tumour development in a host.

❒ viral pathogen

A disease-causing virus.

❒ viral vaccine

Vaccine consisting of live viruses, genetically engineered to avoid causing the disease itself.

❒ virion

A complete infectious virus particle.

❒ viroid

A plant pathogenic agent, composed of an infectious single-stranded low molecular weight RNA, and no coat protein.

❒ virulence

The degree of ability of an organism to cause disease. The relative infectiousness of a bacterium or virus, or its ability to overcome the resistance of the host metabolism.

❒ virulent phage

A phage that destroys its host bacterium.

❒ viruliferous

A vector (usually insect) organism that carries virions and spreads the virus from host to host by mechanical means.

❒ virus

An infectious particle composed of a protein capsule and a nucleic acid core (DNA or RNA), which is dependent on a host organism for replication.

❒ virus-free

Plant, animal, cell, tissue or meristem which exhibits no viral symptoms or contains no identifiable virus particles.

❒ virus-tested

Description of a organism or a cell stock certified as being free of certain specified viruses following recognized procedures of virus diagnosis.

❒ vitamin

Naturally occurring organic substance required by living organisms in small amounts to maintain normal health.

❒ vitrified

Cultured tissue having leaves and sometimes stems with a glassy, transparent or wet and often swollen appearance. The process of vitrification is a general term for a variety of physiological disorders that lead to shoot tip and leaf necrosis.

❒ **vivipary**

1. A form of reproduction in animals in which the developing embryo obtains its nourishment directly from the mother via a placenta or by other means.
2. A form of asexual reproduction in certain plants, in which the flower develops into a bud-like structure that forms a new plant when detached from the parent.
3. The development of young plants in the inflorescence of the parent plant.

❒ **V_{max}**

The maximal rate of an enzyme-catalysed reaction. V_{max} is the product of E_o (the total amount of enzyme) and K_{cat} (the catalytic rate constant).

❒ **vntr**

Abbreviation for variable number tandem repeat.

❒ **volatilization**

The conversion of a solid or liquid into a gas or vapour.

❒ **v region**

Variable region in antibodies.

❒ **v/v**

Abbreviation for volume per volume. The relative proportion of each liquid in a mixture.

❒ **vsg**

Abbreviation for variable surface glycoprotein.

W

❐ wall pressure

Pressure that a cell wall exerts against the turgor of the cell contents. Wall pressure is equal and opposite to the turgor potential.

❐ wash-out

The loss of the slower growing micro-organism when two organisms are being grown together.

❐ water potential

The pressure gradient that induces the flow of water, particularly with reference to plant water uptake from the soil, comprising the net effects of suction, solutes and matric forces.

❐ water stress

Occurs when plants are unable to absorb enough water to replace that lost by transpiration. Short-term water stress leads to turgor loss (wilting). Prolonged stress leads to cessation of growth, and eventually plant death.

❐ wax

Water-insoluble esters of long-chain acids with long-chain alcohols. Waxes form protective waterproof layers on leaves, stems, fruits, animal fur and integuments of insects.

❐ weed

A plant growing where it is not wanted. Generally used to describe plants which colonize readily, and can compete for resources with a planted crop.

❐ weediness

The ability of a plant to colonize a disturbed habitat and compete with cultivated species.

❐ western blot

A technique whereby a complex mixture of size-separated proteins is fixed to a solid support, and then probed with a labelled antibody. Useful, for example, for the measurement of levels of production of a specific protein in a particular tissue or at particular developmental stage.

❐ **wetting agent**

A substance (usually a detergent) that improves the contact of a liquid to a solid surface by reducing its surface tension.

❐ **wild type**

The most frequent allele or genotype found in nature, or a specified organism against which mutants are defined.

❐ **wilt**

Drooping of stems and foliage due to loss of cell turgor. May be caused by water stress or by disease.

❐ **wilting point**

The moisture content of soil at which plants start to wilt, but not to the extent that they fail to recover when placed in a humid atmosphere.

❐ **wobble hypothesis**

An explanation of how one tRNA may recognize more than one codon. The first two bases of the mRNA codon and anticodon pair properly, but the third base in the anticodon has some flexibility that permits it to pair with either the expected base or an alternative.

❐ **w/v**

Abbreviation for weight per volume. The relative proportions of solid and liquid in a solution.

X

❐ x

The basic number of chromosomes in a polyploid series, monoploid/haploid = x; diploid = 2x; triploid = 3x; etc.

❐ xanthophyll

A yellow oxygen-containing carotenoid, present in chloroplasts.

❐ xenia

The immediate effect of pollen on some characters of the endosperm.

❐ xenobiotic

A chemical compound that is not produced by, and often cannot be degraded by, living organisms.

❐ xenogeneic

Refers to organs, genetically engineered ("humanized") todecrease the chance of rejection, that have been grown in an animal of another species for potential transplant to humans.

❐ xenotransplantation

The transplantation of tissue or organs from one species to another species, typically from pigs to humans. Zoonoses are an important issue.

❐ xerophyte

A plant very resistant to drought, typically adapted to extremely dry environments.

❐ x-linked disease

A genetic disease caused by an allele at a locus on the X-chromosome.

❐ x-linked inheritance: hemophilia

Hemophilia is a sex linked trait in humans, inherited in the same way white eye color in Drosophila.

Males are hemizygous, receiving their only X chromosome from their mother. Females are heterozygous, inheriting X chromosomes from both parents.

If a female has a defective gene on one of her two X chromosomes, she will be protected from its effects by the normal gene on her second X chromosome. If a male has a mutant X and a normal Y chromosome, he will be affected by a X-linked disease.

A son, whose mother has two normal alleles, will not be affected by hemophilia even if the father has the disease and the defective gene. A daughter of the same parents will be a heterozygous carrier.

A heterozygous carrier mother and a normal father pass the gene for hemophilia on to possibly one- half of their children. Half the daughters will be carriers and half the sons will be hemophilic. The rest of the siblings will be normal. Daughters, as long as one parent is genotypically normal, can only be carriers. The normal gene on the second X chromosome counteracts the defect and the daughters do not suffer from the trait. If a son receives the defective gene from his mother, he will be hemophilic because the Y chromosome can not counteract the defective gene located on his X chromosome.

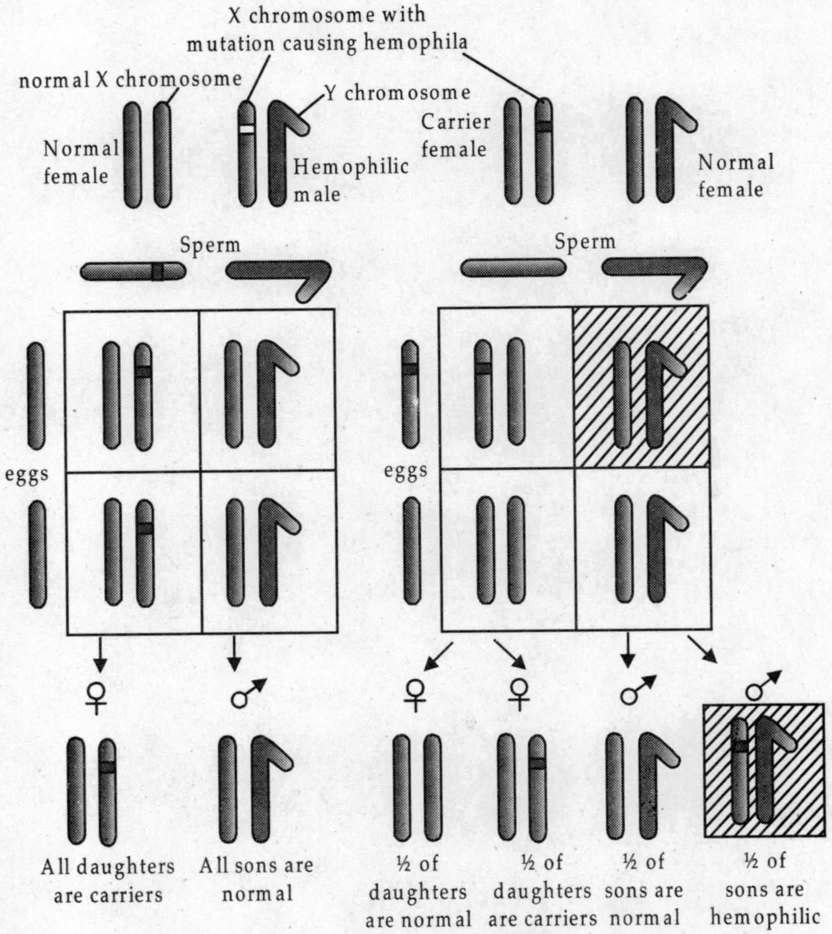

Inheritance of Hemophilia

x-linked

The presence of a gene on the X-chromosome.

xylem

A complex tissue specialized for the conduction of water and mineral nutrients in solution. Xylem may also function as a supporting tissue, particularly secondary xylem.

Y

❑ yac

Abbreviation for yeast artificial chromosome.

❑ yeast

A unicellular ascomycete fungus, commonly found as a contaminant in plant tissue culture.

❑ yeast artificial chromosome

A vector which can be propagated in budding yeast (*Saccharamyces pombe*), consisting of the minimal elements required for a chromosome to replicate, and allowing for the cloning of large DNA fragments (hundreds of kilobase pairs).

❑ yeast episomal vector

A cloning plasmid vector for the yeast *Saccharomyces cerevisiae* maintained as an extrachromosomal nuclear DNA molecule.

❑ yeast extract

A mixture of substances from yeast.

Z

❐ z-dna

A form of DNA, in which the double helix is wound in a left-hand, instead of a right-hand, manner. DNA adopts the Z conformation when purines and pyrimidines alternate on each strand, e.g. 5′CGCGCGCG 3′ or 3′GCGCGCGC5′.

❐ zinc finger

A DNA-binding protein motif, characterized by two closely spaced cysteine and two histidine residues that serve as ligands for a single Zn^{2+} ion. When bound, the structure takes on a conformation in which amino acid side chains protrude in a way that allows interaction with the DNA major groove.

❐ zone of elongation

The section of the young root or shoot just behind the apical meristem, in which the cells are enlarging and elongating rapidly.

❐ zoo blot

Hybridization of cloned DNA from one species to DNA from a range of other organisms to determine the extent to which the cloned DNA is evolutionarily conserved.

❐ zoo FISH

Fluorescence *in situ* hybridization technique, probing metaphase chromosomes of one species with DNA from another species. The technique allows inferences to be made regarding the evolutionary relationships between species.

❐ zoonosis

A disease that is communicable from animals to humans.

❐ zoospore

A spore that possesses flagella and is therefore motile.

❐ zygonema

Stage of meiotic prophase during which chromosome synapsis occurs.

❐ zygospore

A thick-walled resistant spore developing from a zygote resulting from the fusion of gametes in the course of isogamy.

❐ zygote

The diploid cell formed by the fusion of two haploid gametes during fertilization in eukaryotic organisms with sexual reproduction.

❐ zymogen

Inactive enzyme precursor that after secretion is chemically altered to the active form of the enzyme.